Natural Compounds as Antimicrobial Agents, 2nd Edition

Natural Compounds as Antimicrobial Agents, 2nd Edition

Guest Editors

Carlos M. Franco
Beatriz Vázquez Belda

MDPI

Basel • Beijing • Wuhan • Barcelona • Belgrade • Novi Sad • Cluj • Manchester

Guest Editors

Carlos M. Franco
Analytical Chemistry,
Nutrition and Bromatology
Santiago de Compostela
Lugo
Spain

Beatriz Vázquez Belda
Analytical Chemistry
Nutrition and Bromatology
Santiago de Compostela
Lugo
Spain

Editorial Office
MDPI AG
Grosspeteranlage 5
4052 Basel, Switzerland

This is a reprint of the Special Issue, published open access by the journal *Antibiotics* (ISSN 2079-6382), freely accessible at: www.mdpi.com/journal/antibiotics/special_issues/agent_2nd.

For citation purposes, cite each article independently as indicated on the article page online and using the guide below:

Lastname, A.A.; Lastname, B.B. Article Title. *Journal Name* **Year**, *Volume Number*, Page Range.

ISBN 978-3-7258-3616-1 (Hbk)
ISBN 978-3-7258-3615-4 (PDF)
https://doi.org/10.3390/books978-3-7258-3615-4

Contents

About the Editors

Carlos M. Franco

Carlos M. Franco graduated with a degree in veterinary medicine and finished his Ph.D. at the University of Santiago de Compostela, Spain, in 1994, with a thesis on the existence of L. monocytogenes in food, as well as its resistance to several antimicrobials. He completed his posdoctoral studies at the Veterinary Faculty of the Complutense University of Madrid, completing research in several areas of food microbiology, as well as at the Laboratory of Chimie Analityque II of the University of Paris 11, where he studied several analytical methods. Subsequently, he was interested in antimicrobial resistance, the effect of the use of antibiotics in animal production, and antimicrobial resistance in food from ecological or conventional origins. He has researched the detection of antimicrobials and other drugs in food of animal origins, as well as the effect of essential oils on the inhibition of bacteria. He has published more than 140 peer-reviewed papers. Currently, he is researching bacterial biofilms and their elimination or control, as well as other food science topics.

Beatriz Vázquez Belda

Beatriz Vázquez Belda holds degrees in biology and food technology and obtained her Ph.D. in 1997 at the Santiago de Compostela University (USC), Spain. She completed her posdoctoral studies at the Ecole Nationale d'Industrie Laitiére et des Industries Agro-alimentaires (Surgéres, France) and at the Faculte des Sciences Pharmaceutiques et Biologiques (Paris, France). She specializes in food micology, studying fungal contamination on Spanish cheeses, as well as the development of fast detection techniques using microbiological and instrumental methods, taking advantage of the luminescence properties of aflatoxins and other micotoxins. After two years as a lecturer at the Cardenal Herrera-CEU University in Valencia, she returned to the Veterinary Faculty (USC) with a Ramon y Cajal national grant. She continued her interest in researching the use of natural compounds such as essential oils to decrease mycotoxin production in dairy products. She is also the Technical Responsible for physico-chemical analysis at the Laboratory of Hygiene, Inspection and Food Control (LHICA-USC), which is accredited by the National Entity of Accreditation for testing agroalimentary products. Her research has mainly focused on the development of extraction and detection methods for antimicrobials, β-agonists, and corticosteroid drugs that are used for illicit purposes in animal production, methods that are used in routine control analyses of feed and food samples. She has co-authored more than 100 international publications in peer-reviewed journals and other publications.

Preface

The aim of the present reprint is to continue delving into the use of natural compounds as antimicrobial agents in a wide range of applications. The main motivation for this work is the fact that numerous syntetic substances are being continuosly used and disseminated around the world, many of which very little is known about, that may be substituted—at least in part—by these natural compounds. Some of the authors of the chapters in this reprint have worked in this area for several years, such as Prof. Durant Archibold's group, Raffaele Marrone's group, and the Mickymaray group; meanwhile, others focus on other applications, such as the research group of Prof. Viana Pontual and Simone Haberlein, who focuses on the fight against parasites. This reprint is aimed at anyone with an interest in the use of natural compounds against bacteria, as well as other living organisms.

Carlos M. Franco and Beatriz Vázquez Belda
Guest Editors

antibiotics

Editorial

Natural Compounds as Antimicrobial Agents—2nd Edition

Carlos M. Franco * and Beatriz I. Vázquez *

Hygiene, Inspection and Food Control Laboratory, Analytical Chemistry, Nutrition and Bromatology Department, Faculty of Veterinary Medicine, University of Santiago de Compostela, 27002 Lugo, Spain
* Correspondence: carlos.franco@usc.es (C.M.F.); beatriz.vazquez@usc.es (B.I.V.)

This monograph first covers two review papers from Prof. Mickymaray on the mechanisms and efficacy of flavonoids as antimicrobial compounds [1] against opportunistic nontuberculous mycobacteria. Mickymaray and coworkers reviewed the conventional treatments for these mycobacteria and then reviewed the newly discovered mechanisms of anti-nontuberculous mycobacteria, such as the inhibition of cell wall formation, the inhibition of biofilm formation, or even the inhibition of bacterial DNA synthesis, as well as the synergies with other antimycobacterial agents. The second review covers some antimicrobial animal substances, in this case, the secretions of toad skin (Anura, *Bufonidae*) against human pathogens [2]. These authors reported MIC values as low as 4 µg/mL against some Gram-positive bacteria, such as *E. faecalis*, or even less for the secretion of some toad species. Finally, the effectiveness of these secretions as antifungal or antiprotozoal agents is also revealed.

This research begins with a study in our laboratory with respect to the antimicrobial activity of several apitoxins. The apitoxin was directly collected from five apiaries in Ecuador, and the effects of these compounds on different *Salmonella* strains, including not only different serotypes from *Salmonella enterica* subsp. *Enterica*, but also *Salmonella enterica* subsp. arizonae and *Salmonella enterica* subsp. salamae isolates, were studied. Specifically, the relatively high number of different isolates assayed from this genus proved to be one of the most important aspects of this research; previously, the effects of Apitoxin on motility, biofilm formation and gene expression were reported [3]. There is an interesting paper by Brazilian researchers directed by Prof. Viana Pontual regarding the in vitro immune response of peripheral blood mononuclear cells infected with *Trypanosoma cruzi*, the etiological agent of Chagas disease [4], that uses a protein with activity and toxicity to *T. cruzi* from *Moringa oleifera*, a pantropical tree, that is more toxic to the parasite than to immune human cells. This issue also addresses the activity of insect compounds, specifically those of the assassin bug (*Rhynocoris iracundus*) against *Schistosoma mansoni* [5], the venom of *Rhynocoris iracundus*, which inhibits the motility of adult worms of *Schistosoma*, as well as their ability to produce eggs; thus, this venom contains potential antischistosomal compounds. In another extraordinary paper, Shcherbakova et al. studied the sensitization of fungi to 6-demethylmevinolin, a putative natural sensitizer of microbial origin that is able to suppress the biosynthesis of aflatoxin B1 [6], to avoid the frequent increase in fungicide resistance or the use of higher dosages of fungicides in the fight against the fungi cited below. By using this sensitizer, it was possible to show how the former effects colony growth as well as the conidial germination of several vegetal pathogens, such as some *Alternaria*, *Parastagonospora*, *Rhizoctonia*, and *Fusarium* species. These authors confirmed the possibility of using the sensitizer mentioned above instead of using triazole- and strobilurin-based fungicides. In addition, there is one more paper from Prof. Durant-Archibold's group regarding the antibacterial activity of volatile organic compounds produced by the octocoral-associated *Bacillus* sp. BO53 and *Pseudomonas* GA327, thus, revealing for the first time the

check for updates

Received: 6 February 2025
Accepted: 4 March 2025
Published: 6 March 2025

Citation: Franco, C.M.; Vázquez, B.I. Natural Compounds as Antimicrobial Agents—2nd Edition. *Antibiotics* **2025**, *14*, 268. https://doi.org/10.3390/antibiotics14030268

effects of volatile organic compounds produced by these agents associated with corals against important human pathogens, such as methicillin-resistant *S. aureus*, *Acinetobacter baumanni*, and *Pseudomonas aeruginosa*, which in turn are ESKAPE bacteria, due to their easy presentation of resistance. However, these bacteria have previously been shown to have antimicrobial properties [7]. The following paper is a very interesting one, dealing with L-carnitine effects used to grow rabbits, presenting the growth efficiency, hematological, biochemical, and carcass aspects. This work is unique from my point of view because of the nature of carnitine, which is naturally synthesized in the liver, kidneys, or brain, from lysine and methionine, and has shown interesting antimicrobial properties not known until now [8]. Zhang and coworkers studied the inhibition of *Clostridium perfringens* by *Polygonum hydropiper* compound extract, a known plant with various therapeutic properties that previously exhibited activity against other pathogens [9]. These authors also elucidated the effect on the intestinal inflammatory response by modulating NLRP3 inflammasome signaling with various flavonoids as key active compounds. The last published paper is related to the effect of fennel byproduct extracts applicable to meat products. This Italian study by Raffaele Marrone explored the use of fennel waste extracts with both antioxidant and antimicrobial properties [10] against several pathogens, such as *Salmonella*, *E. coli*, and *S. aureus*. They conclude that the cited extracts may be used to preserve the quality of beef burgers. In summary, this Special Issue contains eight research papers, two of which focus on several aspects of the fight against parasites, one of which focuses on the fight against fungi, four of which focus on the fight against bacteria, and one that focuses on the effects of carnitine on several productive parameters during rabbit growth.

Author Contributions: Both authors contributed equally to this editorial. All authors have read and agreed to the published version of the manuscript.

Funding: This research received no external funding.

Conflicts of Interest: The authors declare no conflict of interest.

References

1. Cushnie, T.P.T.; Lamb, A.J. Antimicrobial activity of flavonoids. *Int. J. Antimicrob. Agents* **2005**, *26*, 343–356. [CrossRef] [PubMed]
2. Rodriguez, C.; Ibáñez, R.; Rollins-Smith, L.A.; Gutiérrez, M.; Durant-Archibold, A.A. Antimicrobial Secretions of toads (*Anura, Bufonidae*): Bioactive Extracts and Isolated Compounds against Human Pathogens. *Antibiotics* **2020**, *9*, 843. [CrossRef] [PubMed]
3. Arteaga, V.; Lamas, A.; Regal, P.; Vázquez, B.; Miranda, J.M.; Cepeda, A.; Franco, C.M. Antimicrobial activity of apitoxin from *Apis mellifera* in *Salmonella enterica* strains isolated from poultry and its effects on motility, biofilm formation and gene expression. *Microb. Pathog.* **2019**, *137*, 103771. [CrossRef]
4. Molina, I.; Salvador, F.; Sánchez-Montalvá, A. Actualización en enfermedad de Chagas, Update Chagas disease. *Enfermedades Infecc. Y Microbiol. Clínica* **2016**, *34*, 132–138. [CrossRef]
5. Anderson, T.J.C.; Enabulele, E.E. *Schistosoma mansoni*. *Trends Parasitol.* **2021**, *37*, 176–177. [CrossRef] [PubMed]
6. Mikityuk, O.D.; Voinova, T.M.; Statsyuk, N.V.; Dzhavakhiya, V.G. Suppression of Sporulation, Pigmentation and Zearalenone production in *Fusarium culmorum* by 6-Demethylmevinolin, an Inhibitor of the Aflatoxin B1 Biosynthesis. *AIP Conf. Proc.* **2022**, *2390*, 030058.
7. Radjasa, O.K.; Wiese, J.; Sabdono, A.; Imhoff, J.F. Corals as source of bacteria with antimicrobial activity. *J. Coast. Dev.* **2008**, *11*, 121–130.
8. Olgun, A.; Kista, O.; Yildiran, S.T.; Tezcan, S.; Akman, S.; Erbil, M.K. Antimicrobial efficacy of L-carnitine. *Ann. Microbiol.* **2004**, *54*, 95–101.
9. Ayaz, M.; Junaid, M.; Ullah, F.; Sadiq, A.; Ovais, M.; Ahmad, W.; Ahmad, S.; Zeb, A. Chemical profiling, antimicrobial and insecticidal evaluations of *Polygonum hydropiper* L. *BMC Complement. Altern. Med.* **2016**, *16*, 502. [CrossRef] [PubMed]
10. Hamdy Roby, M.H.; Sarhan, M.A.; Selim, K.A.-H.; Khalel, K.I. Antioxidant and antimicrobial activities of essential oil and extracts of fennel (*Foeniculum vulgare* L.) and chamomile (*Matricaria chamomilla* L.). *Ind. Crops Prod.* **2013**, *44*, 437–445. [CrossRef]

antibiotics

MDPI

Review

Efficacy and Mechanisms of Flavonoids against the Emerging Opportunistic Nontuberculous Mycobacteria

Suresh Mickymaray [1,*], Faiz Abdulaziz Alfaiz [1] and Anand Paramasivam [2]

1 Department of Biology, College of Science, Al-Zulfi, Majmaah University, Majmaah 11952, Riyadh Region, Saudi Arabia; f.alfaiz@mu.edu.sa
2 Department of Basic Medical Sciences, College of Dentistry, Al-Zulfi, Majmaah University, Majmaah 11952, Riyadh Region, Saudi Arabia; anand.p@mu.edu.sa
* Correspondence: s.maray@mu.edu.sa

Received: 4 June 2020; Accepted: 21 July 2020; Published: 27 July 2020

Abstract: Nontuberculous mycobacteria (NTM) are the causative agent of severe chronic pulmonary diseases and is accountable for post-traumatic wound infections, lymphadenitis, endometritis, cutaneous, eye infections and disseminated diseases. These infections are extremely challenging to treat due to multidrug resistance, which encompasses the classical and existing antituberculosis agents. Hence, current studies are aimed to appraise the antimycobacterial activity of flavonoids against NTM, their capacity to synergize with pharmacological agents and their ability to block virulence. Flavonoids have potential antimycobacterial effects at minor quantities by themselves or in synergistic combinations. A cocktail of flavonoids used with existing antimycobacterial agents is a strategy to lessen side effects. The present review focuses on recent studies on naturally occurring flavonoids and their antimycobacterial effects, underlying mechanisms and synergistic effects in a cocktail with traditional agents.

Keywords: nontuberculous mycobacteria; flavonoids; synergistic action; underlying mechanisms

1. Introduction

Mycobacteria belong to Mycobacteriaceae and genus *Actinobacteria*, are slow-growing, immobile, Gram-neutral or weakly Gram-positive thin rod-shaped to filamentous bacteria and can be categorized into three key groups for the determination of diagnosis and therapy. (a) The complex of *Mycobacterium tuberculosis* is the primary causative pathogens of tuberculosis (TB) that consists of a group of organisms' viz., *M. tuberculosis*, *M. caprae*, *M. bovis*, *M. africanum*, *M. pinnipedii*, *M. microti*, *M. mungi*, *M. orygis*, *M.pinnipedii* and *M. surricatae* and *M. canetti*. (b) *M. leprae* and *M. lepromatosis* are the causative pathogens of leprosy. (c) Nontuberculous mycobacteria (NTM) are the additional opportunistic pathogenic mycobacterial complex groups that consists of *M. avium*, *M. marinum*, *M. hemophilum*, *M. kansasii*, *M. scrofulaceum*, *M. gordonae*, *M. abscessus*, *M. fortuitum* and *M. chelonae*. They do not cause TB; however, they can produce pulmonary infections, lymphadenitis, skin disease, endometritis and disseminated disease. Thus, NTM are denoted by other names such as environmental mycobacteria or mycobacteria other than tuberculosis (MOTT) and atypical mycobacteria (ATM) [1–4].

More than 200 different species of NTM have been identified in nature (https://www.bacterio.net/genus/mycobacterium), and among them; about 95% are environmental bacteria with maximum existence as saprophytes, opportunistic pathogens or nonpathogenic to humans and animals [5]. NTM are generally found in the environment, mostly in wet soil, rivers, streams, estuaries, marshland and hospital settings. They are less pathogenic when compared to tuberculous mycobacteria, however they can cause illness to immunocompromised or pulmonary infected individuals [6]. Among NTM pathogens, *M. avium* complex are the most significant and recurrent pathogenic organisms that causes pulmonary and extrapulmonary infections. In addition, M. xenopi, M. kansasii, M. malmoense are the most causative agents for pulmonary infections. Skin and cutaneous tissue infections are also caused by *M. ulcerans* and M. marinum [7]. *M. abscessus*, *M. fortuitum*, *M. chelonae*, *M. chimaera* are the infectious agents accountable for most soft tissue infections [8].

According to the Runyon classification (Figure 1), mycobacteria have broad categories based on phenotypic factors including pigmentation and the frequency of bacterial growth [9]. They are classified as rapidly growing mycobacteria-RGM (visible colonies appear within seven days) and slow-growing mycobacteria-SGM (visible colonies appear after seven days). Most pathogenic mycobacteria are associated with the SGM, due to their virulence and growth rate. The members of the *M. chelonae–M. abscessus* complex and *M. fortuitum* complex are classified under the RGM family (Figure 1). The classification of mycobacteria remains greatly active and is continually developing, owing to the available technological progressions including sequencing of bacterial isolates. However, this improvement provides only taxonomy of the evolving novel mycobacteria, and still, their documentation remains uncertain and is obligatory to find the potential phenotypic and genetic polymorphisms of the *M. abscessus* complex.

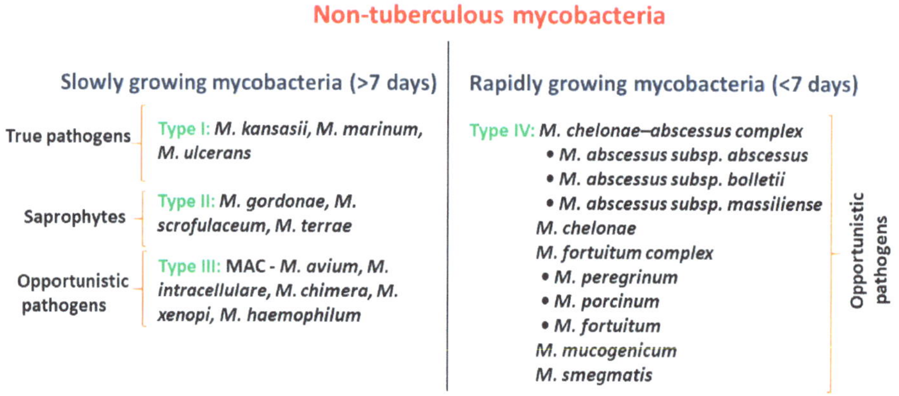

Non-tuberculous mycobacteria
Slowly growing mycobacteria (>7 days)
Rapidly growing mycobacteria (<7 days)
True pathogens
Type I: M. kansasii, M. marinum, M. ulcerans
Saprophytes
Type II: M. gordonae, M. scrofulaceum, M. terrae
Opportunistic pathogens
Type III: MAC - M. avium, M. intracellulare, M. chimera, M. xenopi, M. haemophilum
Type IV: M. chelonae–abscessus complex
• M. abscessus subsp. abscessus
• M. abscessus subsp. bolletii
• M. abscessus subsp. massiliense
M. chelonae
M. fortuitum complex
• M. peregrinum
• M. porcinum
• M. fortuitum
M. mucogenicum
M. smegmatis
Opportunistic pathogens

Figure 1. Classification of nontuberculous mycobacteria.

RGM, *M. chelonei*, *M. fortuitum* and *M. abscessus* complex are well-renowned pathogens that often occur in cutaneous infections related to plastic surgery and cosmetic techniques. They appear widely in different pathologic conditions viz., cellulitis, superficial lymphadenitis, chronic nodular lesions, abscesses, nonhealing ulcers, verrucous lesions and commonly occur in the subcutaneous tissue and skin [10].

M. abscessus is often misidentified as *M. chelonae*. It is documented that *M. chelonae* is seldom accountable for lung disease [11]. In addition, *M. chelonae* fails to develop in the culture at 37 °C when compared to *M. abscessus*. *M. chelonae* is abundant in aquatic systems that can cause infection in immunocompromised hosts [12]. Hence, this inappropriate identification of *M. abscessus* is highly possible in several pilot trials specifically in pulmonary contagions, consequently flouting the significance of this mycobacterium. Notably, the augmented occurrence of *M. abscessus* in the individual with cystic fibrosis directs that this pathogenic organism has developed progressively to become widespread in the past decade [13,14]. The cultures of *M. abscessus* grow in less than seven days using agar medium (the combination of Bactec 12B and Middlebrook 7H10/7H11) and the strains of *M. chelonae* can be cultivated at 30 °C. Most of the NTM species can grow in the RGM culture medium at 30 °C, and *M. xenopi* can grow in the Lowenstein–Jensen (LJ) medium at 36 °C [15].

The RGM organism *M. abscessus* possesses a high level of heterogeneity in the genotype and is capable of rapid evolution by phage mediated gene transfer [16,17]. There are three subtypes in the complex of *M. abscessus*, namely, *M. abscessus*, *M. bolletii* and *M. massiliense* [5]. *M. abscessus* possesses diverse structures in the cell wall due to the occurrence or absence of glycopeptidolipids (GPL) [18]. Similarly, other NTM species have also shown structural variations. The colony morphology and GPL arrangements in *M. abscessus* are normally responsible for interactions with the host and regulating the environment of biofilm development and intracellular survival, which results in disease manifestations and clinical outcomes [19]. The most common point of entry of NTM into the host occurs via direct invasion including trauma, iatrogenic acquisition or postsurgical infections [20]. These bacteria can invade soft tissues and skin in immunodeficient patients during systemic dissemination [21,22]. Shreds of evidence show that the possible human transmission of *M. abscessus* subsp. *massiliense* may occur among cystic fibrosis patients [23,24]. To date, few publications have addressed novel approaches to deal with extensive antimicrobial resistance among the NTM organisms, and thus, the current review aims to appraise the antimycobacterial activity of flavonoids against NTM, its capacity to synergize with existing pharmacological agents and its antivirulence effects.

2. Clinical Epidemiology of NTM

The diseases of NTM are often found in developed nations, where the peak occurrence rates was 10.6 cases per 100,000 individuals in 2000 [25]. Based on pulmonary research by various experts, the respiratory NTM are projected to be at least 15 times more common than TB with at least 200,000 cases per year in the USA [25]. In South Korea, the occurrence of NTM infections have been augmented to 39.6 cases/100,000 people in 2016 and yearly occurrence could be 19.0 cases/100,000 people. An investigation led in Germany described a growing incidence of NTM in 2009 from 2.3 cases/100,000 people to 3.3 cases/100,000 populace in 2014 [26]. Shreds of evidence associated with the occurrence of the disease of NTM and elevation levels are greater in Europe [26], the United States [27–29] and Japan [30]. The higher rates of NTM infection have been reported in East Asian inhabitants particularly China, Vietnam, Hawaii, Philippines, Japan and Korea [27,28]. The individuals with NTM in Japan and the Philippines were at higher risk for *M. abscessus* infection whereas Vietnam and Korean patients were often affected by *M. fortuitum* group infection [27]. *M. avium* complex (MAC) and RGM including *M. abscessus* and *M. chelonae* have been attributed to 85% of pulmonary cases in the United States [31]. Pulmonary diseases are strongly associated with advanced age and more often in women than men [10].

The NTM diseases are generally caused by *M. abscessus*, *M. fortuitum*, MAC and *M. chelonae*. Among them, *M. abscessus* is often found with rising frequency and is most challenging to treat [32]. The swiftly increasing NTMs are normally associated with catheter infections, post-cosmetic surgery of the soft tissue and skin and pulmonary infections [28]. The clinical implications and location of infection of NTM are listed in Table 1. Several investigations have established that the incidence of NTM diseases are greatly escalating in numerous clinical conditions [21,33–35]. The clinical range of the infections is highly connected based on the entry to the host and host susceptibility factors and these infections are multisystem and multigenic-based diseases [21,34]. Disseminated NTM infections typically impact severely immunocompromised patients with primary immunodeficiencies, via inherited or acquired deficiency of the IL-12-IFN-γ pathway, HIV/AIDS, transplant-linked immunosuppression and anti-TNF-α receptor blockers treatment [34,36].

Table 1. Clinical significance and site of infection of nontuberculous mycobacteria (NTM).

List of NTM Species	Clinical Relevance and Possible Site of Infection	Reference
M. abscessus	Peripheral blood, peritoneal biopsy, pulmonary and permanent catheter tip.	
M. asiaticum	Pulmonary	
M. avium	Pulmonary	
M. celatum	Pulmonary	
M. chelonae	Breast abscesses, blood and peritoneal fluid, pleural fluid	
M. flavescens	Pulmonary	
M. fortuitum	Ascetic fluid, peritoneal dialysis fluid, pulmonary, lipoid pneumonia, mediastinal infection, a myocardial and abdominal abscess.	[2,3,37–45]
M. gastri	Pulmonary	
M. gordonae	Urinary tract and rarely liver biopsies	
M. intracellulare	Pulmonary and extrapulmonary	
M. kansasii	Appendiceal abscess	
M. lentiflavum	Extrapulmonary	
M. marinum	Wound-elbow and nasal cavity	
M. riyadhense	Pulmonary infection, sclerotic lesions, maxillary sinus, dural lesion	
M. scrofulaceum	Extrapulmonary	
M. simiae	Pulmonary	
M. smegmatis	Pulmonary	
M. szulgai	Joints/synovial aspiration	
M. terrae	Pulmonary	
M. xenopi	Pulmonary	

3. Challenges in Diagnosing and Treatment of NTM Diseases

RGM are usually isolated from blood, sputum or tissues for diagnosis and are often misidentified as diphtheroids. RGM species normally cultivate as routine culture in liquid broth blood culture medium or on solid agars that can grow quickly within seven days. These strains relatively stain with Gram stain not with Ziehl–Neelsen stain to demonstrate the acid-fast characteristics. A fresh young culture of RGM may not constantly show branching or beaded structures and exhibit weakly Gram-positive bacilli, thus misleading the diagnosis and often incorrectly concluded as diphtheroids [46]. NTM in tissue specimens can also be identified based on the molecular method of determination, which includes, 16S rRNA gene sequencing, PCR analysis and HPLC. The diagnosis of NTM often fails to recognize the species and subspecies of the different samples from the affected individual. Most NTM microscopically appears similar to *Mycobacterium tuberculosis* (MTB), and the colony morphology varies in culture. The culture difference and microscopic appearance are shown in Figure 2. A total of 16S ribosomal RNA sequencing aids in individual NTM species identification [20]. Diagnosis is generally completed by recurrent isolation accompanied by certain clinical and radiological features. There is no explicit treatment of NTM infections and therapy depends upon the particular species and its resistance to antibiotics [47].

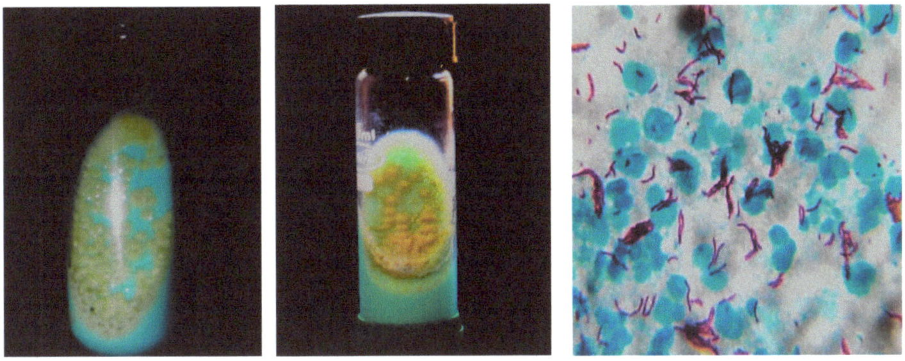

(a) (b) (c)

Figure 2. NTM and *Mycobacterium tuberculosis* (MTB) culture and microscopy. (**a**) NTM grown at 48 h of incubation in LJmedia with typical characteristics of moist, smooth glistening yellow colonies; (**b**) MTB grown at six weeks of incubation in LJ media with typical characteristics of rough, buff yellow-colored cauliflower-like colonies; (**c**) Long and slender pink-colored acid-fast tuberculous mycobacteria by Ziehl–Neelsen stain (100×). The above culture images differentiate the NTM and MTB with almost similar microscopical image.

The diagnosis of NTM are difficult to confirm using acid-fast microscopy, which is the primary diagnostic tool for TB in numerous developing nations. As an outcome, most cases of NTM causing pulmonary infections are not recognized and eventually treated with traditional anti-TB medications. These treatments often fail because NTM are mostly resistant to anti-TB therapy [48]. Hence, in developed nations, caseloads of 8.6/100,000 total population and 20.4/100,000 population over 50 years old are typical [49]. In developing nations, the occurrence rate and diagnosis of NTM cannot be observed due to the lack of laboratory arrangement and identification of mycobacteria. Hence, the escalating rate of pathogenic NTM in developing nations has been greater particularly with the advent of HIV/AIDS patients. Normally, HIV/AIDS individuals with severe immunosuppression are at high risk of NTM infections, which often cause localized or disseminated infections [50]. In addition, the failure of NTM treatment can frequently occur due to resistance to some of the available antibiotics (Table 2).

In addition, using these chemical agents produce various complications including, diarrhea, headache, renal failure and colitis. Mycobacteriosis is an acute/chronic, systemic, granulomatous disease caused by NTM, which is extremely challenging in selecting effective antimicrobial therapy based on the antimicrobial resistance [53]. The RGM involves individualized treatment according to the outcomes found in vitro vulnerability tests for cefoxitin, amikacin, clarithromycin, sulfamethoxazole, ciprofloxacin, imipenem and doxycycline [54]. *The M. fortuitum* and *M. chelonae are* members of *M. abscessus* complex and *M. massiliense*, *M. abscessus* and *M. bolletii* are subspecies, which are the chief NTM related to cutaneous tissue involvement [55]. All these mycobacteria are regularly found with several skin lesions, however *M. fortuitum* is often found in a sole lesion [33]. The susceptibility to antimicrobials generally depends upon the individual species. *M. abscessus* complex is likely to be vulnerable to the cocktail of amikacin, azithromycin, imipenem and cefoxitin, since, it is known that clarithromycin resistance due to the occurrence of the *erm41* gene [56].

Table 2. Various treatment recommendations for NTM [51,52].

Mycobacterium Species	Established Regimens	Additional or Suggested Agents
M. avium complex	rifampin, ethambutol, isoniazid, streptomycin or amikacin	clarithromycin (azithromycin), ciprofloxacin, clofazimine
M. scrofulaceum	-	clarithromycin (azithromycin), ciprofloxacin, clofazimine
M. kansasii	rifampin, ethambutol, isoniazid	streptomycin, ciprofloxacin, clarithromycin
M. marinum	rifampin, ethambutol, doxycycline or trimethoprim-sulfamethoxazole	streptomycin, ciprofloxacin
M. xenopi	rifampin, ethambutol, isoniazid	streptomycin
M. malmoense	-	clarithromycin (azithromycin), ciprofloxacin, clofazimine
M. simiae	-	clarithromycin (azithromycin), ciprofloxacin, clofazimine
M. szulgai	-	streptomycin, ciprofloxacin, clarithromycin
M. hemophilum	-	rifampin, cefoxitin, doxycycline, trimethoprim-sulfamethoxazole
M. fortuitum	amikacin, ciprofloxacin, sulfonamides	clofazimine, cefoxitin, imipenem, a cocktail of azithromycin or clarithromycin, doxycycline, fluoroquinolones, trimethoprim-sulfamethoxazole
M. abscessus	amikacin, streptomycin, cefoxitin	clofazimine, clarithromycin, a cocktail of azithromycin, imipenem, clarithromycin,
M. chelonae	tobramycin, amikacin	clofazimine, clarithromycin, doxycycline, a cocktail of azithromycin, imipenem, cefoxitin, clarithromycin, fluoroquinolones

In vivo study demonstrates that NTM isolates show resistance to azithromycin or clarithromycin [56,57]. Azithromycin is normally the desired antibiotic for *M. abscessus* infections, while azithromycin or clarithromycin is highly efficient in the cases of *M. massiliense* [56,57]. *M. fortuitum*, *M. abscessus* and *M. chelonae* are resistant to all of the existing anti-TB agents [10,56,57]. *M. fortuitum* is highly susceptible to amikacin, trimethoprim-sulfamethoxazole, azithromycin or clarithromycin, fluoroquinolones and doxycycline. *M. chelonae* is also often susceptible to azithromycin or clarithromycin, tobramycin, fluoroquinolones and cefoxitin [55]. The guideline of the therapy recommends performing susceptibility testing of NTM to enhance the option of a cocktail of the antimycobacterial drug relates clinically in vivo trials to antimicrobial treatment for various species of NTM. From the microbiologic perspective, heterogeneity of NTM needs sophisticated and rapid laboratory techniques. Since the present pharmacological treatment of NTM diseases are tricky, and often fails to scope the long-term removal of pathogens. Moreover, it is obligatory to hunt novel agents or treatment and dosage regimens for effective treatment of these NTM diseases, specifically serious in immunocompromised individuals. Hence, it is necessary to find alternative remedial regimens. One of the alternative resources is traditional medicinal plants or their derivatives, which are well-known for their therapeutic properties. Most of the researchers have a positive approach toward natural products due to their natural origin and low noxious with fewer side effects [3,58–66]. A trial of anti-Mycobacterial effects of these medicinal plants, particularly those that are conventionally used for pulmonary infections is significant.

Natural products as a source of medicine are potentially valuable due to their natural origin and low toxicity with lesser side effects. Medicinal herbs with the traditional practice of crude extracts or active principles have been widely used for treating and averting human illnesses for many centuries. These ethnopharmacological techniques have been reinforced to yield bioactive compounds that support to improve modern medicine as beneficial tools [67–70]. Bioactive compounds often contribute a noteworthy function in drug finding by helping as a novel drug of interest and templates for synthetic agents [71–73]. Copious investigations have established that natural bioactive compounds have possible antimycobacterial activities [2,60,74,75]. The single-handed practice of bioactive compounds or cocktails with classical antibiotics signifies a greater alternative treatment. Additionally, the cocktails of those antimicrobial agents often require only a minor amount. Therefore, this smaller amount may provide less toxicity to the host, ensuring great lenience to the antibacterial drugs. Grounded on the existing information, there has been inadequate literature regarding antimycobacterial phytocompounds [76–79].

Thus, the present review aims to emphasize the antimycobacterial effects of flavonoids and their underlying mechanisms.

The literature of flavonoids and antimycobacterial effects were obtained in electronic search using Google Scholar, Science Direct and PubMed The following keywords were used in the Title/Abstract/Keywords: "flavonoids" and "antimycobacterial" or "Nontuberculous mycobacteria" or "*M. fortuitum* or *M. abscessus* or *M. chelonae*," and checking all available findings of clinical, in vivo and in vitro connection among flavonoids and their antimycobacterial effects. The underlying antimycobacterial mechanism was composed and organized in a suitable place.

4. Flavonoids

Most commonly the flavonoids are the secondary metabolites of the plant kingdom with well-known wide-ranging classes of polyphenols. They normally exist in all kinds of vegetables, fruits and beverages [80–84]. WHO estimated that 25% of existing drugs are derived from plants used in folk medicine [85,86]. Besides the long-established clinical use, the plant-derived compounds display good tolerance and acceptance among patients and seem like a credible source of antimicrobial compounds. Among 109 new antibacterial drugs, approved in the period 1981–2006, 69% originated from natural products [87]. One of the major groups of phytochemicals that has been studied extensively for their antimicrobial properties are flavonoids [66,88]. Flavonoids are organized with the structure of two phenyl rings fixed with the heterocyclic ring as C6-C3-C6 and arranged up to a skeleton of 15-carbon. They are classified into many subclasses based on variation in the central carbon ring viz., flavanones, flavonols, flavones, flavan, isoflavones and anthocyanidins [89]. There has been accumulating scientific interest in the study range of flavonoids that demonstrate the following pharmacological functions: antioxidant [90,91], antidiabetic and anti-obesity [92,93], hypolipidemic [94], anti-inflammatory [95], antimicrobial [96–98], anticancer [99–101], anti-aging [102], antiallergic and antithrombotic [103], hepatoprotective [104–107], cardioprotective [108], neuroprotective [109], nephroprotective [110], protect from lung injury [111] and improving endothelial function, adjourning age-related cognitive and neurodegenerative diseases [112,113]. The evidence has validated that the prolonged consumption of dietary flavonoids at higher quantity has also produced minor side effects, which may arise due to the shortage of bioavailability and gut permeability as well as the greater metabolic rate [114]. Moreover, the intake of flavonoids produces a poor absorption coefficient, which may cause only minor toxicity to animals and humans [115,116]. All of these data support investigations to discover and inspect the attractive healing indices of Flavonoids concerning human wellbeing. The daily intake of dietary flavonoids is estimated to be about 1–2.5 g; flavonols and flavones have been found to be 23 mg [114,117]. Hence, regular intake of flavonoids could be favorable in preventing or treating various illnesses and improving health outcomes.

5. Anti-Nontuberculous Mycobacterial Efficacy and Mechanisms

Flavonoids have been used in the treatment of the wide spectrum of human illnesses since time immemorial [118–120]. Flavonoids may inhibit NTM growth with various underlying mechanisms, including inhibiting cell wall formation, biofilm formation, bacterial DNA synthesis and efflux mediated pumping systems. In addition, the mixture of flavonoids with antimycobacterial agents may be a greater approach to combat mycobacterial infections and microbial resistance.

5.1. Inhibition of Cell Wall Formation

Flavonoids inhibit bacterial growth, microbial adhesions and cell wall or transport proteins [121]. Some anti-NTM drugs normally damage the cell membrane's integrity that leads to the leakage of intracellular components, which leads to alterations in membrane permeability. Flavonoids can also damage the cell wall of bacteria [121]. Body cells and tissues are continuously threatened by the injury caused by free radicals and reactive oxygen species (ROS) which are produced during normal oxygen metabolism or are induced by exogenous damage [122]. Eventually, these excess ROS can produce unadorned oxidative stress to the bacterial cell membrane leads to increased permeability, nucleic acid damage and oxidation of protein and fatty acids in the membrane (Figure 3) [123–125]. Unfortunately, these free radicals can attract various inflammatory mediators in the host, contributing to a general inflammatory response and host tissue damage. These elevated ROS species cause depletion of the endogenous scavenging compounds and reduced the levels of antioxidant equilibrium. Flavonoids may have an additive effect on the endogenous scavenging compounds and abolish the effect of the free radical causing inflammatory response and combat to regulate antioxidant levels in the host [126]. Flavonoids are measured as effective ROS scavengers however, the level of flavonoid in human plasma and most tissues is too little to effectively reduce ROS [127]. Moreover, flavonoid as ROS scavenger usage should be carefully measured, since low levels of ROS are, on the contrary, beneficial for bacteria and can persuade resistance. Therefore, the function of flavonoids as an antimicrobial potentiator should rather be related to the regulation of the activities of different proteins and molecular processes, and there is a need for further investigations, specifically regarding their synergistic action.

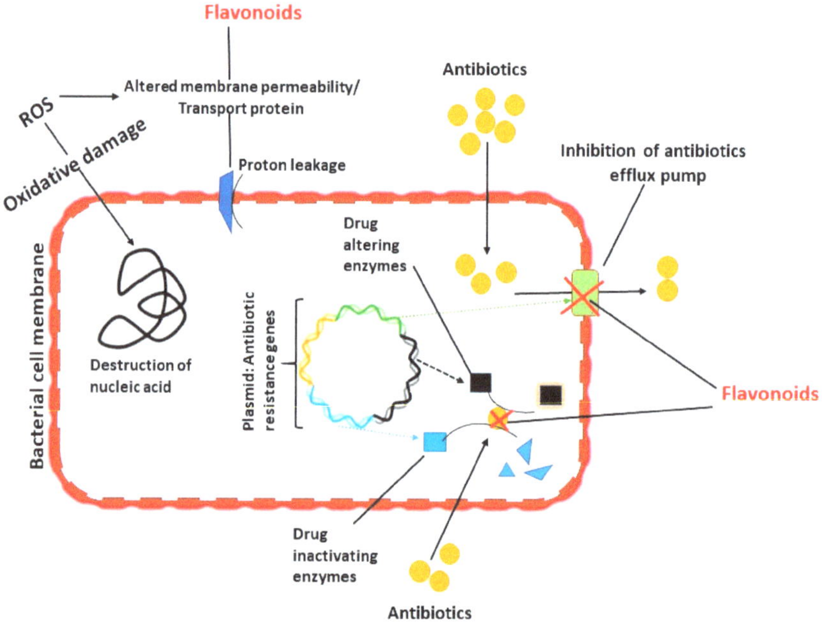

Figure 3. Mechanism of antimycobacterial activity of flavonoids.

Fathima and Rao [128] described that the flavonoid catechin plays a bactericidal action through the oxidative burst and generation of ROS that causes a change in the membrane permeability and membrane injury. Similarly, liposome studies also confirmed membrane disruption during oxidative stress which occurs only at high concentrations of epigallocatechin gallate [129]. Quercetin from propolis (natural resinous mixture produced by honey bees, that have potential antimicrobial applications: upper respiratory tract infections, common cold, wound healing, treatment of burns, acne, herpes simplex and genitalis and neurodermatitis) causes a decrease of proton-motive force and increased membrane permeability in the bacterium which has been employed by the synergistic activity of quercetin with antibiotics, including ampicillin and tetracycline [130,131]. Additionally, flavones- acacetin and apigenin, as well as flavonols morin and rhamnetin caused destabilization of the membrane structure by disordering and disorientation of the membrane lipids and induced leakage from the vesicle [132]. Lipid peroxidation has been shown to destroy the bacterial cell wall and alter membrane potential, ensuing augmented permeability, decreased fluidity and disruption of phospholipids [76]. The connection between the lipid bilayer and production of ROS is often linked in the malondialdehyde production that is the key marker of lipid peroxidation. This lipid peroxidation is not only harmful to the bacterial lipid bilayer, but also affects the host cell membrane. Ethyl acetate leaves extract of *Aegle tamilnadensis* and *Schkuhria pinnata* and their active principles of flavonoids have exerted antioxidant and antimycobacterial activity against *M. smegmatis* with MIC range of 0.01 to 2.50 mg/mL [76,133]. Four well-known testing systems were carried out in this study to assess the antioxidant potential viz., lipid peroxidation inhibition, nitric oxide radical inhibition, ferric thiocyanate and ABTS radical scavenging assay. Based on the findings, ethyl acetate extract demonstrated a noteworthy antioxidant activity and significant antimycobacterial activity [76,133].

Further several research groups have investigated either isolated or identified the structure of flavonoids that possess antibacterial activity and quantified the activity of commercially available flavonoids. For instances, flavonoids such as apigenin [134], galangin [135], pinocembrin [136], ponciretin [137], genkwanin [138], sophoraflavanone G [139], naringin and naringenin [140,141], epigallocatechin gallate and its derivatives [129], luteolin and luteolin 7-glucoside [142–145], quercetin [130,131], 3-*O*-methylquercetin and various quercetin glycosides and kaempferol and its derivatives [85,86,146]. Other flavones [147], flavone glycosides [148], isoflavones [149], flavanones [150], isoflavanones [146], isoflavans [151], flavonols [152], flavonol glycosides and chalcones [152] have potential antibacterial activities.

Heritiera littoralis Dryand mangrove flora produces novel flavonoids; tribuloside, afzelin, and astilbin that were revealed to possess antimycobacterial activity against the various species of NTM with a minimum inhibitory concentration (MIC) of 5.0 mg/mL. All these flavonoids exhibited growth inhibition of NTM while co-administered with standard anti-TB drugs [153]. 2,3,4-trihydroxy-5-methylacetophenone obtained from palmyra palm (*Borassus flabellifer* Linn.) showed potential antimycobacterial activity against *M. smegmatis* with MIC of 10.0 μg/mL [154]. Another study in 2014 showed that total flavonoid contents obtained from fourteen edible plants possess a potent antioxidant (IC_{50} values of DPPH: 8.15 μg/mL; ABTS: 9.16 μg/mL and TEAC: 0.75), antimycobacterial (*M. smegmatis* and *M. fortuitum*: MIC value of 78 μg/mL) and the cytotoxic activities (LC_{50} values stretching from 33 to 102 μg/mL) [155]. Lipophilic flavonoids which are highly hydroxylated can be more disruptive for membrane structure [156,157]. Hence, it is worth observing that the flavonoids decrease the bacterial toxin secretion by damaging the membrane [158,159].

Amikacin is a semi-synthetic aminoglycoside extensively used to treat disease caused by NTM and gentamicin resistant Gram-negative bacterium. Conversely, the clinical use of drugs regularly causes ototoxicity due to the generation of ROS. A natural flavonoid, galangin pretreatment demonstrated to provide defensive functions against amikacin-provoked mitochondrial dysfunction by decreasing ROS generation [160]. The antioxidant properties of quercetin-3-*O*-β-D-glucoside prevent the formation of biofilm and encourage membrane disturbances, ensuing shrinkage of size and outflow of intracellular constituents of *M. smegmatis* [161]. In addition, quercetin accelerates the inhibition of mycobacterial glutamine synthetase. Glutamine synthetase is the key enzyme involved in virulence factors, as well as pathogenesis that had been recognized as a possible antibiotic target [162,163]. This enzyme is normally found in the outer membrane of pathogenic mycobacteria that crucially involves in the synthesis of poly-L-glutamate–glutamine. quercetin plays a key function in regulating the cellular levels of NH_3 in the infected host and eliminate the pathogen through phagosome acidification and phagosome-lysosome fusion [161].

Fatty acid synthase II (FAS-II) is a key enzyme, requires endogenous fatty acid synthesis in the bacterial membrane, represents a possible target for novel antimycobacterial agents [164]. FAS-I is accountable for de novo fatty acid (FA) synthesis to form FA chain elongation (16–24 carbons) and then lengthened by the FAS-II monofunctional enzymes to yield long-chain fatty acids (36–48 carbons) and mycolic acids. Mutation of monofunctional enzymes often provides drug resistance to the mycobacteria [165]. Flavonoids such as isoliquiritigenin, butein, fisetin and 2,2′,4′-trihydroxychalcone prevent the growth of *M. smegmatis* by targeting the dehydratase enzyme of

FAS-II [164]. D-alanine-d-alanine ligase is an enzyme involved in cell wall synthesis. Another study has also confirmed that quercetin and apigenin (4′,5,7-trihydroxyflavone) inhibit ATP binding pocket of D-alanine-D-alanine ligase and prevent bacterial peptidoglycan synthesis [166].

5.2. Inhibition of Biofilm Formation

The biofilm formation is normally associated with virulence, pathogenicity, resistance to antibacterial substances and survival in the environment [167]. Antibacterial resistance of biofilm-developing mycobacteria may cause the failure of the treatment, and biofilms must be materially exterminated to resolve the infection. The formation of biofilms provides relationships among microbial populations with a high spectrum of colonization and functional activities. They form on many surfaces including, human tissue, medical equipment, plumbing pipes and drinking water systems [168]. In hospital wards, the development of biofilms on ventilators and hospital apparatus that permits pathogens to continue as pools which may freely spread to patients. After invading into the host, these biofilms let pathogens disrupt the host immune systems and can persist for a long-time [169]. Studies have also supported that tap water functions as a primary source for human colonization and/or infection outbreak of NTM [169,170]. The developed biofilms often contain *M. fortuitum*, which produce biofilm-dispersing agents such as biosurfactant. Moreover, *M. chelonae* and *M. fortuitum* developed thick biofilms with asymmetrical forms that were comparatively resistant to available antibiotics even at 10× MIC [169].

The hydrophobicity and metal resistance of mycobacteria often permits adhesion of cells and the successive development of biofilms on aquatic surface later. In addition, NTM in tap water are normally able to survive and are often resistant to the chemicals glutaraldehyde and chlorine [169,170]. The proliferation of these NTM from standing biofilms that can aid the spread of infections to individuals, demonstrates a noteworthy health risk in hospital environments [171]. Novel approaches with potential antibiofilm agents that improve treatment efficacy must be developed which is urgently necessary for the suitable therapy of NTM infected patients.

Flavonoids are well recognized as anti-NTM agents and prevent biofilm developments. Research in this area has generated interest in the ability of flavonoids to enhance the outcomes of untreatable infections, especially on antibiotic-resistant bacteria like NTM. Several researchers have confirmed that the structure-relationship of flavonoids enhances the bactericidal actions and demonstrated as antibacterial agents [141,146,172–174]. The anti-NTM activity and inhibition of biofilm effects of flavones and flavanones are usually based on the hydrophobic compounds on one aromatic ring and a hydrogen-bonding group on another aromatic ring [175]. These biofilm developments can be inhibited by the hydrophobic substituents of flavonoids, which comprises various heterocyclic moieties including, alkyl, prenyl, nitrogen or oxygen-containing heterocyclic and alkylamino chains [141, 172]. This structural activation of flavonoids can directly kill the bacteria in the biofilm formation, synergistically activate with the antibiotics and weaken the bacterial pathogenic effects [141,172]. Few recent studies showed a series of flavonoid derivatives significantly exhibited their antimycobacterial activity against various NTM species through inhibition of biofilm formation [176,177]. Apigenin normally has a cyclic or aliphatic chain at the 8-C position that enhanced the antimycobacterial activities and prevents biofilm formation [178]. Few supporting studies demonstrated that C-benzylated dihydrochalcone and the dihydrochalcone dimer have shown significant antibacterial activity against *M. chelonae* and *M. fortuitum* [179]. An active flavanone compound, Platyisoflavanone obtained from *Platycelphium voense* revealed antimycobacterial activity using microplate alamar blue assay against *M. chelonae* with MIC of 23.7 mmol/L [180].

Another study demonstrates that synergistic combinations of amikacin and curcumin (compound isolated from *Curcuma longa*), employs antimycobacterial activity against *M. abscessus* clinical strain with MIC of 128 mg/L. Furthermore, curcumin induced an over-all decrease in microbial masses in the biofilm and considerable loss in cell viability [123]. Two methoxylated flavonoids, flavonoid 7-methylquercetagetin and 7-methylquercetagetin-4′-*O*-β-D-glucopyranoside were extracted from

Paepalanthus latipes which showed significant antimycobacterial activity against NTM species with MIC ranged from 1–2 mg/L [181].

5.3. Inhibition of Efflux Mediated Pumping System

Efflux pumps are well-recognized proteins and protein complexes that provide antibiotic resistance in bacteria, including mycobacteria [182]. Hence, the finding of efflux pump inhibitors is a fascinating target in antimycobacterial treatment. Plant-derived natural bioactive compounds are potent inhibitors of an efflux pump that may capable adjunct to traditional chemotherapy by improving mycobacterial vulnerability to antibiotics. Flavonoids exert noteworthy antimycobacterial activities and exhibited considerable outcomes as antimycobacterial agents [183]. A study showed that the inhibition of the efflux pump has been performed using flavonoid, pinocembrin isolated from *Alpinia katsumadai*, which showed antimycobacterial activities against *M. smegmatis* using MIC: 64 mg/L, further the antimycobacterial activity was synergistically significant in combination with rifampicin [184]. Similarly, the isoflavone biochanin A exhibited significant efflux pump inhibiting activity against *M. smegmatis* that has evoked much attention as promising novel targets in antimycobacterial treatment [144].

A recent study showed that two polymethoxyflavones, Skullcapflavone II (5,2′-dihydroxy-6,7,8,6′-tetramethoxyflavone) and Nobiletin (5,6,7,8,3′,4′-hexamethoxyflavone) exerted as effective antimycobacterial activity and antibiotic resistance modulating activities against *M. smegmatis* [185]. In this study, the efflux inhibitory activity was studied using an ethidium bromide-based fluorometric assay. Conversely, an association between potent modulatory and putative efflux activity of the skullcapflavone II and Nobiletin was not described in this study. However, the outcome has highly emphasized that two polymethoxyflavones are valuable adjuvants in anti-mycobacterial treatments [185]. Nine novel paradol- and gingerol-related compounds known as putative efflux pump inhibitors extracted from *Aframomum melegueta* seeds, which were also possessed significant antimycobacterial activities against *M. smegmatis* [186]. Three novel phenylpropanoids (1′-*S*-1′-acetoxychavicol acetate, trans-p-coumaryl diacetate and 1′-*S*-1′-acetoxyeugenol acetate) isolated from the rhizome of *Alpinia galanga* showed that effective antimycobacterial activity and antibiotic resistance modulating activities against the isolates of *M. smegmatis* with MIC value of 2.5, 6.25 and 5.0 mg/L [187].

Similarly, the function of efflux pumps in clarithromycin resistance with nine clinical isolates of *M. abscessus* subsp. *abscessus* or *bolletii* complex was studied. Based on the findings, the team has highlighted the requirement for additional investigation on *M. abscessus* efflux response to implement more efficient alternative antimicrobial beneficial regimens and direction in the improvement of novel drugs against mycobacterium [77]. In search of efflux pump inhibitors, flavonoids are a promising therapy for potent antimycobacterial activity and antibiotic resistance modulating activities (Figure 3).

5.4. Inhibition of Bacterial DNA Synthesis

Flavonoids are well-known topoisomerases inhibitors, contributes to antimycobacterial activity. DNA topoisomerase is a key enzyme for DNA replication that contribute to a central target for antimycobacterial agents [78]. Earlier, in silico analysis study has confirmed that quercetin is a significant DNA topoisomerase inhibitor at B subunit of the enzyme and prevents the growth of *M. smegmatis* [188]. This statement was further established using different DNA topoisomerase subunits that also showed quercetin binding to the B subunit of topoisomerase and parallel obstruction of ATP binding pocket by the development of H-bonds in the amino acid residues of DNA topoisomerase [78]. Previously, several molecular docking studies suggested that quercetin inhibits DNA topoisomerase and DNA supercoiling, which competitively interacts with the ATP binding site in the B subunit of DNA topoisomerase [78,189,190]. Finally, quercetin binds with DNA that alleviates the DNA topoisomerase complex leads to the breakdown of bacterial DNA [189]. The binding of flavonoids with DNA topoisomerase usually favored by the active groups positioned in the flavonoids viz., 4-carbonyl, 3-hydroxyl, 5-hydroxyl and 7-hydroxyl groups [78,189].

5.5. Synergistic Action of Flavonoids with Antimycobacterial Agents

This synergistic effect of flavonoids with conventional agents is often effective and beneficial for both the proportion and degree of bacterial destructions and microbial resistance modulating activities [191]. The available conventional agents have a spectrum of underlying modes of action, and the combination of two or more agents can contribute diverse targets, ensuing multi-targeting. The implementation of the multi-targeting policy usually eases drug resistance [192]. These synergistic approaches largely evade toxicity and intolerance of the drug [79]. Previously, various in vitro investigations have been studied and reduce the minimum inhibitory concentration of bioactive compounds with conventional antimycobacterial agents (Table 3) [123–125,144,153,181,193,194].

Several studies have demonstrated that the bactericidal antibiotics such as β-lactams, aminoglycosides, and fluoroquinolones induced oxidative stress, regardless of their specific targets, and involved in the ROS-antibiotic bacteria-killing [195,196]. Conversely, other reports failed to indicate the connection between ROS and antibiotic-mediated killing [197]. These varying data may have resulted from the generation of ROS, which is produced through the hyperactivation of normal cell metabolism, as well as the related difficulty or even the impossibility to completely separate the effects of reduced levels of ROS and ROS production as a consequence of the action of antibiotics [195–197]. Flavonoids are synergistic potentiators with conventional agents in improving the antibiotic efficiency against NTM [194]. Flavonoids generally protect the cells from the harmful effects of ROS generation [198,199]. Markedly, Brynildsen et al. [200] suggested that to enhance the antibiotic efficiency not by damaging the bacterial ROS defense systems by flavonoids, but by increasing the endogenous ROS generation in the host, which could negate its capacity to manage with oxidative stress from the available antibiotics. Bactericidal antibiotics such as quinolones, β-lactams and aminoglycosides often induced Fenton reaction resulting in the production of OH• radical [201]. These OH• radicals lead to bactericidal antibiotic-mediated cell loss. Flavonoids play as iron-chelating agents and quenching the hydroxyl radical that attenuate killing by bactericidal drugs [201]. Additionally, the practice of aminoglycoside antibiotics (AGs) such as amikacin, gentamycin, spectinomycin, neomycin, streptomycin and tobramycin, which is driven through the proton motive force and abolished as soon as ROS levels are augmented [202,203]. Flavonoids are iron chelators that protect against AGs by blocking the intake of AGs through the damage of Fe-S cluster synthesis ensuring the impendence of the proton motive force [202]. Co-administration of inhibitory concentrations of resveratrol increased the activity of aminoglycosides, including gentamicin, kanamycin, neomycin, streptomycin and tobramycin, up to 32-fold against various Gram-positive pathogens. Eventually, resveratrol increases the efficacy of aminoglycosides appears to be unrelated to membrane hyperpolarization and disruption of membrane integrity, which have been related with increased aminoglycoside susceptibility [204].

The most common mechanism of AGs resistance is a chemical modification by bacterial aminoglycoside-modifying enzymes: phosphotransferases, acetyltransferases and nucleotidyltransferase [205]. Flavonoids are documented as aminoglycoside-modifying enzyme inhibitors. quercetin and apigenin have recommended as phosphotransferases inhibitor, which occupies the ATP binding site and interacts with the enzyme through a series of hydrogen bonds [206]. Therefore, flavonoids play as chelators that could be employed as potential inhibitors of aminoglycoside-modifying enzymes. However, such a flavonoid application still requires a prospect investigation. To date, many flavonoids were characterized by the antibacterial activities against human pathogens, which play in different mechanisms than those of conventional drugs, and thus could be of significance in the enhancement of antimycobacterial therapy [85]. Important virulence factors, such as bacterial hyaluronidases (produced by both Gram-positive and Gram-negative bacteria), directly interact with host tissues or mask the bacterial surface from host's defense mechanisms. In the bacterial pathogenesis, hyaluronidase-mediated degradation of hyaluronan increases the permeability of connective tissues and decreases the viscosity of body fluids [207]. Notably, flavonols, such as myricetin and quercetin have been identified as hyaluronic acid lyase (Hyal B) inhibitors. Plants have a limitless ability to synthesize aromatic substances, most of which are secondary metabolites. The inhibitory effect of the flavonoids increased with the number of hydroxyl groups present in the flavonoid structure [208].

Table 3. Anti-nontuberculous mycobacterial effects of flavonoids.

Class of Flavonoids	Plant Source (Family)	Compounds	Chemical Structure	NTM	MIC (mg/L)	References
Flavonoid	*Euphorbia paralias* L (Euphorbiaceaea)	quercetin-3-*O*-β-D-glucoside	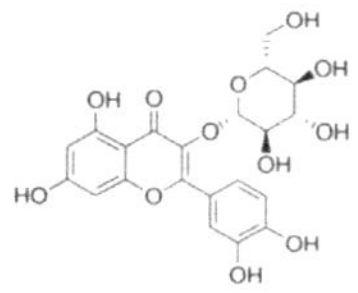	*M. fortuitum* and *M. chelonae*	3.13	[161]
Flavonoid	*Adonis dentate* (Delile) (Ranunculaceae)	quercetin-3-*O*-β-D-glucoside	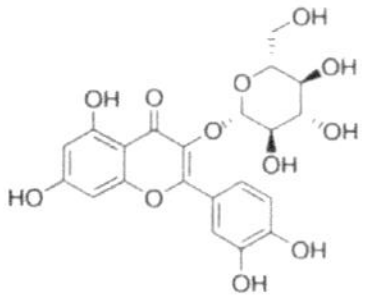	*M. abscessus*	5	[161]
Flavonoid	*Jasoniac andicans* (Delile) Botsch (Asteraceae)	quercetin-3-*O*-β-D-glucoside	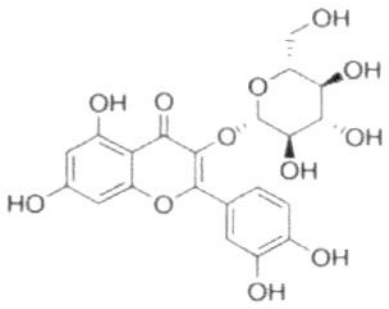	*M. fortuitum* and *M. chelonae*	6.25	[161]
Flavone	*Galenia africana* (Aizoaceae)	5,7,2′-trihydroxyflavone	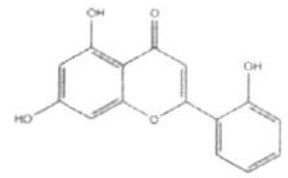	*M. abscessus*	10	[209]

Table 3. *Cont.*

Class of Flavonoids	Plant Source (Family)	Compounds	Chemical Structure	NTM	MIC (mg/L)	References
Flavonoid	*Moltkiopsis ciliate* (Forssk.) I.M (Boraginaceae)	quercetin-3-*O*-β-D-glucoside	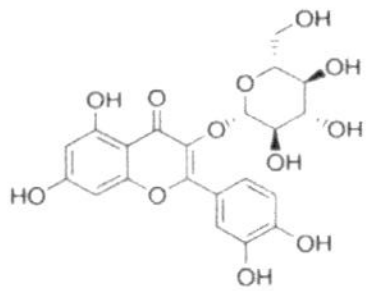	*M. fortuitum* and *M. chelonae*	10	[161]
Flavonoid	*Terminalia albida* (Combretaceae)	gallic acid, flavogallonic acid isomer i, gallagic acid	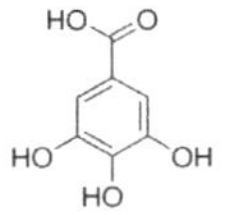	*M. chelonae*	11.81	[193]
Flavonoids	*Pelargonium reniforme* (Geraniaceae)	myricetin and quercitin-3-*O*-β-D-glucoside	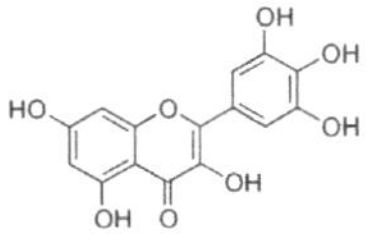	*M. fortuitum*	12.5	[124]
Flavonoid	*Eremophila sturtii* (Myoporaceae)	8,19-dihydroxyserrulat-14-ene and 8-hydroxyserrulat-14-en-19-oic acid	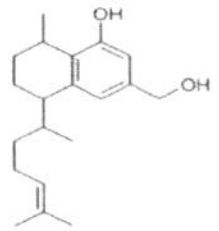	*M. fortuitum* and *M. chelonae*	12.5	[210]
Flavonoid	*Isatis microcarpa* J. Gay ex Boiss. (Brassicaceae)	quercetin-3-*O*-β-D-glucoside	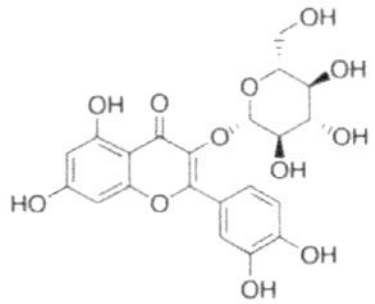	*M. fortuitum* and *M. chelonae*	12.5	[161]

Table 3. *Cont.*

Class of Flavonoids	Plant Source (Family)	Compounds	Chemical Structure	NTM	MIC (mg/L)	References
Flavonoid	*Piper nigrum* L (Piperaceae)	quercetin-3-*O*-β-D-glucoside	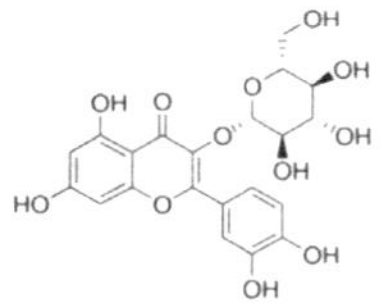	*M. smegmatis*	12.5	[161]
Eugenol	*Alpinia galanga* (Zingiberaceae)	1′-s-1′-acetoxychavicol acetate, trans-p-coumaryl diacetate and 1′-s-1′-acetoxyeugenol acetate	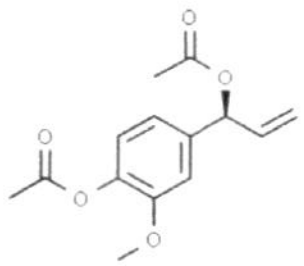	*M. smegmatis*	2.5, 6.25 and 5.0	[187]
Flavonoid	*Rhynchosia precatoria* (Willd.) DC. (Fabaceae)	β-sitosterol, daucosterol, tricin, gallic acid, daidzein, 5,7,3′-trihydroxy-4′-methoxyisoflavo epicatechin, stigmast-5-ene-3β,7α-diol, quercetin, apigenin-7-*O*-β-D-glucoside, luteolin-7-*O*-β-D-glucoside, and calycosin	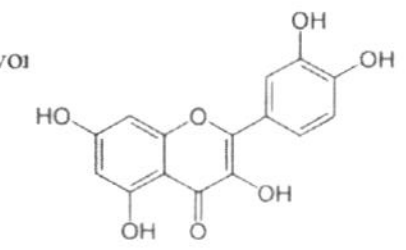	*M. fortuitum* and *M. chelonae*	15.6	[142–145]
Flavonoid	*Lawsonia inermis* (Lythraceae)	lawsonicin	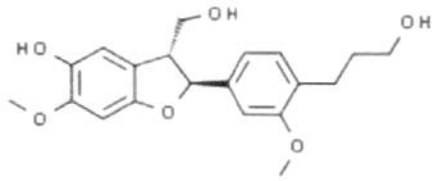	*M. chelonae*	16	[193]
Flavonoid	*Zingiber officinale* Rosc. (*Zingiberaceae*) and *Curcuma longa* L. (*Zingiberaceae*)	flavonoid	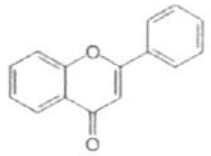	*M. abscessus*	25	[181]

Table 3. *Cont.*

Class of Flavonoids	Plant Source (Family)	Compounds	Chemical Structure	NTM	MIC (mg/L)	References
Flavonoid	*Combretum hereroense*, *C. apiculatum* and *C. collinum* (Combretaceae)	pinocembrin	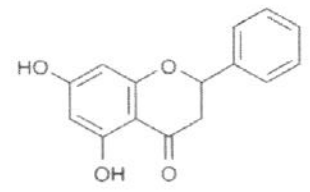	*M. fortuitum*	25	[194]
Flavonoid	*Cistanche tubulosa* (Schrenk) Hoof.f (Orobanchaceae)	quercetin-3-*O*-β-D-glucoside	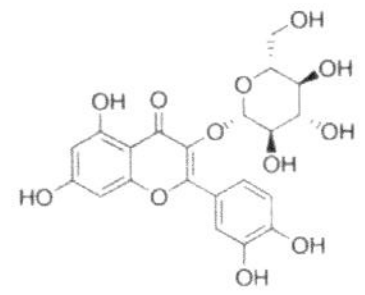	*M. fortuitum* and *M. chelonae*	25	[161]
Flavonoid	*Morcandias nites* (Viv) E.A. Durand & Barratte (Brassicaceae)	quercetin-3-*O*-β-D-glucoside	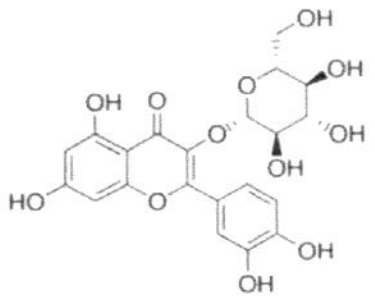	*M. fortuitum* and *M. chelonae*	25	[161]
Flavonoid	*Onopordum acanthium* L (Asteraceae)	quercetin-3-*O*-β-D-glucoside	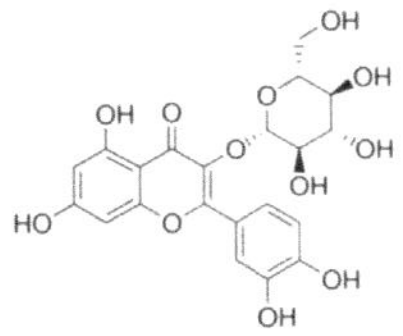	*M. smegmatis*	25	[161]

Table 3. *Cont.*

Class of Flavonoids	Plant Source (Family)	Compounds	Chemical Structure	NTM	MIC (mg/L)	References
Flavonoid	*Phlomis fraticosa* L (Lamiaceae)	quercetin-3-*O*-β-D-glucoside	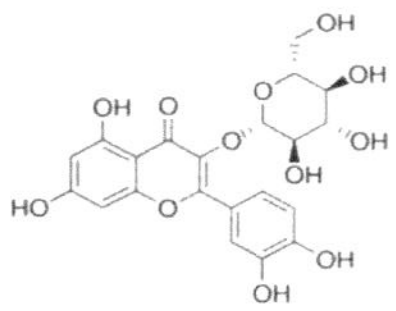	*M. smegmatis*	25	[161]
O-Methylated isoflavone	*Trifolium pretense* (Fabaceae)	biochanin A	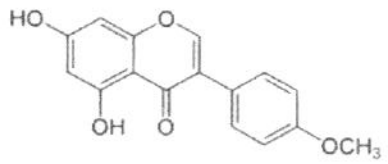	*M. smegmatis*	32	[144]
Stilbene	*Vatica oblongifolia* ssp. Oblongifolia (Dipterocarpaceae)	resveratrol hopeaphenol A, isohopeaphenol A, vaticaphenol A	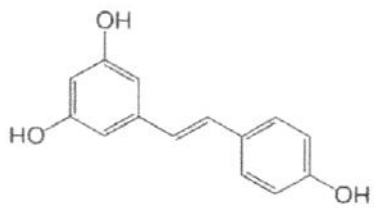	*M. abscessus*	32	[211]
Flavone	–	luteolin	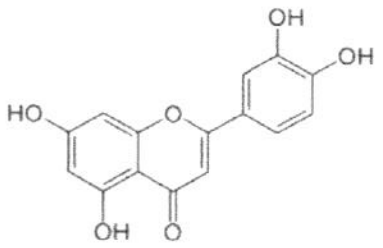	*M. smegmatis*	32	[144]
Flavonoid	–	myricetin	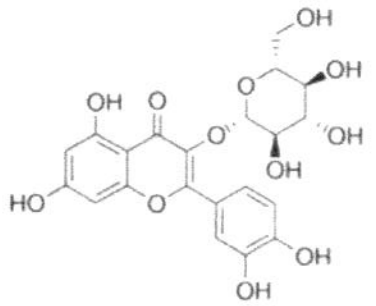	*M. smegmatis*	32	[144]

Table 3. *Cont.*

Class of Flavonoids	Plant Source (Family)	Compounds	Chemical Structure	NTM	MIC (mg/L)	References
Flavonoid	*Thymelea hirsute* L (Thymelaeaceae)	quercetin-3-*O*-β-D-glucoside	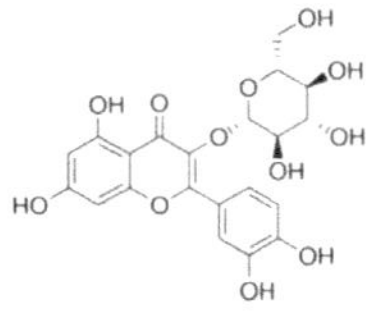	*M. smegmatis*	40	[161]
Methoxylated Flavonoid	*Paepalanthus Latipes* (Eriocaulaceae)	7-methyl quercetagetin-4′-*O*-β-D-glucopyranosid 7-methylquercetagetin	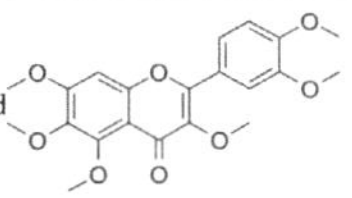	*M. abscessus*	50	[181]
Flavonoid	*Nasturtium africanum* (Braun-Blanq) (Brassicaceae)	quercetin-3-*O*-β-D-glucoside	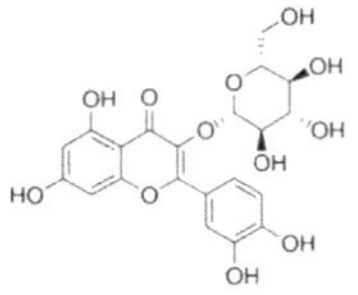	*M. smegmatis*	50	[161]
Flavonoid	*Cesalpinia digyna* (Fabaceae)	Bonducellin	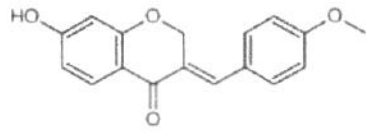	*M. abscessus*	62.5	[212]
Flavonoid	-	carvacrol	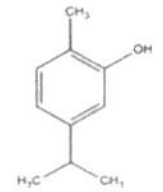	*M. abscessus, M. chelonae, M. fortuitum, M. mucogenicum, M. smegmatis*	64	[125]
Isoflavones	*Iris adriatica* (Iridaceae)	Irigenin, irilone, methoxylated benzophenone	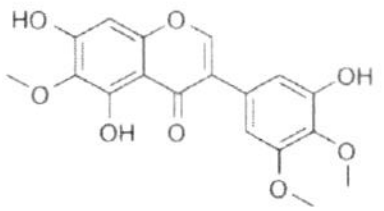	*M. abscessus*	64	[209]

Table 3. *Cont.*

Class of Flavonoids	Plant Source (Family)	Compounds	Chemical Structure	NTM	MIC (mg/L)	References
Flavone	-	baicalein	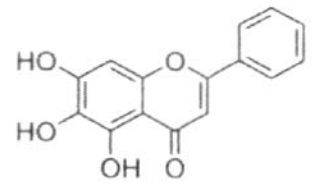	*M. abscessus*	64	[144]
Stilbenoid	-	resveratrol	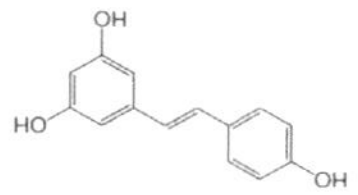	*M. smegmatis*	64	[144]
Flavonoid	*Alpinia katsumadai* (Zingiberaceae)	pinocembrin	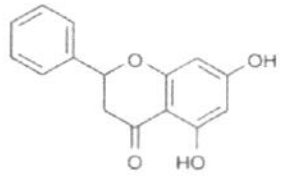	*M. abscessus*	≥ 64	[184]
Flavonoid	*Curcuma longa* L. (Zingiberaceae)	curcumin	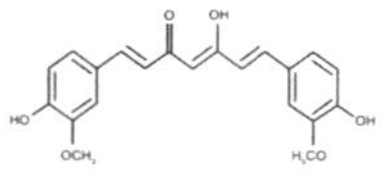	*M. abscessus,*	128	[123]
Flavonoid	*Aloe secundiflora* Engl. (Asphodelaceae)	Flavonoids	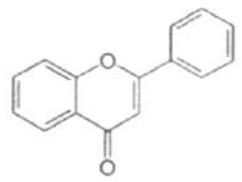	*M. fortuitum* and *M. smegmatis*	150	[213]
Flavonoid	*Colletotrichum tofieldiae* and *Magnaporthe grisea*	2,4-diacetyl phloroglucinol, phloretin	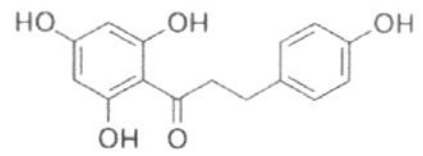	*M. abscessus*	100, 150	[193]
Flavonoid	*Entada abysinnica* steudel ex. A. Rich (Fabaceae)	Flavonoids	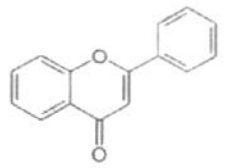	*M. fortuitum* and *M. smegmatis*	250	[214]

Table 3. *Cont.*

Class of Flavonoids	Plant Source (Family)	Compounds	Chemical Structure	NTM	MIC (mg/L)	References
Flavonoid	*Euphorbia albomarginata* Torr. (*Euphorbiaceae*)	Gallic acid methylester, 7-*O*-galloylcatechin, 1,6-di-*O*-galloylglucose, 1-*O*-galloylglucose, trigalloylgallic acid and gallic acid	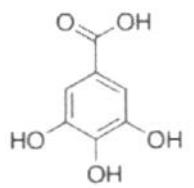	*M. fortuitum* and *M. chelonae*	250	[142,215,216]
Flavonoid	*Helianthus annuus* L. (Asteraceae)	Gallic acid, daidzein and calycosin	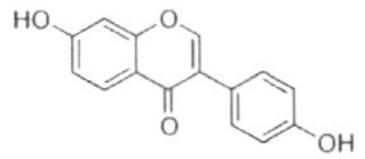	*M. fortuitum* and *M. chelonae*	250	[142,217]
Cinnamolyglico flavonoids	*Heritiera littoralis* (Sterculiaceae)	3-cinnamoyl tribuloside	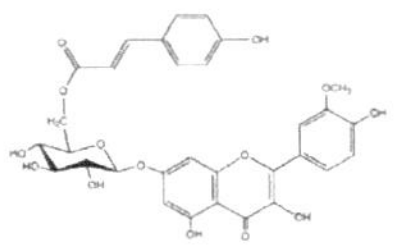	*M. fortuitum*	256	[189]
Flavonoid	*Dorstenia barteri* (Moraceae)	Isobavachalcone, kanzanol C, 4-hydroxylonchocarpin, stipulin, amentoflavone	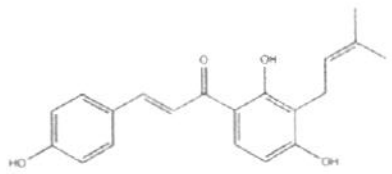	*M. smegmatis*	256	[214]
Flavone glycoside	-	Baicalin	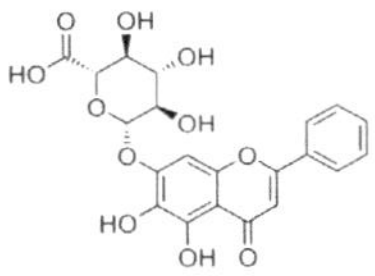	*M. abscessus*	256	[144]
O-methylated isoflavone	-	biochanin A	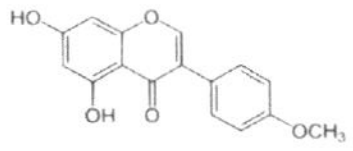	*M. abscessus*,	256	[144]

Table 3. *Cont.*

Class of Flavonoids	Plant Source (Family)	Compounds	Chemical Structure	NTM	MIC (mg/L)	References
Isoflavone	-	Daidzein	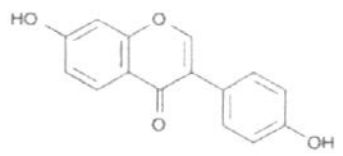	*M. smegmatis*	>256	[144]
O-methylated isoflavone	-	Formononetin	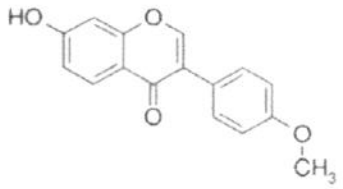	*M. smegmatis*	256	[144]
Isoflavone	-	Genistein	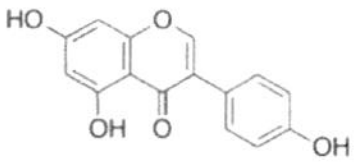	*M. smegmatis*	256	[144]
Flavonoid	*Pelargonium reniforme* and *Pelargonium sidoides* (Geraniaceae)	Gallic acid, methyl gallate, myricetin and quercitin-3-*O*-beta-D-glucoside, 1-*O*-(2-(4-methoxyphenyl)ethyl-6-*O*-galloyl-g	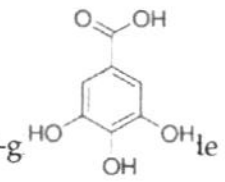	*M. fortuitum*	250, 150	[115]
Polymethoxy flavones	-	Skullcapflavone II and nobiletin, tangeretin, baicalein and wogonin.	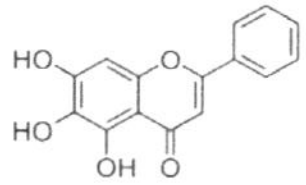	*M. fortuitum* and *M. chelonae*	128, 128, 128, 32, 128	[177]

6. Future Directions and Remarks

Studies on synergistic relations between natural products and synthetic drugs are very limited. Hence, urgent studies are required for a better understanding of synergistic behavior and the underlying mechanisms of action of flavonoids-drug combinations against NTM. This attempt may accelerate the discovery of novel drugs that are effective against antibiotic resistance targets of NTM and reduce the global occurrence of severe chronic pulmonary and extrapulmonary infections. To date, the favorite strategy for the treatment of multidrug resistance is to simultaneously inhibit multiple targets such as the inhibition of DNA gyrase activity and cell wall synthesis. However, in future studies on the synergistic relations between flavonoids and synthetic drugs would be greater effects than treating conventional drugs alone. There are various motives to investigate a novel class of antimicrobial drugs and the flavonoids represent a novel set of opportunities. Based on the chemical profile of the flavonoids, the outcomes can be analyzed to show the target sites of novel drugs against extensively multidrug-resistant NTM. These new classes of drugs may be effective on NTM, which brings about better understandings of flavonoids and structure–activity relationships. Therefore, these plant-derived novel compounds could be useful to cope with the resistance problem. Although these efforts are implemented earlier in the pharma industries and being conducted on NTM drug development projects, the current progress is still inadequate to overwhelm the subject of multidrug resistance. The primary reason for ineffectiveness is based on bacterial resistance and the demands which are not gratified in terms of the requirements for the combinations of novel agents. Novel targets among the bacterial resistance mechanisms and investigation on novel molecules are vital for developing innovative anti-NTM drugs. Further, in vitro, in vivo and clinical, and pharmacokinetics studies and chemical relationship are mandatory to analyze the synergistic relations between flavonoids and synthetic drugs, which may provide the state-of-the-art and translate bench to bed treatments.

7. Conclusions

Recently, NTM have developed into significant bacterial pathogens for both animals and humans. In particular, the concern is the high level of antimicrobial resistance displayed by these organisms, which complicates treatment and possible effective outcomes. The state of the existing antimycobacterial agents and their hitches is relatively serious. In developing nations, the incidence rate and diagnosis of NTM have often not been noticed as a deficiency of laboratory settings and mycobacteria identification. The escalating rate of pathogenic NTM in developing nations is significantly greater in HIV/AIDS patients, which leads to high levels of morbidity and mortality globally. Furthermore, there are restrictions evident by antimycobacterial drugs: the lower bactericidal ability, multidrug usage, high resistance and toxicity and organ damage. Hence, it is imperative to find new drugs as alternative therapies in which flavonoids are promising to be safe for usage, endowed with abundant pharmacological roles that are potentially active against NTM. Several flavonoids have been used in connotation with their antimycobacterial activities and can be potential and cost-effective. They have possible antimycobacterial effects at minor quantities by themselves or in synergistic combinations. A cocktail of flavonoids used with existing antimycobacterial agents is a proposal of a novel strategy to lessen side effects. They often prevent bacterial growth in several underlying mechanisms by increasing the disturbance of the plasma membrane, inhibiting cell wall development, efflux-mediated pumping system and DNA synthesis. These flavonoids are potential in synergetic combination treatment with available conservative pharmacological agents, which can be very suitable and supportive in the search for novel drug treatment against mycobacterial pathogens.

Author Contributions: S.M., F.A.A. and A.P. equally conceived, designed, wrote, revised and improved the review. All authors have read and agreed to the published version of the manuscript

Funding: The authors would like to thank the Deanship of Scientific Research, Majmaah University, Kingdom of Saudi Arabia for academic support under the project number: R-1441-146.

Conflicts of Interest: The authors declare no conflict of interest.

References

1. Peyrani, P.; Ramirez, J.A. Nontuberculous mycobacterial pulmonary infections. In *Pulmonary Complications of HIV*; European Respiratory Society: Sheffield, UK, 2014; pp. 128–137. [CrossRef]
2. Suresh, M.; Rath, P.K.; Panneerselvam, A.; Dhanasekaran, D.; Thajuddin, N. Anti-mycobacterial effect of leaf extract of Centella asiatica. *Res. J. Pharm. Technol.* **2010**, *3*, 872–876.
3. Lim, S.S.; Selvaraj, A.; Ng, Z.Y.; Palanisamy, M.; Mickmaray, S.; Cheong, P.C.H.; Lim, R.L.H. Isolation of actinomycetes with antibacterial activity against multi-drug resistant bacteria. *Malays. J. Microbiol.* **2018**, *14*, 293–305. [CrossRef]
4. Devi, C.A.; Dhanasekaran, D.; Suresh, M.; Thajuddin, N. Diagnostic value of real time PCR and associated bacterial and fungal infections in female genital tuberculosis. *Biomed. Pharm. J.* **2015**, *3*, 73–79.
5. Tortoli, E.; Fedrizzi, T.; Meehan, C.J.; Trovato, A.; Grottola, A.; Giacobazzi, E.; Serpini, G.F.; Tagliazucchi, S.; Fabio, A.; Bettua, C.; et al. The new phylogeny of the genus mycobacterium: The old and the news. *Infect. Genet. Evol.* **2017**, *56*, 19–25. [CrossRef]
6. Catherinot, E.; Roux, A.-L.; Vibet, M.-A.; Bellis, G.; Ravilly, S.; Lemonnier, L.; Le Roux, E.; Bernède-Bauduin, C.; Le Bourgeois, M.; Herrmann, J.-L.; et al. Mycobacterium avium and mycobacterium abscessus complex target distinct cystic fibrosis patient subpopulations. *J. Cyst. Fibros.* **2013**, *12*, 74–80. [CrossRef] [PubMed]
7. Baldwin, S.L.; Larsen, S.E.; Ordway, D.; Cassell, G.; Coler, R.N. The complexities and challenges of preventing and treating nontuberculous mycobacterial diseases. *PLoS Negl. Trop. Dis.* **2019**, *13*, e0007083. [CrossRef]
8. Johansen, M.D.; Herrmann, J.-L.; Kremer, L. Non-tuberculous mycobacteria and the rise of mycobacterium abscessus. *Nat. Rev. Microbiol.* **2020**, *18*, 392–407. [CrossRef]
9. Turenne, C.Y. Nontuberculous mycobacteria: Insights on taxonomy and evolution. *Infect. Genet. Evol.* **2019**, *72*, 159–168. [CrossRef]
10. Franco-Paredes, C.; Marcos, L.A.; Henao-Martínez, A.F.; Rodríguez-Morales, A.J.; Villamil-Gómez, W.E.; Gotuzzo, E.; Bonifaz, A. Cutaneous mycobacterial infections. *Clin. Microbiol. Rev.* **2018**, *32*, e00069-18. [CrossRef]
11. Kim, B.-J.; Kim, B.-R.; Jeong, J.; Lim, J.-H.; Park, S.H.; Lee, S.-H.; Kim, C.K.; Kook, Y.-H.; Kim, B.-J. A description of mycobacterium chelonae subsp. gwanakae subsp. nov., a rapidly growing mycobacterium with a smooth colony phenotype due to glycopeptidolipids. *Int. J. Syst. Evol. Microbiol.* **2018**, *68*, 3772–3780. [CrossRef]
12. Jankovic, M.; Sabol, I.; Zmak, L.; Jankovic, V.K.; Jakopovic, M.; Obrovac, M.; Ticac, B.; Bulat, L.K.; Grle, S.P.; Marekovic, I.; et al. Microbiological criteria in non-tuberculous mycobacteria pulmonary disease: A tool for diagnosis and epidemiology. *Int. J. Tuberc. Lung Dis.* **2016**, *20*, 934–940. [CrossRef] [PubMed]
13. Olivier, K.N.; Weber, D.J.; Wallace, R.J.; Faiz, A.R.; Lee, J.-H.; Zhang, Y.; Brown-Elliot, B.A.; Handler, A.; Wilson, R.W.; Schechter, M.S.; et al. Nontuberculous Mycobacteria. *Am. J. Respir. Crit. Care Med.* **2003**, *167*, 828–834. [CrossRef] [PubMed]
14. Roux, A.L.; Catherinot, E.; Ripoll, F.; Soismier, N.; Macheras, E.; Ravilly, S.; Bellis, G.; Vibet, M.A.; Le Roux, E.; Lemonnier, L.; et al. Multicenter study of prevalence of nontuberculous Mycobacteria in patients with cystic fibrosis in France. *J. Clin. Microbiol.* **2009**, *47*, 4124–4128. [CrossRef] [PubMed]
15. Stephenson, D.; Perry, A.; Appleby, M.R.; Lee, D.; Davison, J.; Johnston, A.; Jones, A.L.; Nelson, A.; Bourke, S.J.; Thomas, M.F.; et al. An evaluation of methods for the isolation of nontuberculous mycobacteria from patients with cystic fibrosis, bronchiectasis and patients assessed for lung transplantation. *BMC Pulm. Med.* **2019**, *19*, 19. [CrossRef] [PubMed]
16. Choo, S.W.; Wee, W.Y.; Ngeow, Y.F.; Mitchell, W.; Tan, J.L.; Wong, G.J.; Zhao, Y.; Xiao, J. Genomic reconnaissance of clinical isolates of emerging human pathogen mycobacterium abscessus reveals high evolutionary potential. *Sci. Rep.* **2014**, *4*, 4061. [CrossRef] [PubMed]
17. Sapriel, G.; Konjek, J.; Orgeur, M.; Bouri, L.; Frézal, L.; Roux, A.-L.; Dumas, E.; Brosch, R.; Bouchier, C.; Brisse, S.; et al. Genome-wide mosaicism within mycobacterium abscessus: Evolutionary and epidemiological implications. *BMC Genom.* **2016**, *17*, 118. [CrossRef]
18. Viljoen, A.; Gutiérrez, A.V.; Dupont, C.; Ghigo, E.; Kremer, L. A simple and rapid gene disruption strategy in mycobacterium abscessus: On the design and application of glycopeptidolipid mutants. *Front. Cell Infect. Microbiol.* **2018**, *8*, 69. [CrossRef]

19. Gutiérrez, A.V.; Viljoen, A.; Ghigo, E.; Herrmann, J.-L.; Kremer, L. Glycopeptidolipids, a double-edged sword of the mycobacterium abscessus complex. *Front. Microbiol.* **2018**, *9*, 1145. [CrossRef]
20. Wallace, J.R.; Mangas, K.M.; Porter, J.L.; Marcsisin, R.; Pidot, S.J.; Howden, B.; Omansen, T.F.; Zeng, W.; Axford, J.K.; Johnson, P.D.R.; et al. Mycobacterium ulcerans low infectious dose and mechanical transmission support insect bites and puncturing injuries in the spread of Buruli ulcer. *PLoS Negl. Trop. Dis.* **2017**, *11*, e0005553. [CrossRef]
21. Wu, U.-I.; Holland, S.M. Host susceptibility to non-tuberculous mycobacterial infections. *Lancet Infect. Dis.* **2015**, *15*, 968–980. [CrossRef]
22. Zhang, M.; Feng, M.; He, J.-Q. Disseminated mycobacterium kansasii infection with cutaneous lesions in an immunocompetent patient. *Int. J. Infect. Dis.* **2017**, *62*, 59–63. [CrossRef] [PubMed]
23. Aitken, M.L.; Limaye, A.; Pottinger, P.; Whimbey, E.; Goss, C.H.; Tonelli, M.R.; Cangelosi, G.A.; Dirac, M.A.; Olivier, K.N.; Brown-Elliott, B.A.; et al. Respiratory outbreak of mycobacterium abscessus subspecies massiliense in a lung transplant and cystic fibrosis center. *Am. J. Respir. Crit. Care Med.* **2012**, *185*, 231–232. [CrossRef] [PubMed]
24. Bryant, J.M.; Grogono, D.M.; Greaves, D.; Foweraker, J.; Roddick, I.; Inns, T.; Reacher, M.; Haworth, C.S.; Curran, M.D.; Harris, S.R.; et al. Whole-genome sequencing to identify transmission of mycobacterium abscessus between patients with cystic fibrosis: A retrospective cohort study. *Lancet* **2013**, *381*, 1551–1560. [CrossRef]
25. Pedrero, S.; Tabernero, E.; Arana-Arri, E.; Urra, E.; Larrea, M.; Zalacain, R. Changing epidemiology of nontuberculous mycobacterial lung disease over the last two decades in a region of the Basque country. *ERJ Open Res.* **2019**, *5*, 00110–02018. [CrossRef] [PubMed]
26. Ringshausen, F.C.; Wagner, D.; de Roux, A.; Diel, R.; Hohmann, D.; Hickstein, L.; Welte, T.; Rademacher, J. Prevalence of nontuberculous Mycobacterial pulmonary disease, Germany, 2009–2014. *Emerg. Infect. Dis.* **2016**, *70*, 1102. [CrossRef] [PubMed]
27. Adjemian, J.; Olivier, K.N.; Seitz, A.E.; Holland, S.M.; Prevots, D.R. Prevalence of Nontuberculous Mycobacterial lung disease in U.S. medicare beneficiaries. *Am. J. Respir. Crit. Care Med.* **2012**, *185*, 881–886. [CrossRef]
28. Adjemian, J.; Olivier, K.N.; Seitz, A.E.; Falkinham, J.O.; Holland, S.M.; Prevots, D.R. Spatial clusters of Nontuberculous Mycobacterial lung disease in the United States. *Am. J. Respir. Crit. Care Med.* **2012**, *186*, 553–558. [CrossRef]
29. Henkle, E.; Hedberg, K.; Schafer, S.; Novosad, S.; Winthrop, K.L. Population-based incidence of pulmonary nontuberculous mycobacterial disease in Oregon 2007 to 2012. *Ann. Am. Thorac. Soc.* **2015**, *12*, 642–647. [CrossRef]
30. Morimoto, K.; Iwai, K.; Uchimura, K.; Okumura, M.; Yoshiyama, T.; Yoshimori, K.; Ogata, H.; Kurashima, A.; Gemma, A.; Kudoh, S. A steady increase in nontuberculous mycobacteriosis mortality and estimated prevalence in Japan. *Ann. Am. Thorac. Soc.* **2014**, *11*, 1–8. [CrossRef]
31. Cassidy, P.M.; Hedberg, K.; Saulson, A.; McNelly, E.; Winthrop Kevin, L. Nontuberculous mycobacterial disease prevalence and risk factors: A changing epidemiology. *Clin. Infect. Dis.* **2009**, *49*, e124–e129. [CrossRef]
32. Larsson, L.-O.; Polverino, E.; Hoefsloot, W.; Codecasa, L.R.; Diel, R.; Jenkins, S.G.; Loebinger, M.R. Pulmonary disease by non-tuberculous mycobacteria—Clinical management, unmet needs and future perspectives. *Expert Rev. Respir. Med.* **2017**, *11*, 977–989. [CrossRef]
33. Wang, S.-H.; Pancholi, P. Mycobacterial skin and soft tissue infection. *Curr. Infect. Dis. Rep.* **2014**, *16*, 438. [CrossRef] [PubMed]
34. Szymanski, E.P.; Leung, J.M.; Fowler, C.J.; Haney, C.; Hsu, A.P.; Chen, F.; Duggal, P.; Oler, A.J.; McCormack, R.; Podack, E.; et al. Pulmonary nontuberculous mycobacterial infection. A multisystem, multigenic disease. *Am. J. Respir. Crit. Care Med.* **2015**, *192*, 618–628. [CrossRef] [PubMed]
35. Ryu, Y.J.; Koh, W.-J.; Daley, C.L. Diagnosis and treatment of nontuberculous mycobacterial lung disease: Clinicians' perspectives. *Tuberc. Respir. Dis.* **2016**, *79*, 74. [CrossRef] [PubMed]
36. Henkle, E.; Winthrop, K.L. Nontuberculous mycobacteria infections in immunosuppressed hosts. *Clin. Chest Med.* **2015**, *36*, 91–99. [CrossRef]
37. Mokaddas, E.; Ahmad, S. Species spectrum of nontuberculous mycobacteria isolated from clinical specimens in Kuwait. *Curr. Microbiol.* **2008**, *56*, 413–417. [CrossRef]

38. Al-Mahruqi, S.H.; van Ingen, J.; Al-Busaidy, S.; Boeree, M.J.; Al-Zadjali, S.; Patel, A.; Dekhuijzen, P.N.R.; van Soolingen, D. Clinical relevance of nontuberculous mycobacteria, Oman. *Emerg. Infect. Dis.* **2009**, *15*, 292–294. [CrossRef]
39. Varghese, B.; Memish, Z.; Abuljadayel, N.; Al-Hakeem, R.; Alrabiah, F.; Al-Hajoj, S.A. Emergence of clinically relevant non-tuberculous mycobacterial infections in Saudi Arabia. *PLoS Negl. Trop. Dis.* **2013**, *7*, e2234. [CrossRef]
40. Al-Harbi, A.; Al-Jahdali, H.; Al-Johani, S.; Baharoon, S.; Bin Salih, S.; Khan, M. Frequency and clinical significance of respiratory isolates of non-tuberculous mycobacteria in Riyadh, Saudi Arabia. *Clin. Respir. J.* **2014**, *10*, 198–203. [CrossRef]
41. Russell, C.D.; Claxton, P.; Doig, C.; Seagar, A.L.; Rayner, A.; Laurenson, I.F. Non-tuberculous mycobacteria: A retrospective review of Scottish isolates from 2000 to 2010. *Thorax* **2014**, *69*, 593–595. [CrossRef]
42. Jankovic, M.; Samarzija, M.; Sabol, I.; Jakopovic, M.; Katalinic Jankovic, V.; Zmak, L.; Ticac, B.; Marusic, A.; Obrovac, M.; van Ingen, J. Geographical distribution and clinical relevance of non-tuberculous mycobacteria in Croatia. *Int. J. Tuberc. Lung Dis.* **2013**, *17*, 836–841. [CrossRef] [PubMed]
43. Albayrak, N.; Simşek, H.; Sezen, F.; Arslantürk, A.; Tarhan, G.; Ceyhan, I. Evaluation of the distribution of non-tuberculous mycobacteria strains isolated in National Tuberculosis Reference Laboratory in 2009–2010, Turkey. *Mikrobiyol. Bul.* **2012**, *46*, 560–567. [PubMed]
44. Simons, S.; van Ingen, J.; Hsueh, P.R.; Van Hung, N.; Dekhuijzen, P.N.; Boeree, M.J.; van Soolingen, D. Nontuberculous mycobacteria in respiratory tract infections, eastern Asia. *Emerg. Infect. Dis.* **2011**, *17*, 343–349. [CrossRef] [PubMed]
45. Lai, C.C.; Hsueh, P.R. Diseases caused by nontuberculous mycobacteria in Asia. *Future Microbiol.* **2014**, *9*, 93–106. [CrossRef]
46. Baron, E.J. Clinical Microbiology in Underresourced Settings. *Clin. Lab. Med.* **2019**, *39*, 359–369. [CrossRef]
47. Wu, M.-L.; Aziz, D.B.; Dartois, V.; Dick, T. NTM drug discovery: Status, gaps and the way forward. *Drug Discov. Today* **2018**, *23*, 1502–1519. [CrossRef]
48. Fleshner, M.; Olivier, K.N.; Shaw, P.A.; Adjemian, J.; Strollo, S.; Claypool, R.J.; Folio, L.; Zelazny, A.; Holland, S.M.; Prevots, D.R. Mortality among patients with pulmonary non-tuberculous mycobacteria disease. *Int. J. Tuberc. Lung Dis.* **2016**, *20*, 582–587. [CrossRef]
49. Winthrop, K.L.; McNelley, E.; Kendall, B.; Marshall-Olson, A.; Morris, C.; Cassidy, M.; Saulson, A.; Hedberg, K. Pulmonary nontuberculous mycobacterial disease prevalence and clinical features. *Am. J. Respir. Crit. Care Med.* **2010**, *182*, 977–982. [CrossRef]
50. Martínez González, S.; Cano Cortés, A.; Sota Yoldi, L.A.; García García, J.M.; Alba Álvarez, L.M.; Palacios Gutiérrez, J.J. Non-tuberculous mycobacteria. An emerging threat? *Arch. Bronconeumol.* **2017**, *53*, 554–560. [CrossRef]
51. Novosad, S.A.; Beekmann, S.E.; Polgreen, P.M.; Mackey, K.; Winthrop, K.L. Treatment of Mycobacterium abscessus Infection. *Emerg. Infect. Dis.* **2016**, *22*, 511–514. [CrossRef]
52. Wolinsky, E. Mycobacterial Diseases Other Than Tuberculosis. *Clin. Infect. Dis.* **1992**, *15*, 1–12. [CrossRef] [PubMed]
53. Redelman-Sidi, G.; Sepkowitz, K.A. Rapidly growing mycobacteria infection in patients with cancer. *Clin. Infect. Dis.* **2010**, *51*, 422–434. [CrossRef] [PubMed]
54. Esteban, J.; García-Coca, M. Mycobacterium biofilms. *Front. Microbiol.* **2017**, *8*, 2651. [CrossRef]
55. Forbes, B.A.; Hall, G.S.; Miller, M.B.; Novak, S.M.; Rowlinson, M.-C.; Salfinger, M.; Somoskövi, A.; Warshauer, D.M.; Wilson, M.L. Practice guidelines for clinical microbiology laboratories: Mycobacteria. *Clin. Microbiol. Rev.* **2018**, *31*. [CrossRef] [PubMed]
56. Mougari, F.; Guglielmetti, L.; Raskine, L.; Sermet-Gaudelus, I.; Veziris, N.; Cambau, E. Infections caused by mycobacterium abscessus: Epidemiology, diagnostic tools and treatment. *Expert Rev. Anti-Infect. Ther.* **2016**, *14*, 1139–1154. [CrossRef] [PubMed]
57. Marion, E.; Song, O.-R.; Christophe, T.; Babonneau, J.; Fenistein, D.; Eyer, J.; Letournel, F.; Henrion, D.; Clere, N.; Paille, V.; et al. Mycobacterial toxin induces analgesia in buruli ulcer by targeting the angiotensin pathways. *Cell* **2014**, *157*, 1565–1576. [CrossRef]
58. Suresh, M.; Rath, P.K.; Panneerselvam, A.; Dhanasekaran, D.; Thajuddin, N. Antifungal activity of selected Indian medicinal plant salts. *J. Glob. Pharm. Technol.* **2010**, *2*, 71–74.

59. Prabakar, K.; Sivalingam, P.; Mohamed Rabeek, S.I.; Muthuselvam, M.; Devarajan, N.; Arjunan, A.; Karthick, R.; Suresh, M.M.; Wembonyama, J.P. Evaluation of antibacterial efficacy of phyto fabricated silver nanoparticles using Mukia scabrella (Musumusukkai) against drug resistance nosocomial gram negative bacterial pathogens. *Colloids Surf. B Biointerfaces* **2013**, *104*, 282–288. [CrossRef]
60. Mickymaray, S.; Al Aboody, M.S.; Rath, P.K.; Annamalai, P.; Nooruddin, T. Screening and antibacterial efficacy of selected Indian medicinal plants. *Asian Pac. J. Trop. Biomed.* **2016**, *6*, 185–191. [CrossRef]
61. Moorthy, K.; Punitha, T.; Vinodhini, R.; Mickymaray, S.; Shonga, A.; Tomass, Z.; Thajuddin, N. Efficacy of different solvent extracts of Aristolochia krisagathra and Thottea ponmudiana for potential antimicrobial activity. *J. Pharm. Res.* **2015**, *9*, 35–40.
62. Mickymaray, S.; Alturaiki, W. Antifungal Efficacy of Marine Macroalgae against Fungal Isolates from Bronchial Asthmatic Cases. *Molecules* **2018**, *23*, 3032. [CrossRef] [PubMed]
63. Kannaiyan, M.; Manuel, V.N.; Raja, V.; Thambidurai, P.; Mickymaray, S.; Nooruddin, T. Antimicrobial activity of the ethanolic and aqueous extracts of Salacia chinensis Linn. against human pathogens. *Asian Pac. J. Trop. Dis.* **2012**, *2*, S416–S420. [CrossRef]
64. Mickymaray, S. Efficacy and mechanism of traditional medicinal plants and bioactive compounds against clinically important pathogens. *Antibiotics* **2019**, *8*, 257. [CrossRef] [PubMed]
65. Mickymaray, S.; Al Aboody, M.S. In vitro antioxidant and bactericidal efficacy of 15 common spices: Novel therapeutics for urinary tract infections? *Medicina* **2019**, *55*, 289. [CrossRef]
66. Suresh, M.; Alfonisan, M.; Alturaiki, W.; Aboody, M.S.A.; Alfaiz, F.A.; Premanathan, M.; Vijayakumar, R.; Umamagheswari, K.; Ghamdi, S.A.; Alsagaby, S.A. Investigations of bioactivity of *Acalypha indica* (L.), *Centella asiatica* (L.) and croton bonplandianus (Baill) against multidrug resistant bacteria and cancer cells. *J. Herb. Med.* **2020**, 100359. [CrossRef]
67. Kumar, G.; Murugesan, A.G.; Rajasekara Pandian, M. Effect of Helicteres isora bark extract on blood glucose and hepatic enzymes in experimental diabetes. *Pharmazie* **2006**, *61*, 353–355.
68. Ganesan, K.; Xu, B. Ethnobotanical studies on folkloric medicinal plants in Nainamalai, Namakkal District, Tamil Nadu, India. *Trends Phytochem. Res.* **2017**, *1*, 153–168.
69. Kumar, G.; Banu, G.S.; Murugesan, A.G.; Pandian, M.R. Hypoglycaemic effect of Helicteres isora bark extract in rats. *J. Ethnopharmacol.* **2006**, *107*, 304–307. [CrossRef]
70. Sinaga, M.; Ganesan, K.; Kumar Nair, S.K.P.; Gani, S.B. Preliminary phytochemical analysis and in vitro antibacterial activity of bark and seeds of Ethiopian neem (Azadirachta indica A. Juss). *World J Pharm. Pharma. Sci.* **2016**, *5*, 1714–1723. [CrossRef]
71. Zhang, T.; Jayachandran, M.; Ganesan, K.; Xu, B. Black truffle aqueous extract attenuates oxidative stress and inflammation in STZ-induced hyperglycemic rats via Nrf2 and NF-κB pathways. *Front. Pharmacol.* **2018**, *9*, 1257. [CrossRef]
72. Jayachandran, M.; Wu, Z.; Ganesan, K.; Khalid, S.; Chung, S.M.; Xu, B. Isoquercetin upregulates antioxidant genes, suppresses inflammatory cytokines and regulates AMPK pathway in streptozotocin-induced diabetic rats. *Chem. Biol. Interact.* **2019**, *303*, 62–69. [CrossRef] [PubMed]
73. Sukalingam, K.; Ganesan, K.; Xu, B. *Trianthema portulacastrum* L. (giant pigweed): Phytochemistry and pharmacological properties. *Phytochem. Rev.* **2017**, *16*, 461–478. [CrossRef]
74. Vijayakumar, R.; Sandle, T.; Al-Aboody, M.S.; AlFonaisan, M.K.; Alturaiki, W.; Mickymaray, S.; Premanathan, M.; Alsagaby, S.A. Distribution of biocide resistant genes and biocides susceptibility in multidrug-resistant Klebsiella pneumoniae, Pseudomonas aeruginosa and Acinetobacter baumannii—A first report from the Kingdom of Saudi Arabia. *J. Infect. Public Health* **2018**, *11*, 812–816. [CrossRef] [PubMed]
75. Vinodhini, R.; Moorthy, K.; Suresh, M. Incidence and virulence traits of Candida dubliniensis isolated from clinically suspected patients. *Asian J. Pharm. Clin. Res.* **2016**, *9*, 77. [CrossRef]
76. Chandran, R.P.; Kumar, S.N.; Manju, S.; Kader, S.A.; Dileep Kumar, B.S. In vitro α-glucosidase inhibition, antioxidant, anticancer, and antimycobacterial properties of ethyl acetate extract of Aegle tamilnadensis Abdul Kader (Rutaceae) leaf. *Appl. Biochem. Biotechnol.* **2015**, *175*, 1247–1261. [CrossRef]
77. Vianna, J.S.; Machado, D.; Ramis, I.B.; Silva, F.P.; Bierhals, D.V.; Abril, M.A.; von Groll, A.; Ramos, D.F.; Lourenço, M.C.S.; Viveiros, M.; et al. The Contribution of Efflux Pumps in Mycobacterium abscessus Complex Resistance to Clarithromycin. *Antibiotics* **2019**, *8*, 153. [CrossRef]
78. Górniak, I.; Bartoszewski, R.; Króliczewski, J. Comprehensive review of antimicrobial activities of plant flavonoids. *Phytochem. Rev.* **2019**, *18*, 241–272. [CrossRef]

79. Talevi, A. Multi-target pharmacology: Possibilities and limitations of the "skeleton key approach" from a medicinal chemist perspective. *Front. Pharmacol.* **2015**, *6*, 205. [CrossRef]
80. Ganesan, K.; Xu, B. Anti-diabetic effects and mechanisms of dietary polysaccharides. *Molecules* **2019**, *24*, 2556. [CrossRef]
81. Ganesan, K.; Xu, B. Polyphenol-rich dry common beans (*Phaseolus vulgaris* L.) and their health benefits. *Int. J. Mol. Sci.* **2017**, *18*, 2331. [CrossRef]
82. Ganesan, K.; Xu, B. Polyphenol-rich lentils and their health promoting effects. *Int. J. Mol. Sci.* **2017**, *18*, 2390. [CrossRef] [PubMed]
83. Ganesan, K.; Xu, B. A critical review on polyphenols and health benefits of black soybeans. *Nutrients* **2017**, *9*, 455. [CrossRef] [PubMed]
84. Ganesan, K.; Xu, B. A critical review on phytochemical profile and health promoting effects of mung bean (Vigna radiata). *Food Sci. Hum. Wellness* **2018**, *7*, 11–33. [CrossRef]
85. Cushnie, T.P.; Lamb, A.J. Recent advances in understanding the antibacterial properties of flavonoids. *Int. J. Antimicrob. Agents* **2011**, *38*, 99–107. [CrossRef] [PubMed]
86. Cushnie, T.P.; Taylor, P.W.; Nagaoka, Y.; Uesato, S.; Hara, Y.; Lamb, A.J. Investigation of the antibacterial activity of 3-*O*-octanoyl-(−)-epicatechin. *J. Appl. Microbiol.* **2008**, *105*, 1461–1469. [CrossRef] [PubMed]
87. Newman, D.J. Natural products as leads to potential drugs: An old process or the new hope for drug discovery? *J. Med. Chem.* **2008**, *51*, 2589–2599. [CrossRef] [PubMed]
88. Aboody, M.S.A.; Mickymaray, S. Anti-fungal efficacy and mechanisms of flavonoids. *Antibiotics* **2020**, *9*, 45. [CrossRef]
89. Ganesan, K.; Jayachandran, M.; Xu, B. Diet-derived phytochemicals targeting colon cancer stem cells and microbiota in colorectal cancer. *Int. J. Mol. Sci.* **2020**, *21*, 3976. [CrossRef]
90. Islam, T.; Ganesan, K.; Xu, B. New insight into mycochemical profiles and antioxidant potential of edible and medicinal mushrooms: A review. *Int. J. Med. Mushrooms* **2019**, *21*, 237–251. [CrossRef]
91. Kumar, G.; Sharmila Banu, G.; Ganesan Murugesan, A. Effect of Helicteres isora bark extracts on heart antioxidant status and lipid peroxidation in streptozotocin diabetic rats. *J. Appl. Biomed.* **2008**, *6*, 89–95. [CrossRef]
92. Ganesan, K.; Xu, B. Anti-obesity effects of medicinal and edible mushrooms. *Molecules* **2018**, *23*, 2880. [CrossRef] [PubMed]
93. Kumar, G.; Sharmila Banu, G.; Ganesan Murugesan, A.; Pandian, M.R. Antihyperglycaemic and antiperoxidative effect of Helicteres isora L. bark extracts in streptozotocin-induced diabetic rats. *J. Appl. Biomed.* **2007**, *5*, 97–104. [CrossRef]
94. Kumar, G.; Murugesan, A.G. Hypolipidaemic activity of Helicteres isora L. bark extracts in streptozotocin induced diabetic rats. *J. Ethnopharmacol.* **2008**, *116*, 161–166. [CrossRef] [PubMed]
95. Xu, B.; Ganesan, K.; Mickymaray, S.; Alfaiz, F.A.; Thatchinamoorthi, R.; Aboody, M.S.A. Immunomodulatory and antineoplastic efficacy of common spices and their connection with phenolic antioxidants. *Bioact. Compd. Health Dis.* **2020**, *3*, 15. [CrossRef]
96. Pandian, M.R.; Banu, G.S.; Kumar, G.; Smila, K.H. Screening of antibacterial activity of fruit extract of citrus medica against bacteria involved in typhoid fever. *Nat. J. Life Sci.* **2006**, *3*, 289–292.
97. Ke, Y.; Al Aboody, M.S.; Alturaiki, W.; Alsagaby, S.A.; Alfaiz, F.A.; Veeraraghavan, V.P.; Mickymaray, S. Photosynthesized gold nanoparticles from Catharanthus roseus induces caspase-mediated apoptosis in cervical cancer cells (HeLa). *Artif. CellsNanomed. Biotechnol.* **2019**, *47*, 1938–1946. [CrossRef] [PubMed]
98. Pandian, M.R.; Banu, G.S.; Kumar, G. A study of the antimicrobial activity of Alangium salviifolium. *Indian J. Pharmacol.* **2006**, *38*, 203. [CrossRef]
99. Ganesan, K.; Xu, B. Telomerase inhibitors from natural products and their anticancer potential. *Int. J. Mol. Sci.* **2017**, *19*, 13. [CrossRef]
100. Ganesan, K.; Xu, B. Molecular targets of vitexin and isovitexin in cancer therapy: A critical review. *Ann. N. Y. Acad. Sci.* **2017**, *1401*, 102–113. [CrossRef]
101. Ganesan, K.; Guo, S.; Fayyaz, S.; Zhang, G.; Xu, B. Targeting Programmed Fusobacterium nucleatum Fap2 for Colorectal Cancer Therapy. *Cancers* **2019**, *11*, 1592. [CrossRef]
102. Chedraui, P.; Pérez-López, F.R. Nutrition and health during mid-life: Searching for solutions and meeting challenges for the aging population. *Climacteric* **2013**, *16*, 85–95. [CrossRef] [PubMed]

103. Bojić, M.; Maleš, Ž.; Antolić, A.; Babić, I.; Tomičić, M. Antithrombotic activity of flavonoids and polyphenols rich plant species. *Acta Pharm.* **2019**, *69*, 483–495. [CrossRef] [PubMed]
104. Ganesan, K.; Jayachandran, M.; Xu, B. A critical review on hepatoprotective effects of bioactive food components. *Crit. Rev. Food Sci. Nutr.* **2017**, *58*, 1165–1229. [CrossRef] [PubMed]
105. Sukalingam, K.; Ganesan, K.; Xu, B. Protective effect of aqueous extract from the leaves of justicia tranquebariesis against thioacetamide-induced oxidative stress and hepatic fibrosis in rats. *Antioxidants* **2018**, *7*, 78. [CrossRef]
106. Ganesan, K.; Sukalingam, K.; Xu, B. Solanum trilobatum L. Ameliorate thioacetamide-induced oxidative stress and hepatic damage in albino rats. *Antioxidants* **2017**, *6*, 68. [CrossRef]
107. Kumar, G.; Banu, G.S.; Pappa, P.V.; Sundararajan, M.; Pandian, M.R. Hepatoprotective activity of Trianthema portulacastrum L. against paracetamol and thioacetamide intoxication in albino rats. *J. Ethnopharmacol.* **2004**, *92*, 37–40. [CrossRef]
108. Gabrielová, E.; Bartošíková, L.; Nečas, J.; Modrianský, M. Cardioprotective effect of 2,3-dehydrosilybin preconditioning in isolated rat heart. *Fitoterapia* **2019**, *132*, 12–21. [CrossRef]
109. Braidy, N.; Behzad, S.; Habtemariam, S.; Ahmed, T.; Daglia, M.; Nabavi, S.M.; Sobarzo-Sanchez, E.; Nabavi, S.F. Neuroprotective effects of citrus fruit-derived flavonoids, nobiletin and tangeretin in alzheimer's and parkinson's disease. *CNS Neurol. Disord. Drug Targets* **2017**, *16*. [CrossRef]
110. Kumar, G.; Banu, G.S.; Murugesan, A.G.A. Influence of Helicteres isora L. bark extracts on glycemic control and renoprotective activity in streptozotocin-induced diabetic rats. *Int. J. Pharma Sci. Nanotechnol.* **2008**, *1*, 275–280.
111. Ye, J.; Guan, M.; Lu, Y.; Zhang, D.; Li, C.; Li, Y.; Zhou, C. Protective effects of hesperetin on lipopolysaccharide-induced acute lung injury by targeting MD2. *Eur. J. Pharmacol.* **2019**, *852*, 151–158. [CrossRef]
112. Murphy, K.J.; Walker, K.M.; Dyer, K.A.; Bryan, J. Estimation of daily intake of flavonoids and major food sources in middle-aged Australian men and women. *Nutr. Res.* **2019**, *61*, 64–81. [CrossRef]
113. McGrattan, A.M.; McGuinness, B.; McKinley, M.C.; Kee, F.; Passmore, P.; Woodside, J.V.; McEvoy, C.T. Diet and inflammation in cognitive ageing and alzheimer's disease. *Curr. Nutr. Rep.* **2019**, *8*, 53–65. [CrossRef] [PubMed]
114. Ziberna, L.; Fornasaro, S.; Čvorović, J.; Tramer, F.; Passamonti, S. Bioavailability of flavonoids. In *Polyphenols in Human Health and Disease*; Watson, R.R., Preedy, V.R., Zibadi, S., Eds.; Academic Press, Elsevier: Cambridge, MA, USA, 2014; pp. 489–511. [CrossRef]
115. Kumar, G.; Sharmila Banu, G.; Murugesan, A.G. Attenuation of Helicteres isora L. bark extracts on streptozotocin-induced alterations in glycogen and carbohydrate metabolism in albino rats. *Hum. Exp. Toxicol.* **2009**, *28*, 689–696. [CrossRef] [PubMed]
116. Kumar, G.; Banu, G.S.; Pandian, M.R. Biochemical activity of selenium and glutathione on country made liquor (CML) induced hepatic damage in rats. *Indian J. Clin. Biochem.* **2007**, *22*, 105–108. [CrossRef] [PubMed]
117. Kumar, G.; Banu, G.S.; Kannan, V.; Pandian, M.R. Antihepatotoxic effect of beta-carotene on paracetamol induced hepatic damage in rats. *Indian J. Exp. Biol.* **2005**, *43*, 351–355. [PubMed]
118. Yadav, S.S.; Singh, M.K.; Singh, P.K.; Kumar, V. Traditional knowledge to clinical trials: A review on therapeutic actions of Emblica officinalis. *Biomed. Pharm.* **2017**, *93*, 1292–1302. [CrossRef] [PubMed]
119. Zhang, S.; Chen, C.; Lu, W.; Wei, L. Phytochemistry, pharmacology, and clinical use of Panax notoginseng flowers buds. *Phytother. Res.* **2018**, *32*, 2155–2163. [CrossRef]
120. Chan, E.W.; Lye, P.Y.; Wong, S.K. Phytochemistry, pharmacology, and clinical trials of Morus alba. *Chin. J. Nat. Med.* **2016**, *14*, 17–30. [CrossRef]
121. Yusook, K.; Weeranantanapan, O.; Hua, Y.; Kumkrai, P.; Chudapongse, N. Lupinifolin from Derris reticulata possesses bactericidal activity on Staphylococcus aureus by disrupting bacterial cell membrane. *J. Nat. Med.* **2017**, *71*, 357–366. [CrossRef]
122. Kumar, G.; Banu, G.; Pandian, M. Evaluation of the antioxidant activity of Trianthema portulacastrum L. *Indian J. Pharmacol.* **2005**, *37*, 331–333. [CrossRef]
123. Marini, E.; Di Giulio, M.; Magi, G.; Di Lodovico, S.; Cimarelli, M.E.; Brenciani, A.; Nostro, A.; Cellini, L.; Facinelli, B. Curcumin, an antibiotic resistance breaker against a multiresistant clinical isolate of mycobacterium abscessus. *Phytother. Res.* **2017**, *32*, 488–495. [CrossRef] [PubMed]

124. Kim, C.E.; Griffiths, W.J.; Taylor, P.W. Components derived fromPelargoniumstimulate macrophage killing of mycobacterium species. *J. Appl. Microbiol.* **2009**, *106*, 1184–1193. [CrossRef] [PubMed]
125. Marini, E.; Di Giulio, M.; Ginestra, G.; Magi, G.; Di Lodovico, S.; Marino, A.; Facinelli, B.; Cellini, L.; Nostro, A. Efficacy of carvacrol against resistant rapidly growing mycobacteria in the planktonic and biofilm growth mode. *PLoS ONE* **2019**, *14*, e0219038. [CrossRef] [PubMed]
126. Sharmila Banu, G.; Kumar, G.; Murugesan, A.G. Effect of ethanolic leaf extract of Trianthema portulacastrum L. on aflatoxin induced hepatic damage in rats. *Indian J. Clin. Biochem.* **2009**, *24*, 414–418. [CrossRef]
127. Brunetti, C.; Di Ferdinando, M.; Fini, A.; Pollastri, S.; Tattini, M. Flavonoids as antioxidants and developmental regulators: Relative significance in plants and humans. *Int. J. Mol. Sci.* **2013**, *14*, 3540–3555. [CrossRef]
128. Fathima, A.; Rao, J.R. Selective toxicity of Catechin-a natural flavonoid towards bacteria. *Appl. Microbiol. Biotechnol.* **2016**, *100*, 6395–6402. [CrossRef]
129. Sirk, T.W.; Brown, E.F.; Friedman, M.; Sum, A.K. Molecular binding of catechins to biomembranes: Relationship to biological activity. *J. Agric. Food. Chem.* **2009**, *57*, 6720–6728. [CrossRef]
130. Stepanović, S.; Antić, N.; Dakić, I.; Svabić-Vlahović, M. In vitro antimicrobial activity of propolis and synergism between propolis and antimicrobial drugs. *Microbiol. Res.* **2003**, *158*, 353–357. [CrossRef]
131. Wagh, V.D. Propolis: A wonder bees product and its pharmacological potentials. *Adv. Pharm. Sci.* **2013**, *2013*, 308249. [CrossRef]
132. Ollila, F.; Halling, K.; Vuorela, P.; Vuorela, H.; Slotte, J.P. Characterization of flavonoid-biomembrane interactions. *Arch. Biochem. Biophys.* **2002**, *399*, 103–108. [CrossRef]
133. Masoko, P.; Masiphephethu, M.V. Phytochemical investigation, antioxidant and antimycobacterial activities of schkuhria pinnata (Lam) thell extracts against mycobacterium smegmatis. *J. Evid. Based Integr. Med.* **2019**, *24*, 2515690x19866104. [CrossRef] [PubMed]
134. Liu, R.; Zhang, H.; Yuan, M.; Zhou, J.; Tu, Q.; Liu, J.-J.; Wang, J. Synthesis and biological evaluation of apigenin derivatives as antibacterial and antiproliferative agents. *Molecules* **2013**, *18*, 11496–11511. [CrossRef] [PubMed]
135. Boonsai, P.; Phuwapraisirisan, P.; Chanchao, C. Antibacterial activity of a cardanol from Thai Apis mellifera propolis. *Int. J. Med. Sci.* **2014**, *11*, 327–336. [CrossRef]
136. Shen, X.; Liu, Y.; Luo, X.; Yang, Z. Advances in Biosynthesis, Pharmacology, and Pharmacokinetics of Pinocembrin, a Promising Natural Small-Molecule Drug. *Molecules* **2019**, *24*, 2323. [CrossRef] [PubMed]
137. Kim, D.H.; Bae, E.A.; Han, M.J. Anti-Helicobacter pylori activity of the metabolites of poncirin from Poncirus trifoliata by human intestinal bacteria. *Biol. Pharm. Bull.* **1999**, *22*, 422–424. [CrossRef]
138. Lucarini, R.; Tozatti, M.G.; Silva, M.L.A.; Gimenez, V.M.M.; Pauletti, P.M.; Groppo, M.; Turatti, I.C.C.; Cunha, W.R.; Martins, C.H.G. Antibacterial and anti-inflammatory activities of an extract, fractions, and compounds isolated from Gochnatia pulchra aerial parts. *Braz. J. Med. Biol. Res.* **2015**, *48*, 822–830. [CrossRef]
139. Edziri, H.; Mastouri, M.; Mahjoub, M.A.; Mighri, Z.; Mahjoub, A.; Verschaeve, L. Antibacterial, antifungal and cytotoxic activities of two flavonoids from retama raetam flowers. *Molecules* **2012**, *17*, 7284–7293. [CrossRef]
140. Xie, Y.; Chen, J.; Xiao, A.; Liu, L. Antibacterial activity of polyphenols: Structure-activity relationship and influence of hyperglycemic condition. *Molecules* **2017**, *22*, 1913. [CrossRef]
141. Xie, Y.; Yang, W.; Tang, F.; Chen, X.; Ren, L. Antibacterial activities of flavonoids: Structure-activity relationship and mechanism. *Curr. Med. Chem.* **2015**, *22*, 132–149. [CrossRef]
142. Coronado-Aceves, E.W.; Sánchez-Escalante, J.J.; López-Cervantes, J.; Robles-Zepeda, R.E.; Velázquez, C.; Sánchez-Machado, D.I.; Garibay-Escobar, A. Antimycobacterial activity of medicinal plants used by the Mayo people of Sonora, Mexico. *J. Ethnopharmacol.* **2016**, *190*, 106–115. [CrossRef]
143. Yeung, M.-F.; Lau, C.B.S.; Chan, R.C.Y.; Zong, Y.; Che, C.-T. Search for antimycobacterial constituents from a Tibetan medicinal plant, Gentianopsis paludosa. *Phytother. Res.* **2009**, *23*, 123–125. [CrossRef]
144. Lechner, D.; Gibbons, S.; Bucar, F. Plant phenolic compounds as ethidium bromide efflux inhibitors in mycobacterium smegmatis. *J. Antimicrob. Chemother.* **2008**, *62*, 345–348. [CrossRef]
145. Kuete, V.; Nono, E.C.N.; Mkounga, P.; Marat, K.; Hultin, P.G.; Nkengfack, A.E. Antimicrobial activities of the CH2Cl2–CH3OH (1:1) extracts and compounds from the roots and fruits ofPycnanthus angolensis(Myristicaceae). *Nat. Prod. Res.* **2011**, *25*, 432–443. [CrossRef]
146. Cushnie, T.P.T.; Lamb, A.J. Antimicrobial activity of flavonoids. *Int. J. Antimicrob. Agents* **2005**, *26*, 343–356. [CrossRef]

147. Hariri, B.M.; McMahon, D.B.; Chen, B.; Adappa, N.D.; Palmer, J.N.; Kennedy, D.W.; Lee, R.J. Plant flavones enhance antimicrobial activity of respiratory epithelial cell secretions against Pseudomonas aeruginosa. *PLoS ONE* **2017**, *12*, e0185203. [CrossRef]
148. Tagousop, C.N.; Tamokou, J.D.; Ekom, S.E.; Ngnokam, D.; Voutquenne-Nazabadioko, L. Antimicrobial activities of flavonoid glycosides from Graptophyllum grandulosum and their mechanism of antibacterial action. *BMC Complementary Altern. Med.* **2018**, *18*, 252. [CrossRef]
149. Křížová, L.; Dadáková, K.; Kašparovská, J.; Kašparovský, T. Isoflavones. *Molecules* **2019**, *24*, 1076. [CrossRef]
150. Bisignano, C.; Filocamo, A.; La Camera, E.; Zummo, S.; Fera, M.T.; Mandalari, G. Antibacterial activities of almond skins on cagA-positive and-negative clinical isolates of Helicobacter pylori. *BMC Microbiol.* **2013**, *13*, 103. [CrossRef]
151. Eerdunbayaer; Orabi, M.A.A.; Aoyama, H.; Kuroda, T.; Hatano, T. Structures of two new flavonoids and effects of licorice phenolics on vancomycin-resistant Enterococcus species. *Molecules* **2014**, *19*, 3883–3897. [CrossRef]
152. Chen, H.; Yu, W.; Chen, G.; Meng, S.; Xiang, Z.; He, N. Antinociceptive and antibacterial properties of anthocyanins and flavonols from fruits of black and non-black mulberries. *Molecules* **2017**, *23*, 4. [CrossRef]
153. Christopher, R.; Nyandoro, S.S.; Chacha, M.; de Koning, C.B. A new cinnamoylglycoflavonoid, antimycobacterial and antioxidant constituents from Heritiera littoralis leaf extracts. *Nat. Prod. Res.* **2014**, *28*, 351–358. [CrossRef]
154. Reshma, M.V.; Jacob, J.; Syamnath, V.L.; Habeeba, V.P.; Dileep Kumar, B.S.; Lankalapalli, R.S. First report on isolation of 2,3,4-trihydroxy-5-methylacetophenone from palmyra palm (*Borassus flabellifer* Linn.) syrup, its antioxidant and antimicrobial properties. *Food Chem.* **2017**, *228*, 491–496. [CrossRef]
155. Dzoyem, J.P.; Kuete, V.; McGaw, L.J.; Eloff, J.N. The 15-lipoxygenase inhibitory, antioxidant, antimycobacterial activity and cytotoxicity of fourteen ethnomedicinally used African spices and culinary herbs. *J. Ethnopharmacol.* **2014**, *156*, 1–8. [CrossRef]
156. Matijašević, D.; Pantić, M.; Rašković, B.; Pavlović, V.; Duvnjak, D.; Sknepnek, A.; Nikšić, M. The antibacterial activity of coriolus versicolor methanol extract and its effect on ultrastructural changes of staphylococcus aureus and salmonella enteritidis. *Front. Microbiol.* **2016**, *7*, 1226. [CrossRef]
157. Mishra, A.K.; Mishra, A.; Kehri, H.K.; Sharma, B.; Pandey, A.K. Inhibitory activity of Indian spice plant Cinnamomum zeylanicum extracts against Alternaria solani and Curvularia lunata, the pathogenic dematiaceous moulds. *Ann. Clin. Microbiol. Antimicrob.* **2009**, *8*, 9. [CrossRef]
158. Shah, S.; Stapleton, P.D.; Taylor, P.W. The polyphenol (-)-epicatechin gallate disrupts the secretion of virulence-related proteins by Staphylococcus aureus. *Lett. Appl. Microbiol.* **2008**, *46*, 181–185. [CrossRef]
159. Lee, J.H.; Regmi, S.C.; Kim, J.A.; Cho, M.H.; Yun, H.; Lee, C.S.; Lee, J. Apple flavonoid phloretin inhibits Escherichia coli O157:H7 biofilm formation and ameliorates colon inflammation in rats. *Infect. Immun.* **2011**, *79*, 4819–4827. [CrossRef]
160. Kim, Y.R.; Kim, M.A.; Cho, H.J.; Oh, S.K.; Lee, I.K.; Kim, U.K.; Lee, K.Y. Galangin prevents aminoglycoside-induced ototoxicity by decreasing mitochondrial production of reactive oxygen species in mouse cochlear cultures. *Toxicol. Lett.* **2016**, *245*, 78–85. [CrossRef]
161. Safwat, N.A.; Kashef, M.T.; Aziz, R.K.; Amer, K.F.; Ramadan, M.A. Quercetin 3-*O*-glucoside recovered from the wild Egyptian Sahara plant, Euphorbia paralias L., inhibits glutamine synthetase and has antimycobacterial activity. *Tuberculosis* **2018**, *108*, 106–113. [CrossRef]
162. Mowbray, S.L.; Kathiravan, M.K.; Pandey, A.A.; Odell, L.R. Inhibition of glutamine synthetase: A potential drug target in mycobacterium tuberculosis. *Molecules* **2014**, *19*, 13161–13176. [CrossRef]
163. Harth, G.; Clemens, D.L.; Horwitz, M.A. Glutamine synthetase of mycobacterium tuberculosis: Extracellular release and characterization of its enzymatic activity. *Proc. Natl. Acad. Sci. USA* **1994**, *91*, 9342–9346. [CrossRef] [PubMed]
164. Brown, A.K.; Papaemmanouil, A.; Bhowruth, V.; Bhatt, A.; Dover, L.G.; Besra, G.S. Flavonoid inhibitors as novel antimycobacterial agents targeting Rv0636, a putative dehydratase enzyme involved in mycobacterium tuberculosis fatty acid synthase II. *Microbiology* **2007**, *153*, 3314–3322. [CrossRef] [PubMed]
165. Duan, X.; Xiang, X.; Xie, J. Crucial components of mycobacterium type II fatty acid biosynthesis (Fas-II) and their inhibitors. *Fems. Microbiol. Lett.* **2014**, *360*, 87–99. [CrossRef] [PubMed]
166. Wu, D.; Kong, Y.; Han, C.; Chen, J.; Hu, L.; Jiang, H.; Shen, X. D-Alanine:D-alanine ligase as a new target for the flavonoids quercetin and apigenin. *Int. J. Antimicrob. Agents* **2008**, *32*, 421–426. [CrossRef]

167. Korablioviené, J.; Mauricas, M.; Ambrozevičiené, Č.; Valius, M.; Kaupinis, A.; Čaplinskas, S.; Korabliov, P. Mycobacteria produce proteins involved in biofilm formation and growth-affecting processes. *Acta Microbiol. Immunol. Hung.* **2018**, *65*, 405–418. [CrossRef]
168. Munayco, C.V.; Grijalva, C.G.; Culqui, D.R.; Bolarte, J.L.; Suárez-Ognio, L.A.; Quispe, N.; Calderon, R.; Ascencios, L.; Del Solar, M.; Salomón, M.; et al. Outbreak of persistent cutaneous abscesses due to mycobacterium chelonae after mesotherapy sessions, Lima, Peru. *Rev. Saude Publica* **2008**, *42*, 146–149. [CrossRef]
169. Kostakioti, M.; Hadjifrangiskou, M.; Hultgren, S.J. Bacterial biofilms: Development, dispersal, and therapeutic strategies in the dawn of the postantibiotic era. *Cold Spring Harb. Perspect. Med.* **2013**, 3, a010306. [CrossRef]
170. Falkinham, J.O., 3rd. Nontuberculous mycobacteria from household plumbing of patients with nontuberculous mycobacteria disease. *Emerg. Infect. Dis.* **2011**, *17*, 419–424. [CrossRef]
171. Williams, M.M.; Yakrus, M.A.; Arduino, M.J.; Cooksey, R.C.; Crane, C.B.; Banerjee, S.N.; Hilborn, E.D.; Donlan, R.M. Structural analysis of biofilm formation by rapidly and slowly growing nontuberculous mycobacteria. *Appl. Env. Microbiol.* **2009**, *75*, 2091–2098. [CrossRef]
172. Farhadi, F.; Khameneh, B.; Iranshahi, M.; Iranshahy, M. Antibacterial activity of flavonoids and their structure-activity relationship: An update review. *Phytother. Res.* **2019**, *33*, 13–40. [CrossRef]
173. Orhan, D.D.; Özçelik, B.; Özgen, S.; Ergun, F. Antibacterial, antifungal, and antiviral activities of some flavonoids. *Microbiol. Res.* **2010**, *165*, 496–504. [CrossRef] [PubMed]
174. Mandalari, G.; Bennett, R.N.; Bisignano, G.; Trombetta, D.; Saija, A.; Faulds, C.B.; Gasson, M.J.; Narbad, A. Antimicrobial activity of flavonoids extracted from bergamot (Citrus bergamia Risso) peel, a byproduct of the essential oil industry. *J. Appl. Microbiol.* **2007**, *103*, 2056–2064. [CrossRef] [PubMed]
175. Lin, Y.-M.; Zhou, Y.; Flavin, M.T.; Zhou, L.-M.; Nie, W.; Chen, F.-C. Chalcones and flavonoids as anti-Tuberculosis agents. *Bioorganic Med. Chem.* **2002**, *10*, 2795–2802. [CrossRef]
176. Bhunu, B.; Mautsa, R.; Mukanganyama, S. Inhibition of biofilm formation in mycobacterium smegmatis by Parinari curatellifolia leaf extracts. *BMC Complementary Altern. Med.* **2017**, *17*, 285. [CrossRef]
177. Nguta, J.M.; Appiah-Opong, R.; Nyarko, A.K.; Yeboah-Manu, D.; Addo, P.G. Current perspectives in drug discovery against tuberculosis from natural products. *Int. J. Mycobacteriol.* **2015**, *4*, 165–183. [CrossRef] [PubMed]
178. Liu, R.; Zhao, B.; Wang, D.E.; Yao, T.; Pang, L.; Tu, Q.; Ahmed, S.M.; Liu, J.J.; Wang, J. Nitrogen-containing apigenin analogs: Preparation and biological activity. *Molecules* **2012**, *17*, 14748–14764. [CrossRef] [PubMed]
179. Prawat, U.; Chairerk, O.; Phupornprasert, U.; Salae, A.W.; Tuntiwachwuttikul, P. Two new C-benzylated dihydrochalcone derivatives from the leaves of Melodorum siamensis. *Planta Med.* **2013**, *79*, 83–86. [CrossRef]
180. Gumula, I.; Heydenreich, M.; Derese, S.; Ndiege, I.O.; Yenesew, A. Four isoflavanones from the stem bark of Platycelphium voënse. *Phytochem. Lett.* **2012**, *5*, 150–154. [CrossRef]
181. Moreira, R.R.D.; Martins, G.Z.; Pietro, R.C.L.R.; Sato, D.N.; Pavan, F.R.; Leite, S.R.A.; Vilegas, W.; Leite, C.Q.F. *Paepalanthus* spp.: Antimycobacterial activity of extracts, methoxylated flavonoids and naphthopyranone fractions. *Rev. Bras. Farmacogn.* **2013**, *23*, 268–272. [CrossRef]
182. Alcalde-Rico, M.; Hernando-Amado, S.; Blanco, P.; Martínez, J.L. Multidrug efflux pumps at the crossroad between antibiotic resistance and bacterial virulence. *Front. Microbiol.* **2016**, *7*, 1483. [CrossRef]
183. Kumar, S.; Varela, M.F. Biochemistry of bacterial multidrug efflux pumps. *Int. J. Mol. Sci.* **2012**, *13*, 4484–4495. [CrossRef]
184. Gröblacher, B.; Kunert, O.; Bucar, F. Compounds of Alpinia katsumadai as potential efflux inhibitors in mycobacterium smegmatis. *Bioorg. Med. Chem.* **2012**, *20*, 2701–2706. [CrossRef]
185. Solnier, J.; Martin, L.; Bhakta, S.; Bucar, F. Flavonoids as novel efflux pump inhibitors and antimicrobials against both environmental and pathogenic intracellular mycobacterial species. *Molecules* **2020**, *25*, 734. [CrossRef]
186. Gröblacher, B.; Maier, V.; Kunert, O.; Bucar, F. Putative mycobacterial efflux inhibitors from the seeds of Aframomum melegueta. *J. Nat. Prod.* **2012**, *75*, 1393–1399. [CrossRef]
187. Roy, S.K.; Pahwa, S.; Nandanwar, H.; Jachak, S.M. Phenylpropanoids of Alpinia galanga as efflux pump inhibitors in mycobacterium smegmatis mc2 155. *Fitoterapia* **2012**, *83*, 1248–1255. [CrossRef]
188. Suriyanarayanan, B.; Sarojini Santhosh, R. Docking analysis insights quercetin can be a non-antibiotic adjuvant by inhibiting Mmr drug efflux pump in mycobacterium sp. and its homologue EmrE in Escherichia coli. *J. Biomol. Struct. Dyn.* **2015**, *33*, 1819–1834. [CrossRef]

189. Plaper, A.; Golob, M.; Hafner, I.; Oblak, M.; Solmajer, T.; Jerala, R. Characterization of quercetin binding site on DNA gyrase. *Biochem. Biophys. Res. Commun.* **2003**, *306*, 530–536. [CrossRef]
190. Fang, Y.; Lu, Y.; Zang, X.; Wu, T.; Qi, X.; Pan, S.; Xu, X. 3D-QSAR and docking studies of flavonoids as potent Escherichia coli inhibitors. *Sci. Rep.* **2016**, *6*, 23634. [CrossRef]
191. Hemaiswarya, S.; Kruthiventi, A.K.; Doble, M. Synergism between natural products and antibiotics against infectious diseases. *Phytomedicine* **2008**, *15*, 639–652. [CrossRef]
192. Wagner, H.; Ulrich-Merzenich, G. Synergy research: Approaching a new generation of phytopharmaceuticals. *Phytomedicine* **2009**, *16*, 97–110. [CrossRef] [PubMed]
193. Sahar Traoré, M.; Aliou Baldé, M.; Camara, A.; Saïdou Baldé, E.; Diané, S.; Telly Diallo, M.S.; Keita, A.; Cos, P.; Maes, L.; Pieters, L.; et al. The malaria co-infection challenge: An investigation into the antimicrobial activity of selected Guinean medicinal plants. *J. Ethnopharmacol.* **2015**, *174*, 576–581. [CrossRef] [PubMed]
194. Katerere, D.R.; Gray, A.I.; Nash, R.J.; Waigh, R.D. Phytochemical and antimicrobial investigations of stilbenoids and flavonoids isolated from three species of Combretaceae. *Fitoterapia* **2012**, *83*, 932–940. [CrossRef] [PubMed]
195. Dwyer, D.J.; Belenky, P.A.; Yang, J.H.; MacDonald, I.C.; Martell, J.D.; Takahashi, N.; Chan, C.T.; Lobritz, M.A.; Braff, D.; Schwarz, E.G.; et al. Antibiotics induce redox-related physiological alterations as part of their lethality. *Proc. Natl. Acad. Sci. USA* **2014**, *111*, E2100–E2109. [CrossRef]
196. Vatansever, F.; de Melo, W.C.; Avci, P.; Vecchio, D.; Sadasivam, M.; Gupta, A.; Chandran, R.; Karimi, M.; Parizotto, N.A.; Yin, R.; et al. Antimicrobial strategies centered around reactive oxygen species–bactericidal antibiotics, photodynamic therapy, and beyond. *Fems. Microbiol. Rev.* **2013**, *37*, 955–989. [CrossRef]
197. Van Acker, H.; Coenye, T. The role of reactive oxygen species in antibiotic-mediated killing of bacteria. *Trends Microbiol.* **2017**, *25*, 456–466. [CrossRef]
198. Jayachandran, M.; Zhang, T.; Ganesan, K.; Xu, B.; Chung, S.S.M. Isoquercetin ameliorates hyperglycemia and regulates key enzymes of glucose metabolism via insulin signaling pathway in streptozotocin-induced diabetic rats. *Eur. J. Pharm.* **2018**, *829*, 112–120. [CrossRef]
199. Kumar, G.; Sharmila Banu, G.; Murugesan, A.G.; Rajasekara Pandian, M. Effect of Helicteres isora. Bark Extracts on Brain Antioxidant Status and Lipid Peroxidation in Streptozotocin Diabetic Rats. *Pharm. Biol.* **2007**, *45*, 753–759. [CrossRef]
200. Brynildsen, M.P.; Winkler, J.A.; Spina, C.S.; MacDonald, I.C.; Collins, J.J. Potentiating antibacterial activity by predictably enhancing endogenous microbial ROS production. *Nat. Biotechnol.* **2013**, *31*, 160–165. [CrossRef]
201. Kohanski, M.A.; Dwyer, D.J.; Hayete, B.; Lawrence, C.A.; Collins, J.J. A common mechanism of cellular death induced by bactericidal antibiotics. *Cell* **2007**, *130*, 797–810. [CrossRef]
202. Ezraty, B.; Vergnes, A.; Banzhaf, M.; Duverger, Y.; Huguenot, A.; Brochado, A.R.; Su, S.Y.; Espinosa, L.; Loiseau, L.; Py, B.; et al. Fe-S cluster biosynthesis controls uptake of aminoglycosides in a ROS-less death pathway. *Science* **2013**, *340*, 1583–1587. [CrossRef]
203. Farha, M.A.; Brown, E.D. Discovery of antibiotic adjuvants. *Nat. Biotechnol.* **2013**, *31*, 120–122. [CrossRef]
204. Nøhr-Meldgaard, K.; Ovsepian, A.; Ingmer, H.; Vestergaard, M. Resveratrol enhances the efficacy of aminoglycosides against Staphylococcus aureus. *Int. J. Antimicrob. Agents* **2018**, *52*, 390–396. [CrossRef]
205. Ramirez, M.S.; Tolmasky, M.E. Aminoglycoside modifying enzymes. *Drug Resist. Updat.* **2010**, *13*, 151–171. [CrossRef]
206. Shakya, T.; Stogios, P.J.; Waglechner, N.; Evdokimova, E.; Ejim, L.; Blanchard, J.E.; McArthur, A.G.; Savchenko, A.; Wright, G.D. A small molecule discrimination map of the antibiotic resistance kinome. *Chem. Biol.* **2011**, *18*, 1591–1601. [CrossRef]
207. Girish, K.S.; Kemparaju, K. The magic glue hyaluronan and its eraser hyaluronidase: A biological overview. *Life Sci.* **2007**, *80*, 1921–1943. [CrossRef]
208. Hertel, W.; Peschel, G.; Ozegowski, J.-H.; Müller, P.-J. Inhibitory effects of triterpenes and flavonoids on the enzymatic activity of hyaluronic acid-splitting enzymes. *Arch. Pharm. Int. J. Pharm. Med. Chem.* **2006**, *339*, 313–318. [CrossRef]
209. Ochensberger, S.; Alperth, F.; Mitić, B.; Kunert, O.; Mayer, S.; Mourão, M.F.; Turek, I.; Luca, S.V.; Skalicka-Woźniak, K.; Maleš, Ž.; et al. Phenolic compounds of Iris adriatica and their antimycobacterial effects. *Acta Pharm.* **2019**, *69*, 673–681. [CrossRef]

210. Anakok, O.F.; Ndi, C.P.; Barton, M.D.; Griesser, H.J.; Semple, S.J. Antibacterial spectrum and cytotoxic activities of serrulatane compounds from the Australian medicinal plant Eremophila neglecta. *J. Appl. Microbiol.* **2011**, *112*, 197–204. [CrossRef]
211. Zgoda-Pols, J.R.; Freyer, A.J.; Killmer, L.B.; Porter, J.R. Antimicrobial Resveratrol Tetramers from the Stem Bark ofVaticaoblongifoliassp.oblongifolia. *J. Nat. Prod.* **2002**, *65*, 1554–1559. [CrossRef]
212. Roy, S.K.; Kumari, N.; Gupta, S.; Pahwa, S.; Nandanwar, H.; Jachak, S.M. 7-Hydroxy-(E)-3-phenylmethylene-chroman-4-one analogues as efflux pump inhibitors against mycobacterium smegmatis mc2 155. *Eur. J. Med. Chem.* **2013**, *66*, 499–507. [CrossRef]
213. Okemo, P.; Kirimuhuzya, C.; Otieno, J.; Magadula, J.; Mariita, R.; Orodho, J. Methanolic extracts of Aloe secundiflora Engl. inhibits in vitro growth of tuberculosis and diarrhea-causing bacteria. *Pharmacogn. Res.* **2011**, *3*, 95. [CrossRef] [PubMed]
214. Okemo, P.; Mbugua, P.; Mariita, R.; Orodho, J. Antifungal, antibacterial and antimycobacterial activity of Entada abysinnica Steudel ex A. Rich (Fabaceae) methanol extract. *Pharmacogn. Res.* **2010**, *2*, 163. [CrossRef] [PubMed]
215. Graham, J.G.; Pendland, S.L.; Prause, J.L.; Danzinger, L.H.; Schunke Vigo, J.; Cabieses, F.; Farnsworth, N.R. Antimycobacterial evaluation of Peruvian plants. *Phytomedicine* **2003**, *10*, 528–535. [CrossRef] [PubMed]
216. Sharma, A.; Dutta, P.; Sharma, M.; Rajput, N.K.; Dodiya, B.; Georrge, J.J.; Kholia, T.; Bhardwaj, A. BioPhytMol: A drug discovery community resource on anti-mycobacterial phytomolecules and plant extracts. *J. Cheminf.* **2014**, *6*, 1–10. [CrossRef]
217. Cantrell, C.; Franzblau, S.; Fischer, N. Antimycobacterial plant terpenoids. *Antimycobact. Planta Med.* **2001**, *67*, 685–694. [CrossRef]

antibiotics

MDPI

Review

Antimicrobial Secretions of Toads (Anura, Bufonidae): Bioactive Extracts and Isolated Compounds against Human Pathogens

Candelario Rodriguez [1,2,3,4], Roberto Ibáñez [5,6], Louise A. Rollins-Smith [7], Marcelino Gutiérrez [1,*] and Armando A. Durant-Archibold [1,2,*]

1 Centro de Biodiversidad y Descubrimiento de Drogas, Instituto de Investigaciones Científicas y Servicios de Alta Tecnología (INDICASAT AIP), Clayton, Panama City 0843-01103, Panama; Crodriguez@indicasat.org.pa
2 Departamento de Bioquímica, Facultad de Ciencias Naturales, Exactas y Tecnología, Universidad de Panamá, Apartado 0824-03366, Panama
3 Department of Biotechnology, Acharya Nagarjuna University, Nagarjuna Nagar, Guntur 522510, India
4 Scientific Station COIBA, (COIBA AIP), Ciudad del Saber, Apartado 0816-02852, Panama
5 Smithsonian Tropical Research Institute (STRI), Balboa 0843-03092, Panama; ibanezr@si.edu
6 Departamento de Zoología, Facultad de Ciencias Naturales, Exactas y Tecnología, Universidad de Panamá, Apartado 0824-03366, Panama
7 Department of Pathology, Microbiology, and Immunology, and Department of Pediatrics, Vanderbilt University School of Medicine, Nashville, TN 37232, USA; louise.rollins-smith@vanderbilt.edu
* Correspondence: mgutierrez@indicasat.org.pa (M.G.); adurant@indicasat.org.pa (A.A.D.-A.)

Received: 29 October 2020; Accepted: 23 November 2020; Published: 26 November 2020

check for updates

Abstract: Species of the family Bufonidae, better known as true toads, are widespread and produce bioactive substances in the secretions obtained from specialized skin macroglands. Some true toads have been employed as a folk remedy to treat infectious diseases caused by microbial pathogens. Recent publications based on in silico analysis highlighted the Bufonidae as promising sources of antimicrobial peptides. A review of the literature reveals that Bufonidae skin secretion extracts show inhibitory activity in vitro against clinical isolates of bacteria, resistant and standard strains of bacterial, and fungal and parasitic human pathogens. Secondary metabolites belonging to the classes of alkaloids, bufadienolides, and peptides with antimicrobial activity have been isolated from species of the genera *Bufo*, *Bufotes*, *Duttaphrynus*, and *Rhinella*. Additionally, some antimicrobial extracts and purified compounds display low cytotoxicity against mammal cells.

Keywords: alkaloids; amphibians; antimicrobial peptides; bufadienolides; Bufonidae; crude extract; parotoid gland; skin secretion; toad

1. Introduction (True Toads, Antimicrobial Resistance of Human Pathogens against Drugs)

Amphibians of the Bufonidae family are distributed worldwide, mainly in arid and glacial regions [1]. More than 600 species of the Bufonidae family have been classified into 50 genera, with the genus *Bufo* being the most abundant. [2]. Toads of the Bufonidae family are known as "true toads" because they possess specialized macroglands, identified as parotoids, behind the eyes [3]. Additionally, some true toads have macroglands on the limbs (Figure 1a,b).

The bufonid secretions contain small molecules such as alkaloids and steroids, as well as larger molecules such as peptides and proteins [4]. Skin gland secretions can be obtained by gentle manual compression, although other techniques have been successfully applied such as mild electrical discharges and chemical stimulation (Figure 1c,d). Amphibian stimulation should be carried out at intervals of 4 weeks in order to avoid harm to the animals [5]. Parotoid and skin glands of amphibians are involved in cutaneous respiration, reproduction, thermoregulation, and defense [6]. In vivo assays, via intraperitoneal injection in mice, revealed that parotoid gland secretions of *Rhinella marina* (Linnaeus, 1758) and *Rhaebo guttatus* (Schneider, 1799) contain metabolites that may act as antipredator agents through toxic and nociceptive effects [7]. True toads biosynthesize peptides and steroids that are able to inhibit the growth of amphibian pathogens. Three steroids, arenobufagin, gamma-bufotalin, and telocinobufagin, of the bufadienolide class were isolated from mucosal and skin gland secretions of the toad *Anaxyrus boreas* (Baird and Girard, 1852). All three bufadienolides showed activity against the lethal fungus of amphibians *Batrachochytrium dendrobatidis* (Bd). Furthermore, arenobufagin enhanced the growth of the anti-Bd bacterium *Janthinobacterium lividum* [8]. Two cathelicidin-derived peptides designated as BG-CATH(37) and BG-CATH(5-37) were identified from the DNA of the Asian toad *Bufo gargarizans* (Cantor, 1842). Both peptides were found to inhibit the aquatic bacteria *Vibrio splendidus*, *Streptococcus iniae*, and *Aeromorus hydrophila* that are usually found in the habitat of this toad [9].

Scientific research focused on the potential use of extracts and compounds isolated from true toads for medical treatment has increased during the last years [10]. Currently, clinical trials are underway to evaluate the anticancer activity of the aqueous extracts obtained from the dried skin of *Bufo gargarizans* [11]. Some evidence supports the use of bufonid secretions, found in true toads, in folk treatments for diseases caused by microbes. In the Colombian forest region of Cundinamarca, species of *Rhinella* toads are employed to treat erysipelas, which is an infection caused by the multidrug-resistant bacteria *Streptococcus pyogenes*. The treatment consists of rubbing the toad skin on the infected areas of the body in order to eradicate the infection [12]. Furthermore, *Rhinella marina* toads, from the Brazilian Amazon, have also been used for treatment of patients with erysipelas [13]. In the Northeastern region of Brazil, the skin and fat tissues from *Rhinella jimi* (Stevaux, 2002) are employed to treat asthma, cancer, and infections [14].

Figure 1. Species of true toads, with white arrows indicating skin macroglands. (**a**) Adult specimen of *Rhinella horribilis* (Wiegmann, 1833). (**b**) Adult toad of *Incilius alvarius* (Girard, 1859); (© 2014 Hunter McCall). (**c**) Mild electrical stimulation of an adult male of *Rhinella centralis* (Narvaes and Rodrigues, 2009) employing a transcutaneous amphibian stimulator [15]. (**d**) Stimulation of *Rhinella horribilis* by gentle compression of the parotoid glands.

Due to the low efficacy of many antimicrobial drugs against many pathogenic microorganisms, the discovery of new antimicrobial compounds, such as peptide substances, has increasingly gained attention [16]. In this regard, amphibians are rich sources of antimicrobial peptides, and more than

100 of these chemical compounds have been discovered. These molecules confer to amphibians immunological, analgesic, antimicrobial, antiparasitic, hormonal regulation, mating, and wound healing properties for survival [17]. Recently, it was found by employing transcriptomic and peptidomic techniques that the deionized-water extract of the parotoid gland secretion from *B. gargarizans* contains a high amount of unique peptides, 23 of which showed defensive properties [18]. Antimicrobial peptides of the buforin, kassinin, temporin, peroniin, and rugosauperoilin families were identified from the DNA of the skin of the toad *Rhinella diptycha* (Cope, 1862). Caseinolytic activity was detected confirming the protease activity of toad skin secretions [19]. Furthermore, eight peptides with potential antimicrobial activities were predicted from the DNA sequence of H2A histones from the toads *Duttaphrynus melanostictus* (Schneider, 1799) and *Phrynoidis asper* (Gravenhorst, 1829). In silico analysis revealed that these peptides show a positively charged face, aggregation of bacterial cell membranes, and absence of cleavage sites for alpha-chymotrypsin, which suggests promising antimicrobial activity [20].

According to the World Health Organization (WHO) three of the top 10 causes of deaths worldwide are bacterial diarrhea, lower respiratory infections, and tuberculosis [21]. Fungal infections (mycosis) are opportunistic and affect mainly asthmatic, human immunodeficiency virus (HIV)-infected, and hospitalized patients. Current statistical data estimate 15,000 and 181,000 new cases worldwide per year of aspergillosis and cryptococosis, respectively [22]. Parasitic tropical diseases remain a challenge. There is a high incidence of leishmaniasis in 90 countries, with around eight million people at risk for American trypanosomiasis, 2000 positive cases between 2017 and 2018 of African trypanosomiasis, and about 1700 diagnosed cases each year of malaria in the United States, mostly in returning travelers [23]. Microbial illnesses that have been in the population but are increasing rapidly in prevalence and geographic localities are defined as reemerging infections [24]. The incidence of these diseases is mainly related to microbial adaptation, partly as a consequence of the inadequate and overuse of antimicrobial drugs by humans. Additionally, pathogenic bacteria, fungi, and parasites are able to develop strategies such as specific gene overexpression, incorporation of new virulence genes, and rapid genetic mutations to overcome biological pressures. These gene adaptions are then translated into proteins that confer microbial resistance [25].

Pathogenic microbes such as bacteria, fungi, and protozoan parasites may resist the bioactivity of drugs due to mechanisms related to membrane permeability, efflux systems, binding target, and drug inactivation [26]. The Gram-negative pathogen *Pseudomonas aeruginosa* changes membrane permeability against the antibiotic imipenem via deletion of D2 pores involved in drug transport across the plasma membrane [27]. The intracellular concentration of a drug decreases with the use of ATP-binding cassette (ABC) efflux pump systems in bacteria and represents a major cause of multidrug resistance [28]. Bacteria are able to develop modification to alter the drug-binding target interaction. Recently, it was reported that the *Mycobaterium tuberculosis* genome encodes 20 different cytochrome p450 (CYP) enzymes, which are involved in the biosynthesis of membrane steroids. Inhibition of CYP51B1 by rifampicin remains among the only few options for treatment of tuberculosis [29]. Gram-negative and Gram-positive pathogens produce enzymes that decrease the activity of antibiotics by modifying their chemical structure. Chloramphenicol-resistant strains of *Escherichia coli* and *Staphylococcus aureus* produce chloramphenicol acetyltransferase, an enzyme that catalyzes the acetylation of chloramphenicol into the inactive metabolite 1,2-di-acetyl-chloramphenicol [30]. Penicillin-resistant *S. aureus* strains deactivate β-lactam antibiotics via hydrolysis of the penam ring. β-Lactamases transform penicillins into penicilloic acids, which are not toxic for bacteria [31].

Fungal pathogens have been shown to produce efflux pumps at the plasma membrane for antimycotic azoles [32]. An ABC transporter was characterized in *Cryptococcus neoformans* and found to confer resistance against fluconazole [33]. Two genes that code for ABC transporters were identified in *Candida krusei*. Both genes display correlation with resistance against miconazole [34]. Alterations in the chemical composition at the membrane level in order to decrease susceptibility against antimycotics have been reported for human pathogenic fungi [32]. An amphotericin-resistant strain of *Aspergillus flavus* was produced experimentally from a susceptible wild-type strain. Chemical analysis of the cell membrane showed significant changes in chemical composition of carbohydrates, suggesting that acquired resistance is associated with modifications in membrane glycans [35]. Clinical isolates of *Candida albicans* that present mutations on the gen *erg3*, which encodes for ergosterol biosynthesis, were found to display resistance against amphotericin. The chemical analysis revealed highly decreased ergosterol content in comparison with polyene-sensitive strains [36].

Trypansoma brucei, the causative agent of African trypanosomiasis, exhibits cross-resistance against the drugs melarsoprol and pentamidine. This adaptation was revealed to be caused by the loss of the aquaporin-2 (AQP2) function at the cell membrane [37]. Genetic mutations on the *Plasmodium falciparum* chloroquine-resistant transporter (PFCRT) were detected in resistant clinical isolates [38]. Induced overexpression of the P-glycoprotein (LtrMDR1) from *Leishmania tropica* revealed that this protein pumps miltefosine out of the parasite cells, conferring resistance by decreasing intracellular drug concentrations [39]. The cellular disruption caused by metabolites produced by drug metabolism in parasites can be neutralized by the catalytic actions of cytosolic oxide-reductase enzymes. In this sense, the gene expression profile among sensitive and resistant strains of *Leishmania braziliensis* clinical isolates to antimony was evaluated. The results showed overexpression of the gene *TRYR*, which encodes the parasitic enzyme thrypanothione reductase [40].

Resistance mechanisms expressed by pathogenic microbes against drugs reveal the need for the development of more efficient treatments ideally focused on different targets not recognized by pathogens. In this sense, therapy based on bacteriophages, also known as phages, which are viruses that invade bacteria and disrupt the metabolism leading to bacterial lysis, has gained attention [41]. Another option is the research for new chemical entities able to inhibit growth and, hence, overcome the resistance that pathogenic microbes display. Bioactive compounds obtained from understudied natural sources such as toxins from amphibians, insects, marine invertebrates, spiders, and reptiles represent a promise for unknown chemical entities [42]. Former review articles described different bioactivities of compounds and extracts, isolated from the skin of amphibians, against several diseases [4,10]. Hence, our present work focused on a description and understanding, for the first time, of the importance of natural products isolated from true toads with potential for the development of antimicrobial drugs.

2. Methodology

This article reviews the literature of antimicrobial activity of molecules and extracts prepared from secretions of toads from the family Bufonidae. Selection of published information from the electronic databases PubMed. Gov, MDPI.com, SciFinder®, and ScienDirect.com, as well as Google Scholar, was carried out employing the key words "amphibians", "antimicrobial secretions", "animal venom", "Bufonidae", and "toads". As result, this literature review covers papers from 1975 to 2020.

3. Antimicrobial Activities

Compounds and crude extracts prepared from skin secretions obtained from toads of the family Bufonidae display inhibitory activity against clinically isolated, drug resistant, and standard pathogenic strains of microbes (Table 1).

3.1. Antibacterial

The aqueous-soluble fraction of the parotoid gland secretion and the methanolic skin extract from the Sudan toad *Duttaphrynus melanostictus* have antibiotic potential against *Bacillus cereus*, *Escherichia coli*, *Klebsiella pneumoniae*, *Salmonella typhimurium*, *Staphylococcus aureus*, and *Staphylococcus epidermidis*; however, no activity was found against a methicillin-resistant strain of *S. aureus* [43,44]. The granular secretion from *Rhinella jimi* was extracted with 70% ethanol and evaluated against drug-resistant bacteria; however, no activity was observed. When tested in association with antibiotics of the aminoglycosides and β-lactam classes, the secretion was able to reduce the minimal inhibitory concentration (MIC) values of antibiotics against clinical isolates of *E. coli*, *Pseudomonas aeruginosa*, and *S. aureus* [45]. The dorsal skin secretion from *Phrynoidis asper* was extracted with deionized water and investigated for antibiotic potential. As a result, it was shown to have antibiotic properties with pronounced activity against the Gram-positive pathogens *S. aureus* and *Bacillus subtilis*. The protein profile of the antimicrobial skin secretion of *P. asper* was analyzed by tandem mass spectrometry (MS^2). The bands from sodium dodecyl sulfate polyacrylamide gel electrophoresis (SDS-PAGE) wells were digested and showed fragments of proteins involved in different cellular processes, such as actins, cathepsins, histones, and synthases [46]. Buforin I, an antimicrobial peptide (AMP) with broad-spectrum activity, was isolated from the stomach of the toad *Bufo gargarizans* (Figure 2). Sequencing by Edman's degradation revealed that Buforin I (MH^+ = 4309 g/mol) and its endoproteinase Lys-C product, named Buforin II (MH^+ = 2432 g/mol), are homologous to the N-terminal region of *Xenopus* histone H2A. Buforin II was found to have more potent bioactivity than Buforin I [47]. Labeling experiments showed that Buforin II is able to penetrate cell membranes of *E. coli* even at doses lower than its MIC value, as well as bind to RNA and DNA and, hence, disrupt cellular metabolism [48]. Structural–activity relationship analysis testing of synthesized analogues of Buforin II revealed that the proline hinge is essential for antimicrobial activity [49]. The skin secretion of the toad *Bufotes sitibundus* (Pallas, 1771) contains the peptides Buforin I, Maximin 1, Alyteserin-1a, and a novel +2 charged hydrophobic peptide named Maximin-Bk. This peptide has a molecular weight of 2012 g/mol (MH^+) and showed antimicrobial activity with very low hemolytic activity against human erythrocytes even at 100 mg/mL [50]. Lectins with antibacterial activity LBP-1 (50 KDa) and LBP-2 (56 KDa) were isolated from skin of *Rhinella arenarum* (Hensel, 1867) via saccharide extraction. Microbial inhibition was determined to be bacteriostatic, and both proteins did not agglutinate human type O, A, and B erythrocytes [51]. Total aqueous crude secretion extracts were prepared by pooling skin (granular gland) and parotoid gland secretions followed by extraction with ultra-pure water. Total aqueous crude extracts obtained from *Bufo bufo* (Linnaeus, 1758), *Bufo verrucosissimus* (Pallas, 1814), and *B. sitibundus* displayed activity against *E. coli*, *S. aureus*, *Enterococcus faecalis*, *Enterococcus faecium*, *S. epidermidis*, and *S. thyphimurium* standard strains without hemolysis between 0.5 and 50 mg/mL [52]. A lysozyme (15 kDa) was isolated from the saline-soluble skin secretion extract of the Asian toad *B. gargarizans*. The protein was named Ba-lysozyme and exhibits potent bactericidal activity. The complete sequence of Ba-lysozyme was deduced by peptide mass fingerprinting and phylogenetic analysis. According to the occurrence in Ba-lysozyme of the amino-acid residues at positions glutamate-35 and aspartate-52, commonly found in other lysozymes and essential for lytic activity, the antibiotic mechanism of Ba-lysozyme is suggested to be via enzymatic degradation; however, the potent activity against both *S. aureus* (Gram+) and *E. coli* (Gram−) points to a different mechanism [53]. Solutions prepared from the parotoid gland secretion of *Rhinella icterica* (Spix, 1824) (no solvent reported) are able to inhibit the growth of *E. coli* and *S. aureus* in 15 and 30 min, respectively. This difference in antimicrobial action has been observed for some drugs, such as lincomycin and erythromycin, and may be related to differences in the structure of their cell walls [54]. Bufotenine, an indole alkaloid widespread among toads, shows antibacterial activity against *B. subtilis*, *E. coli*, *Proteus mirabilis*, and *S. aureus* with inhibition halos of 4, 6, 9, and 5 mm, respectively [55,56]. Two bufadienolides, known as marinobufagin and telocinobufagin, were isolated as major components from the chloroform/methanol (9:1)-soluble parotoid gland secretion extract of the toad *Rhinella rubescens* (Lutz, 1925). Both steroids presented antimicrobial properties with MIC values comparable to commercial antibiotics [57]. More recently, it was revealed that telocinobufagin was able to significantly decrease the bacterial burdens in spleen and enhance the Th1 immune response against the pathogen *S. typhimurium* in mice, as revealed by interferon gamma induction. Furthermore, after co-injection with formalin-inactivated *S. typhimirium*, telocinobufagin promoted the production of immunoglobulin G (IgG) and IgG2a antibodies, making it a promising adjuvant for antibiotic vaccines [58].

Table 1. Antimicrobial activity from true toads against human pathogenic bacteria, fungi, and parasites.

Toad Species [a]	Sample (Solvent Used)	Effect	Bioactivity [b,c,d]	Toxicity (μg/mL) [d,e]	Reference
Bufo bufo	Skin gland + parotoid gland secretion (ultra-pure water)	Antibacterial	MIC (μg/mL): 250 (*S. thyphimirium*); 62.5 (*S. aureus*); 3.9 (*E. Faecalis, E. faecium, S. epidermidis*)	IC_{50}: 0.35 (HEK-293)	[52]
		Antifungal	MIC (μg/mL): 250 (*C. albicans*)		
Bufo gargarizans (*Bufo andrewsi*)	Ba-lysozyme	Antibacterial	MIC (μM): 8 (*E. coli*); 1 (*S. aureus*)	*	[53]
Bufo gargarizans	Buforin-I	Antibacterial	MIC (μg/mL): 8 (*E. coli, Serratia* sp., *S. mutans*); 4 (*B. subtilis, P. putida, S. thyphimurium, S. aureus, S. pneumoniae*)	*	[47]
		Antifungal	MIC (μg/mL): 4 (*C. albicans, C. neoformans*)		
Bufo verrucosissimus	Skin gland + parotoid gland secretion (ultra-pure water)	Antibacterial	MIC (μg/mL): 62.5 (*S. aureus*); 3.9 (*E. faecalis, E. faecium, S. epidermidis*)	IC_{50}: 0.99 (HEK-293)	[52]
		Antifungal	MIC (μg/mL): 125 (*C. albicans*)		
Bufotes sitibundus (*Bufo kavirensis*)	Maximin-Bk	Antibacterial	MIC (μg/mL): 20.78 (*L. mesenteroides*); 19.4 (*B. subtilis*); 18.5 (*B. cereus*); 16.3 (*S. aureus*); 10.3 (*P. aeruginosa*); 8.9 (*K. pneumoniae*); 8.1 (*E. coli*)	*	[50]
		Antifungal	MIC (μg/mL): 35.6 (*A. fumigates*); 32.1 (*C. albicans*); 28.6 (*A. niger*); 25.7 (*P. lilacinum*)		
Bufotes sitibundus (*Bufotes variabilis*)	Skin gland + parotoid gland secretion (ultra-pure water)	Antibacterial	MIC (μg/mL): 125 (*S. aureus, S. thyphimirium*); 7.8 (*S. epidermidis*); 3.9 (*E. faecalis, E. faecium*)	IC_{50}: 1.46 (HEK-293)	[52]
		Antifungal	MIC (μg/mL): 125 (*C. albicans*)		

Table 1. *Cont.*

Toad Species [a]	Sample (Solvent Used)	Effect	Bioactivity [b,c,d]	Toxicity (μg/mL) [d,e]	Reference
Duttaphrynus melanostictus (*Bufo melanostictus*)	Granular gland secretion (0.9% NaCl)	Antifungal	Halo zones (7.5% *w*/*v*): *C. albicans* (20.3 mm); *M. gypsum* (24.5 mm); *T. mentagrophytes* (24.7 mm)	*	[59]
	Parotoid gland secretion (distilled water)	Antibacterial	Halo zones: *E.coli* (7 mm); *S. typhimurium* (7 mm); *S. epidermidis* (8 mm); *K. pneumoniae* (9 mm); *B. cereus* (10 mm); *S. aureus* (11 mm)	*	[43]
	Skin (0.9% NaCl)	Antifungal	Halo zones: *P. notatum* (21 mm); *A. niger* (23 mm)	*	[60]
		Antibacterial	Halo zones: *E. coli* (19 mm); *K. pneumoniae* (24 mm); *S. aureus* (27 mm); *P. vulgaris* (33 mm)	*	[44]
	Skin gland secretion (distilled water)	Antibacterial	Halo zones: *K. pneumoniae* (19 mm); *E. coli* (25 mm); *P. vulgaris* (28 mm); *S. aureus* (30 mm)	*	[44]
		Antifungal	Halo zones: *P. notatum* (23 mm); *A. niger* (25 mm)	*	[60]
Leptophryne cruentata	Skin gland secretion (acetate buffer)	Antifungal	Halo zone: *T. mentagrophytes* (14.5 mm)	*	[61]
Phrynoidis asper (*Bufo asper*)	Skin secretion (deionized water)	Antibacterial	MIC (μg/mL): 100 (*E. coli*); 50 (*B. cereus*, *K. pneumoniae*, *P. aeruginosa*); 25 (*B. subtilis*); 12 (*S. aureus*)	*	[46]
Rhaebo guttatus	Skin gland secretion ($CHCl_3$/MeOH)	Antiparasitic	IC_{50} (μg/mL): 0.05 (*P. falciparum*, ring)	LD_{50}: 34.83 (BGM)	[62]

Table 1. *Cont.*

Toad Species [a]	Sample (Solvent Used)	Effect	Bioactivity [b,c,d]	Toxicity (µg/mL) [d,e]	Reference
Rhinella arenarum (*Bufo arenarum*)	Venom (distilled water)	Antibacterial	MIC (µg/mL): >1250 (*A. baumannii, B. subtilis*); 1250 (*E. coli, K. pneumoniae, S. aureus*); 625 (*A. hydrophila*); 312.5 (*P. aeruginosa*)	*	[63]
		Antifungal	Inactive for *C. albicans* and *A. niger*		
	LBP-1	Antibacterial	Halo zones (25 µg): *E. coli* (16 mm); *E. faecalis* (12 mm); *P. morganii* (20 mm)	*	[51]
	LBP-2	Antibacterial	Halo zones (25 µg): *E. coli* (17.5 mm); *E. faecalis* (12.5 mm); *P. morganii* (19 mm)		
Rhinella centralis	19-Hydroxy-bufalin	Antiparasitic	IC_{50} (µg/mL): 7.81 (*T. cruzi*, tryp)	IC_{50}: 71.58 (Vero)	[64]
Rhinella icterica	Parotoid gland secretion	Antibacterial	medium inhibition at 25 mg/mL for *E. coli; S. aureus*	*	[54]
Rhinella jimi	Parotoid gland secretion (EtOH)	Antibacterial	MIC (µg/mL): ≥2048 (*E. coli; P. aeruginosa; S. aureus*)	LD_{50}: 365.94 (shrimp)	[45]
	Hellebrigenin	Antiparasitic	IC_{50} (µg/mL): 126.2 (*L. chagasi*, prom); 91.75 (*T. cruzi*, tryp)	$IC_{50} > 200$ (Macrophages)	[65]
	Telocinobufagin	Antiparasitic	IC_{50} (µg/mL): 61.2 (*L. chagasi*, prom)		
Rhinella marina	Skin gland secretion (crude)	Antibacterial	MIC (µg/mL): >25 (*E. coli*); 21 (*S. aureus*); 10.79 (*P. aeruginosa*)	$IC_{50} > 100$ (MRC5)	[66]
		Antiparasitic	IC_{50} (µg/mL): 14.82 (*L. braziliensis*, prom); 9.34 (*L. guyanensis*, prom); 2.43 (*P. falciparum*, ring)		
	Skin gland secretion (MeOH)	Antiparasitic	MIC (µg/mL): ≥100 (*L. braziliensis*, prom); 12.04 (*P. falciparum*, ring); 3.99 (*L. guyanensis*, prom)	*	[66]
	Skin gland secretion ($CHCl_3$/MeOH)	Antiparasitic	IC_{50}: 0.534 (*P. falciparium*, ring)	$LD_{50} > 200$ (BGM)	[62]
	16-Desacetil-cinobufagin	Antibacterial	MIC (µg/mL) < 3.12 (*E. coli; P. aeruginosa; S. aureus*)	$IC_{50} > 100$ (MRC5)	[66]
	Marinobufagin	Antibacterial	MIC (µg/mL) < 3.12 (*S. aureus*)	$IC_{50} > 100$ (MRC5)	[66]
	Telocinobufagin	Antiparasitic	IC_{50} (µg/mL): 1.28 (*P. falciparum*, ring)	$IC_{50} > 200$ (BGM)	[62]

Table 1. *Cont.*

Toad Species [a]	Sample (Solvent Used)	Effect	Bioactivity [b,c,d]	Toxicity (μg/mL) [d,e]	Reference
Rhinella rubescens (*Bufo rubescens*)	Marinobufagin	Antibacterial	MIC (μg/mL): 128 (*S. aureus*); 16 (*E. coli*)	*	[57]
	Telocinobufagin	Antibacterial	MIC (μg/mL): 128 (*S. aureus*); 64 (*E. coli*)		
Sclerophrys pantherina (*Amietophrynus pantherinus*)	Skin gland secretion (phosphate buffer)	Antifungal	MIC (mg/mL): 0.39 (*F. verticillioides*); 0.04 (*C. albicans*); 0.02 (*A. flavus*)	*	[67]

[a] Current names, scientific names in the original publications are in parentheses. [b] Abbreviations: Gram—(*A. baumannii—Acinetobacter baumannii; A. hydrophila—Aeromonas hydrophila; E. coli—Escherichia coli; K. Pneumoniae—Klebsiella pneumoniae; P. aeruginosa—Pseudomonas aeruginosa; P. vulgaris—Proteus vulgaris; P. morganii—Proteus morganii; S. thyphimurium—Salmonella thyphimurium*); Gram+ (*B. cereus—Bacillus cereus; B. subtilis—Bacillus subtilis; E. faecalis—Enterococcus faecalis; E. faecium—Enterococcus faecium; L. mesenteroides—Leuconostoc mesenteroides; P. putida—Pseudomonas putida; S. aureus—Staphylococcus aureus; S. epidermidis—Staphylococcus epidermidis; S. mutans—Streptococcus mutans; S. pneumoniae—Streptococcus pneumoniae*); fungi (*A. flavus—Aspergillus flavus; A. fumigatus—Aspergillus fumigatus; A. niger—Aspergillus niger; C. albicans—Candida albicans; C. neoformans—Cryptococcus neoformans; F. verticillioides—Fusarium verticillioides; M. gypsum—Microsporum gypsum; P. lilacinum—Penicillium lilacinum; P. notatum—Penicillium notatum; T. mentagrophytes—Trichophyton mentagrophytes*); protozoas (*L. braziliensis—Leishmania braziliensis; L. chagasi—Leishmania chagasi; L. guyanensis—Leishmania guyanensis; P. falciparum—Plasmodium falciparum; T. cruzi—Trypanosoma cruzi*). Drug-resistant or clinically isolated strains are in bold. [c] Parasitic stages: prom: promastigotes, tryp: trypomastigotes, ring: ring stage; [d] MIC: minimal inhibitory concentration; IC_{50}: half maximal inhibitory concentration; LD_{50}: median lethal dose. [e] Toxicity model description: BGM, kidney glomerular cells; HEK-293, noncancerous kidney cells; macrophages, BALB/C mice macrophages; MRC5, lung human normal fibroblasts; shrimp, Artemia salina larvae; Vero, epithelial kidney monkey cells. * Not evaluated.

a

Buforin I = AGRGKQGGKVRAKAKTRSSRAGLQFPVGRVHRLLRKGNY

Maximin-Bk = ILGPVLGLVGRLAGGLIKRE

b

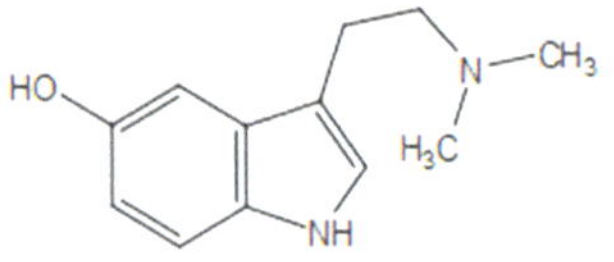

Alkaloid	Formula	MW
Bufotenine	$C_{12}H_{16}N_2O$	204.2

c

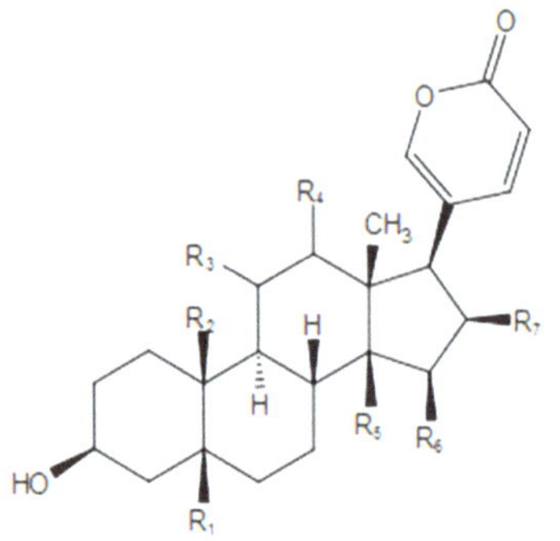

Bufadienolide	R_1	R_2	R_3	R_4	R_5	R_6	R_7	Formula	MW
Arenobufagin	H	CH_3	a-OH	=O	OH	H	H	$C_{24}H_{32}O_6$	416.5
19-hydroxy-Bufalin	H	CH_2OH	H	H	OH	H	H	$C_{24}H_{34}O_5$	402.5
gamma-Bufotalin	H	CH_3	a-OH	H	OH	H	H	$C_{24}H_{34}O_5$	402.5
16-deacetyl-Cinobufagin	H	CH_3	H	H	-O-		OH	$C_{24}H_{32}O_5$	400.5
Hellebrigenin	OH	CHO	H	H	OH	H	H	$C_{24}H_{32}O_6$	416.5
Marinobufagin	OH	CH_3	H	H	-O-		H	$C_{24}H_{32}O_5$	400.5
Telocinobufagin	OH	CH_3	H	H	OH	H	H	$C_{24}H_{34}O_5$	402.5

Figure 2. Chemical structures of antimicrobial molecules isolated from Bufonidae. (**a**) Primary structure of peptides; the underlined sequence in Buforin I represents the primary structure of buforin II. (**b**) Alkaloid. (**c**) Bufadienolides. MW: Molecular weight (g/mol).

3.2. Antifungal

Screenings of crude skin secretions from bufonids have become a recent field of research for novel sources of antimycotics. The aqueous-soluble parotoid gland secretions from the toads *B. bufo*, *B. verrucosissimus*, and *B. variabilis* inhibit the growth of *Candida albicans* ATCC 10239. Additionally, extracts from these toads were evaluated for hemolytic activity on red blood cells from healthy rabbits, and no toxicity was detected [52]. The methanol-soluble skin extract and the water-soluble parotoid gland secretions from the toad *D. melanostictus* were found to be active with a similar inhibition halo against *Aspergillus niger* and *Penicillium notatum* [60]. Micrographs of the fungal morphology after treatment with the *D. melanostictus* secretions prepared by extraction with physiological saline revealed disrupted integrity of the cell wall, which was observed as pore formation and shrinkage of the membrane [40]. The human pathogen *Trichophyton mentagrophytes* is inhibited by the saline phosphate buffer-soluble extract of the skin secretion from the toad *Leptophryne cruentata* (Tschudi, 1838). According to gas chromatography (GC) analysis, *L. cruentata* skin secretions contain amines, fatty acids, and steroids; however, no indole alkaloids were detected [61]. Toads of the family Bufonidae biosynthesize indole alkaloids that may represent as much as 15% of the dried weight of skin gland secretions [68]. The antifungal effects of amphibian alkaloids, such as samandarines and indoles, were evaluated on the nonpathogenic fungus *Saccharomyces cerevisiae*. The results showed that the alkaloids induced disturbances in the cytoplasm and plasma membrane, as observed by the appearance of vacuoles and translucent bodies in the cytoplasm [56]. The skin secretion from *Rhinella jimi* (Stevaux, 2002) was purified by Soxhlet extraction. The fixed oil obtained was evaluated in combination with some antifungal drugs. The oil extract was able to increase the potency (MIC) of amphotericin-B from 512 to 64 µg/mL against *Candida krusei* ATCC 6258. The chemical analysis by GC revealed that the fixed oil of *R. jimi* is composed mainly of the methyl esters of linoleic, oleic, palmitic, and stearic fatty acids [69]. Recently, the antifungal properties of some methyl esters against human pathogenic fungi were revealed [70]. The skin secretion of the South African toad *Sclerophrys pantherina* (Smith, 1828) was extracted with saline buffer phosphate and was found to be bioactive against the pathogens *Aspergillus flavus*, *C. albicans*, and *Fusarium verticillioides*. A temporal analysis of the MIC values suggested that the toad secretion inhibited fungal growth by killing the fungus, as revealed by stable inhibition for 120 h [67]. Two peptides with antifungal properties have been isolated from bufonid toads. Buforin I from *B. gargarizans* was equipotent against standard strains of *C. albicans*, *Cryptococcus neoformans*, and *S. cerevisae* [47]. Maximin-Bk from *B. sitibundus* was able to inhibit *Aspergillus fumigatus*, *A. niger*, *C. albicans*, and *Penicillium lilacinum* at nanomolar levels. On the basis of the observed low MIC values, researchers suggested that the inhibitory activity of Maximin-Bk is carried out via fungicidal effects [50].

3.3. *Antiprotozoal*

The antimicrobial potential of bufadienolides from toads was initially ignored and believed not to possess significant inhibitory activities against human pathogens [71]. Hellebrigenin and telocinobufagin were extracted from the skin secretion of the toad *R. jimi* via bioguided isolation. Both bufadienolides inhibit the growth of *Leishmania chagasi*, but only hellebrigenin was active against *Trypanosoma cruzi*. According to electron microscopy analysis, antileishmanial activity of telocinobufagin is mediated by damage to mitochondria and plasma membranes of the parasites [65]. The parotoid gland secretion of the Panamanian toad *Rhinella centralis* contains 19-hydroxy-bufalin as a major component. In vitro bioassays revealed that 19-hydroxy-bufalin exerted growth inhibition of *Trypanosoma cruzi* with significant selectivity, as its cytotoxicity was limited against normal kidney Vero cells [64]. The chloroform/methanol extract prepared from the parotoid gland secretion of the toads *Rhinella marina* (Linnaeus, 1758) and *Rhaebo guttatus* (Schneider, 1799), as well as telocinobufagin, displays antimalarial activity against the cloroquine-resistant strain W2 of *Plasmodium falciparum*. Remarkably, toad poisons and telocinobufagin presented low cytotoxicity against human cells and high selectivity for parasites [62]. Recently, the crude (no solvent reported) and methanolic extracts from the *R. marina* parotoid secretion showed antileishmanial and antiplasmodial activity. In general, both extracts displayed antiparasitic activity, although the methanolic extracts showed less inhibition against *Leishmania braziliensis* promastigotes. Biological evaluations revealed that *R. marina* crude extract did not induce DNA damage or mutagenesis [66].

4. Concluding Remarks

Bioactive molecules produced by bufonids in skin gland secretions as a defense mechanism against pathogens and predators have gained interest for drug discovery and development. Traditional uses of true toads as medicine for infections caused by microbes highlight the potential of bufonids as source of antimicrobial drugs. Bioguided isolation has allowed purification of alkaloids, peptides, and steroids with activity against human pathogens. Additionally, recent publications using omics technologies of gland and skin secretions have demonstrated that bufonids produce peptides with unknown structures, as well as peptides with known antimicrobial activity. Microbial resistance against drugs continues to be an important factor in the reemergence of infectious diseases. Crude extracts and isolated metabolites such as alkaloids, peptides, and steroids from Bufonidae display potent activity against clinical isolates, resistant and standard strains of bacteria, fungi, and protozoan parasites. An elucidation of biochemical mechanisms involved in the reported microbial inhibition by true toad secondary metabolites is needed. Bufonidae skin gland secretions represent a valuable source of antimicrobial agents with potential for the development of novel therapeutic drugs in studies against pathogens.

Author Contributions: All authors contributed equally to the design of the work, performing the literature research, and writing and editing the manuscript. All authors have read and agreed to the published version of the manuscript.

Funding: This research received no external funding.

Acknowledgments: We acknowledge INDICASAT AIP for funding of the research project IND-JAL-05 and 02-12-H. C.R. thanks the Secretaría Nacional de Ciencia, Tecnología, e Innovación (SENACYT) and the Instituto para la Formación y Aprovechamiento de los Recursos Humanos (IFARHU) for the doctoral scholarship. R.I. acknowledges the Panama Amphibian Rescue and Conservation Project (Cheyenne Mountain Zoo, Houston Zoo, Smithsonian Institution and Zoo New England) and Cobre Panamá (First Quantum Minerals) for financial support. C.R., R.I., M.G., and A.A.D.-A. are thankful to the Sistema Nacional de Investigación (SNI) of the SENACYT for its support. L.A.R.-S. was supported by a research grant (IOS-1557634) from the National Science Foundation of the United States.

Conflicts of Interest: The authors declare no conflict of interest.

References

1. AmphibiaWeb. Available online: amphibiaweb.org (accessed on 29 September 2020).
2. Amphibian Species of the World. Available online: https://amphibiansoftheworld.amnhorg/Amphibia/Anura/Bufonidae (accessed on 29 September 2020).
3. Ódonohoe, M.E.A.; Luna, M.C.; Regueira, E.; Brunetti, A.E.; Basso, N.G.; Lynch, J.D.; Pereyra, M.O.; Hermida, G.N. Diversity and evolution of the parotoid macrogland in true toads (Anura: Bufonidae). *Zool. J. Linn. Soc.* **2019**, *187*, 453–478. [CrossRef]
4. Rodriguez, C.; Rollins-Smith, L.; Ibáñez, R.; Durant-Archibold, A.A.; Gutiérrez, M. Toxins and pharmacologically active compounds from species of the family Bufonidae (Amphibia, Anura). *J. Ethnopharmacol.* **2017**, *198*, 235–254. [CrossRef] [PubMed]
5. Clarke, B.T. The natural history of amphibian skin secretions, their normal functioning and potential medical applications. *Biol. Rev.* **1997**, *72*, 365–379. [CrossRef] [PubMed]
6. Toledo, R.C.; Jared, C. Cutaneous granular glands and amphibian venoms. *Comp. Biochem. Physiol. Part A Physiol.* **1995**, *111*, 1–29. [CrossRef]
7. Mailho-Fontana, P.L.; Antoniazzi, M.M.; Toledo, L.F.; Verdade, V.K.; Sciani, J.M.; Barbaro, K.C.; Pimenta, D.C.; Rodrigues, M.T.; Jared, C. Passive and active defense in toads: The parotoid macroglands in *Rhinella marina* and *Rhaebo guttatus*. *J. Exp. Zool. Part A Ecol. Genet. Physiol.* **2014**, *321*, 65–77. [CrossRef]
8. Barnhart, K.; Forman, M.E.; Umile, T.P.; Kueneman, J.; McKenzie, V.; Salinas, I.; Minbiole, K.P.C.; Woodhams, D.C. Identification of bufadienolides from the boreal toad, *Anaxyrus boreas*, active against a fungal pathogen. *Microb. Ecol.* **2017**, *74*, 990–1000. [CrossRef]
9. Sun, T.; Zhan, B.; Gao, Y. A novel cathelicidin from *Bufo bufo gargarizans* Cantor showed specific activity to its habitat bacteria. *Gene* **2015**, *571*, 172–177. [CrossRef]
10. Qi, J.; Zulfiker, A.H.M.; Li, C.; Good, D.; Wei, M.Q. The development of toad toxins as potential therapeutic agents. *Toxins* **2018**, *10*, 336. [CrossRef]
11. ClinicalTrials.gov. Available online: https://clinicaltrials.gov/ct2/results?cond=&term=cinobufacini&cntry=&state=&city=&dist= (accessed on 30 September 2020).
12. Riós-Orjuela, J.C.; Falcón-Espitia, N.; Arias-Escobar, A.; Espejo-Uribe, M.J.; Chamorro-Vargas, C. Knowledge and interactions of the local community with the herpetofauna in the forest reserve of Quininí (Tibacuy-Cundinamarca, Colombia). *J. Ethnobiol. Ethnomed.* **2020**, *16*, 1–11. [CrossRef]
13. Barros, F.B.; Varela, S.A.M.; Pereira, H.M.; Vicente, L. Medicinal use of fauna by a traditional community in the Brazilian Amazonia. *J. Ethnobiol. Ethnomed.* **2012**, *8*, 37. [CrossRef]
14. Ferreira, F.S.; Albuquerque, U.P.; Melo Coutinho, H.D.; Almeida, W.D.O.; Nóbrega Alves, R.R. The trade in medicinal animals in northeastern Brazil. *Evid.-Based Complement. Altern. Med.* **2012**, *2012*, 126938. [CrossRef] [PubMed]
15. Grant, J.B.; Land, B. Transcutaneous Amphibian Stimulator (TAS): A device for the collection of amphibian skin secretions. *Herpetol. Rev.* **2002**, *33*, 38–41.
16. Koo, H.B.; Seo, J. Antimicrobial peptides under clinical investigation. *Pept. Sci.* **2019**, *111*, 24122. [CrossRef]
17. Xu, X.; Lai, R. The chemistry and biological activities of peptides from amphibian skin secretions. *Chem. Rev.* **2015**, *115*, 1760–1846. [CrossRef] [PubMed]

18. Huo, Y.; Xv, R.; Ma, H.; Zhou, J.; Xi, X.; Wu, Q.; Duan, J.; Zhou, M.; Chen, T. Identification of <10 KDa peptides in the water extraction of Venenum Bufonis from Bufo gargarizans using Nano LC-MS/MS and De novo sequencing. *J. Pharm. Biomed. Anal.* **2018**, *157*, 156–164. [PubMed]
19. Yumi, P.; Shibao, T.; Cologna, C.T.; Morandi-Filho, R.; Wiezel, G.A.; Fujimura, P.T.; Ueira-Viera, C.; Arantes, E.C. Deep sequencing analysis of toad *Rhinella schneideri* skin glands and partial biochemical characterization of its cutaneous secretion. *J. Venom. Anim. Toxins Incl. Trop. Dis.* **2018**, *24*, 1–15.
20. Dailami, M.; Artika, I.M.; Kusrini, M.D. Analysis and prediction of some histone-derived antimicrobial peptides from toads *Duttaphrynus melanostictus* and *Phyrinoidis asper*. *J. Pure Appl. Chem. Res.* **2016**, *5*, 67–76. [CrossRef]
21. World Health Organization (WHO). Available online: https://www.who.int/news-room/fact-sheets/detail/the-top-10-causes-of-death (accessed on 6 September 2020).
22. Centers for Disease Control (CDC)—Fungal Diseases. Available online: https://www.cdc.gov/fungal/index.html (accessed on 15 September 2020).
23. Centers for Disease Control (CDC)—Parasites. Available online: https://www.cdc.gov/parasites/index.html (accessed on 15 September 2020).
24. Morse, S.S. Factors in the emergence of infectious diseases. *Emerg. Infect. Dis.* **1995**, *1*, 7–15. [CrossRef]
25. Smolinski, M.; Hamburg, M.A.; Lederberg, J. *Microbial Threats to Health*; National Academies Press: Cambridge, MA, USA, 2003.
26. Wang, Z.A. Literature review of bacterial drug resistance. *Mater. Sci. Forum* **2020**, *980*, 197–209. [CrossRef]
27. Yoneyama, H.; Nakae, T. Mechanism of efficient elimination of protein D2 in outer membrane of imipenem-resistant *Pseudomonas aeruginosa*. *Antimicrob. Agents Chemother.* **1993**, *37*, 2385–2390. [CrossRef]
28. Ducret, V.; Gonzalez, M.R.; Leoni, S.; Valentini, M.; Perron, K. The CzcCBA efflux system requires the CadA P-type ATPase for timely expression upon zinc excess in *Pseudomonas aeruginosa*. *Front. Microbiol.* **2020**, *11*, 911. [CrossRef]
29. Lockart, M.M.; Butler, J.T.; Mize, C.J.; Fair, M.N.; Cruce, A.A.; Conner, K.P.; Atkins, W.M.; Bowman, M.K. Multiple drug binding modes in *Mycobacterium tuberculosis* CYP51B1. *J. Inorg. Biochem.* **2020**, *205*, 110994. [CrossRef]
30. Shaw, W.V.; Brodsky, R.F. Characterization of chloramphenicol acetyltransferase from chloramphenicol-resistant *Staphylococcus aureus*. *J. Bacteriol.* **1968**, *95*, 28–36. [CrossRef]
31. Knowles, J.R. Penicillin resistance: The chemistry of β-lactamase inhibition. *Acc. Chem. Res.* **1985**, *18*, 97–104. [CrossRef]
32. Berman, J.; Krysan, D.J. Drug resistance and tolerance in fungi. *Nat. Rev. Microbiol.* **2020**, *18*, 319–331. [CrossRef]
33. Posteraro, B.; Sanguinetti, M.; Sanglard, D.; La Sorda, M.; Boccia, S.; Romano, L.; Morace, G.; Fadda, G. Identification and characterization of a *Cryptococcus neoformans* ATP binding cassette (ABC) transporter-encoding gene, CnAFR1, involved in the resistance to fluconazole. *Mol. Microbiol.* **2003**, *47*, 357–371. [CrossRef]
34. Katiyar, S.K.; Edlind, T.D. Identification and expression of multidrug resistance-related ABC transporter genes in *Candida krusei*. *Med. Mycol.* **2001**, *39*, 109–116. [CrossRef]
35. Seo, K.; Akiyoshi, H.; Ohnishi, H. Alteration of cell wall composition leads amphotericin B resistance in *Aspergillus flavus*. *Microbiol. Immunol.* **1999**, *43*, 1017–1025. [CrossRef]
36. Martel, C.M.; Parker, J.E.; Bader, O.; Weig, M.; Gross, U.; Warrilow, A.G.S.; Rolley, N.; Kelly, D.E.; Kelly, S.L. Identification and characterization of four azole-resistant erg3 mutants of *Candida albicans*. *Antimicrob. Agents Chemother.* **2010**, *54*, 4527–4533. [CrossRef]
37. Baker, N.; Glover, L.; Munday, J.C.; Andrés, D.A.; Barret, M.P.; De Koning, H.P.; Horn, D. Aquaglyceroporin 2 controls susceptibility to melarsoprol and pentamidine in African trypanosomes. *Proc. Natl. Acad. Sci. USA* **2012**, *109*, 10996–11001. [CrossRef]
38. Ross, L.S.; Dhingra, S.K.; Mok, S.; Yeo, T.; Wicht, K.J.; Kumpornsin, K.; Takala-Harrison, S.; Witkowski, B.; Fairhurst, R.M.; Ariey, F.; et al. Emerging Southeast Asian PfCRT mutations confer *Plasmodium falciparum* resistance to the first-line antimalarial piperaquine. *Nat. Commun.* **2018**, *9*, 25–28. [CrossRef] [PubMed]

39. Pérez-Victoria, J.M.; Cortés-Selva, F.; Parodi-Talice, A.; Bavchvarov, B.I.; Pérez-Victoria, F.J.; Muñoz-Martínez, F.; Maitrejean, M.; Costi, M.P.; Barron, D.; Di Pietro, A.; et al. Combination of suboptimal doses of inhibitors targeting different domains of LtrMDR1 efficiently overcomes resistance of *Leishmania* spp. to miltefosine by inhibiting drug efflux. *Antimicrob. Agents Chemother.* **2006**, *50*, 3102–3110. [CrossRef] [PubMed]
40. Adaui, V.; Schnorbusch, K.; Zimic, M.; Gutirrez, A.; Decuypere, S.; Vanaerschot, M.; De Donker, S.; Maes, I.; Llanos-Cuentas, A.; Chappuis, F.; et al. Comparison of gene expression patterns among *Leishmania braziliensis* clinical isolates showing a different in vitro susceptibility to pentavalent antimony. *Parasitology* **2011**, *138*, 183–193. [CrossRef] [PubMed]
41. Domingo-Calap, P.; Delgado-Martínez, J. Bacteriophages: Protagonists of a post-antibiotic era. *Antibiotics* **2018**, *7*, 66. [CrossRef] [PubMed]
42. Holford, B.M.; Daly, M.; King, G.F.; Norton, R.S. Venoms to the rescue. *Science* **2018**, *361*, 842–844. [CrossRef]
43. Syafiq, M.; Zahari, A.; Darnis, D.S.; Haziyamin, T.; Abdul, T. Protein profiles and antimicrobial activity of common Sudan toad, *Duttaphrynus melanostictus* paratoid secretions. *Nat. Preced.* **2015**. [CrossRef]
44. Thirupathi, K.; Chandrakala, G.; Krishna, L.; Rao, T.B.; Venkaiah, Y. Antibacterial activity of skin secretion and its extraction from the toad *Bufo melanostictus*. *Eur. J. Pharm. Med. Res.* **2018**, *3*, 283–286.
45. Sales, D.L.; Bezerra Morais-Braga, M.F.; Lucas Dos Santos, A.T.; Targino Machado, A.J.; De Araujo Filho, J.A.; De Queiroz Dias, D.; Bezerra Da Cunha, F.A.; De Aquino Saraiva, R.; Alencar De Menezes, I.R.; Melo Coutinho, H.D.; et al. Antibacterial, modulatory activity of antibiotics and toxicity from *Rhinella jimi* (Stevaux, 2002) (Anura: Bufonidae) glandular secretions. *Biomed. Pharmacother.* **2017**, *92*, 554–561. [CrossRef]
46. Dahham, S.S.; Sen Hew, C.; Jaafar, I.; Gam, L.H. The protein profiling of asian giant toad skin secretions and their antimicrobial activity. *Int. J. Pharm. Pharm. Sci.* **2016**, *8*, 88–95.
47. Park, C.B.; Kim, M.S.; Kim, S.C. A novel antimicrobial peptide from *Bufo bufo gargarizans*. *Biochem. Biophys. Res. Commun.* **1996**, *218*, 408–413. [CrossRef]
48. Park, C.B.; Kim, H.S.; Kim, S.C. Mechanism of action of the antimicrobial peptide buforin II: Buforin II kills microorganisms by penetrating the vell membrane and inhibiting cellular functions. *Biochem. Biophys. Res. Commun.* **1998**, *244*, 253–257. [CrossRef]
49. Park, C.B.; Yi, K.S.; Matsuzaki, K.; Kim, M.S.; Kim, S.C. Structure-activity analysis of buforin II, a histone H2A-derived antimicrobial peptide: The proline hinge is responsible for the cell-penetrating ability of buforin II. *Proc. Natl. Acad. Sci. USA* **2000**, *97*, 8245–8250. [CrossRef]
50. Zare-Zardini, H.; Ebrahimi, L.; Mahdi Ejtehadi, M.; Hashemi, A.; Ghorani Azam, A.; Atefi, A.; Soleimanizadeh, M. Purification and characterization of one novel cationic antimicrobial peptide from skin secretion of *Bufo kavirensis*. *Turk. J. Biochem.* **2013**, *38*, 416–424. [CrossRef]
51. Sánchez Riera, A.; Daud, A.; Gallo, A.; Genta, S.; Aybar, M.; Sánchez, S. Antibacterial activity of lactose-binding lectins from *Bufo arenarum* skin. *Biocell* **2003**, *27*, 37–46. [CrossRef]
52. Nalbantsoy, A.; Karis, M.; Tansel Yalcin, H.; Göçmen, B. Biological activities of skin and parotoid gland secretions of bufonid toads (*Bufo bufo, Bufo verrucosissimus* and *Bufotes variabilis*) from Turkey. *Biomed. Pharmacother.* **2016**, *80*, 298–303. [CrossRef]
53. Zhao, Y.; Jin, Y.; Lee, W.H.; Zhang, Y. Purification of a lysozyme from skin secretions of *Bufo andrewsi*. *Comp. Biochem. Physiol. Part C Toxicol. Pharmacol.* **2006**, *142*, 46–52. [CrossRef]
54. Pinto, E.G.; Felipe, A.C.; Nadaletto, D.; Rall, V.L.M.; Martinez, R.M. Research of the antimicrobial activity of the poison from *Rhinella icterica* (Amphibia, Anura). *Rev. Inst. Adolfo Lutz* **2009**, *68*, 471–475.
55. Cei, J.M.; Erspamer, V.; Roseghini, M. Taxonomic and evolutionary significance of biogenic amines and polypeptides occurring in amphibian skin. II. Toads of the genera *Bufo* and *Melanophryniscus*. *Syst. Zool.* **1968**, *17*, 232–245. [CrossRef]
56. Preusser, H.J.; Habermehl, G.; Sablofski, M.; Schmall-Haury, D. Antimicrobial activity of alkaloids from amphibian venoms and effects on the ultrastructure of yeast cells. *Toxicon* **1975**, *13*, 285–289. [CrossRef]
57. Cunha Filho, G.A.; Alberto, C.; Lemos, S.; Castro, M.S.; Ma, M.; Leite, R.S.; Kyaw, C.; Pires, O.R.; Bloch, C.; Ferroni, E. Antimicrobial activity of the bufadienolides marinobufagin and telocinobufagin isolated as major components from skin secretion of the toad *Bufo rubescens*. *Toxicon* **2005**, *45*, 777–782. [CrossRef]
58. Wu, S.C.; Fu, B.D.; Shen, H.Q.; Yi, P.F.; Zhang, L.Y.; Lv, S.; Guo, X.; Xia, F.; Wu, Y.L.; Wei, X.B. Telocinobufagin enhances the Th1 immune response and protects against *Salmonella typhimurium* infection. *Int. Immunopharmacol.* **2015**, *25*, 353–362. [CrossRef] [PubMed]

59. Barlian, A.; Anggadiredja, K.; Kusumorini, A. Damage in fungal morphology underlies the antifungal effect of lyophilisate of granular gland secretion from *Duttaphrynus melanostictus* frog. *J. Biol. Sci.* **2011**, *11*, 282–287. [CrossRef]
60. Thirupathi, K.; Shankar, C.; Chandrakala, G.; Krishna, L.; Venkaiah, Y. The antifungal activity of skin secretion and its extract of Indian toad *Bufo melanosctictus*. *Int. J. Environ. Ecol. Fam. Urban. Stud.* **2019**, *9*, 7–12.
61. Made Artika, I.; Pinontoan, S.; Kusrini, M.D. Antifungal activity of skin secretion of bleeding toad *Leptophryne cruentata* and javan tree frog *Rhacophorus margaritifer*. *Am. J. Biochem. Biotechnol.* **2015**, *11*, 5–10. [CrossRef]
62. Banfi, F.F.; Guedes, K.S.; Andrighetti, C.R.; Aguiar, A.C.; Debiasi, B.W.; Noronha, J.C.; Rodrigues, D.J.; Vieira Junior, G.M.; Sanchez Marino, B.A. Antiplasmodial and cytotoxic activities of toad venoms from Southern Amazon, Brazil. *Korean J. Parasitol.* **2016**, *54*, 415–421. [CrossRef]
63. Kalayci, S.; Iyigundogdu, Z.U.; Yazici, M.M.; Asutay, B.A.; Demir, O.; Sahin, F. Evaluation of antimicrobial and antiviral activities of different venoms. *Infect. Disord. Targets* **2016**, *16*, 44–53. [CrossRef]
64. Rodriguez, C.; Ibáñez, R.; NG, M.; Spadafora, C.; Durant-Archibold, A.A.; Gutiérrez, M. 19-Hydroxy-bufalin, a major bufadienolide isolated from the parotoid gland secretions of the Panamanian endemic toad *Rhinella centralis* (Bufonidae), inhibits the growth of *Trypanosoma cruzi*. *Toxicon* **2020**, *177*, 89–92. [CrossRef]
65. Tempone, A.G.; Pimenta, D.C.; Lebrun, I.; Sartoreli, P.; Taniwaki, N.M.; De Andrade, H.F.; Antoniazzi, M.M.; Jared, C. Antileishmanial and antitrypanosomal activity of bufadienolides isolated from the toad *Rhinella jimi* parotoid macrogland secretion. *Toxicon* **2008**, *52*, 13–21. [CrossRef]
66. De Medeiros, D.S.S.; Rego, T.B.; Santos, A.A.; Pontes, A.S.; Moreira-Dill, L.S.; Matos, N.B.; Zuliani, J.P.; Stábeli, R.G.; Teles, C.B.G.; Soares, A.M.; et al. Biochemical and biological profile of parotoid zecretion of the Amazonian *Rhinella marina* (Anura: Bufonidae). *Biomed. Res. Int.* **2019**, *2019*, 292315.
67. Katerere, D.R.; Dawood, A.; Esterhuyse, A.J.; Vismer, H.F.; Govender, T. Antifungal activity of epithelial secretions from selected frog species of South Africa. *Afr. J. Biotechnol.* **2013**, *12*, 6411–6418.
68. Erspamer, V.; Vitali, T.; Roseghini, M.; Cei, J.M. 5-Methoxy- and 5-Hydroxyindoles in the skin of *Bufo alvarius*. *Biochem. Pharmacol.* **1967**, *16*, 1149–1164. [CrossRef]
69. Sales, L.D.; Oliveira, O.P.; Santos, C.M.E.; Queiroz, D.D.; Kerntopf, M.R.; Melo, C.H.D.; Martins, C.J.G.; Dias, F.F.R.; Silva, F.F.; Nóbrega, A.R.R.; et al. Chemical identification and evaluation of the antimicrobial activity of fixed oil extracted from *Rhinella jimi*. *Pharm. Biol.* **2014**, *53*, 98–103. [CrossRef] [PubMed]
70. Pinto, M.E.A.; Araújo, S.G.; Morais, M.I.; Sá, N.P.; Lima, C.M.; Rosa, C.A.; Siqueira, E.P.; Johann, S.; Lima, L.A.R.S. Antifungal and antioxidant activity of fatty acid methyl esters from vegetable oils. *An. Acad. Bras. Cienc.* **2017**, *89*, 1671–1681. [CrossRef] [PubMed]
71. Habermehl, G.G. Antimicrobial activity of amphibian venoms. *Stud. Nat. Prod. Chem.* **1995**, *15*, 327–339.

antibiotics
MDPI

Article

Antimicrobial Activity of Five Apitoxins from *Apis mellifera* on Two Common Foodborne Pathogens

Alexandre Lamas [1], Vicente Arteaga [2], Patricia Regal [1], Beatriz Vázquez [1], José Manuel Miranda [1], Alberto Cepeda [1] and Carlos Manuel Franco [1,*]

1 Laboratorio de Higiene Inspección y Control de Alimentos, Departamento de Química Analítica, Nutrición y Bromatología, Universidad de Santiago de Compostela, 27002 Lugo, Spain; alexandre.lamas@usc.es (A.L.); patricia.regal@usc.es (P.R.); beatriz.vazquez@usc.es (B.V.); josemanuel.miranda@usc.es (J.M.M.); alberto.cepeda@usc.es (A.C.)

2 Laboratorio de Microbiología, Escuela de Ciencias Agrícolas y Ambientales (ECAA) Pontificia Universidad Católica del Ecuador, Sede Ibarra, Ibarra 100112, Ecuador; varteaga@pucesi.edu.ec

* Correspondence: carlos.franco@usc.es; Tel.: +34-982-822-407; Fax: +34-982-254-592

Received: 2 June 2020; Accepted: 26 June 2020; Published: 30 June 2020

Abstract: Antimicrobial resistance is one of today's major public health challenges. Infections caused by multidrug-resistant bacteria have been responsible for an increasing number of deaths in recent decades. These resistant bacteria are also a concern in the food chain, as bacteria can resist common biocides used in the food industry and reach consumers. As a consequence, the search for alternatives to common antimicrobials by the scientific community has intensified. Substances obtained from nature have shown great potential as new sources of antimicrobial activity. The aim of this study was to evaluate the antimicrobial activity of five bee venoms, also called apitoxins, against two common foodborne pathogens. A total of 50 strains of the Gram-negative pathogen *Salmonella enterica* and 8 strains of the Gram-positive pathogen *Listeria monocytogenes* were tested. The results show that the minimum inhibitory concentration (MIC) values were highly influenced by the bacterial genus. The MIC values ranged from 256 to 1024 µg/mL in *S. enterica* and from 16 to 32 µg/mL in *L. monocytogenes*. The results of this study demonstrate that apitoxin is a potential alternative agent against common foodborne pathogens, and it can be included in the development of new models to inhibit the growth of pathogenic bacteria in the food chain.

Keywords: apitoxin; antimicrobial resistance; natural antimicrobial compounds; foodborne pathogens; *Salmonella*; *Listeria monocytogenes*

1. Introduction

The discovery and development of antimicrobial agents in the first half of the 20th century created a new paradigm. Since that time, common infections that would have caused death have become treatable with antibiotics, saving millions of lives. At first, the use of antimicrobials was generalized, and they were used to treat both human and animal infections [1]. Antimicrobials were, and still are, used for zootechnical purposes in farm animals [1]. Soon after the discovery of antibiotics, the phenomenon of antimicrobial resistance was addressed. In his 1945 Nobel Prize lecture, Sir Alexander Fleming stated that "there is the danger that the ignorant man may easily under dose himself and by exposing his microbes to non-lethal quantities of the drug make them resistant" [2]. This warning has become a reality; antimicrobial resistance is a global public health problem [3]. It is estimated that antimicrobial resistance in common bacterial infections is responsible for 700,000 deaths worldwide each year, with the potential to reach millions of deaths per year by 2050. In the European Union alone, there are 25,000 deaths each year related to antimicrobial resistance. In addition, antimicrobial resistance causes serious economic damage, estimated at $1.5 trillion in health care costs and lost productivity [4]. Multidrug-resistant bacteria are also a serious problem in the food production chain [5]. Studies in recent years have found a large number of strains of multidrug-resistant foodborne pathogens. The use of antibiotics in production animals and the resultant selective pressure on the environmental microbiota together constitute one of the main causes of the current exponential increase in antimicrobial resistance.

Therefore, a current research priority is the search for and discovery of alternatives to conventional antibiotics. The three principal research strategies can be classified as (i) naturally occurring alternatives, (ii) synthetic designs, and (iii) biotechnology-based strategies [6]. The most common naturally occurring alternatives are bacteriocins, bacteriophages, and antimicrobial peptides (AMPs) [6]. Of these, AMPs have received great attention from the research community in recent years. These naturally derived molecules are part of the innate immune system in both prokaryotic and eukaryotic cells; their main advantages with respect to other natural alternatives are their broad-spectrum activity and lack of susceptibility to resistance development [6–8]. AMPs' mode of action is based on the permeabilization of bacterial membranes and the formation of cytotoxic pores, but they can also inhibit nucleotides, proteins, and cell wall biosynthesis [9]. Practical studies have demonstrated that AMPs are a promising alternative for combating common foodborne pathogens such as *Salmonella*, *L. monocytogenes*, and *Staphylococcus aureus* [8,10,11]. The water-soluble peptide melittin from honeybee venom is one of these promising AMPs. Melittin has demonstrated both antimicrobial and antiviral activity in in vitro studies [12,13]. Melittin is a 26 amino acid cationic linear peptide with an N-terminal hydrophobic region, a C-terminal hydrophilic region, and asymmetrical distribution of polar and nonpolar amino acid residues. This suggests an amphipathic nature in α-helical conformation that makes melittin a membrane-active molecule. Due to its nature, melittin exerts antimicrobial activity by destabilizing the bacterial membrane and causing pore formation, which induces a loss of osmotic balance and, ultimately, cell lysis. Specifically, the perpendicular orientation of melittin to the cell membrane causes its insertion, peptide aggregation, and the bending of lipids, resulting in the leakage of cytoplasmic contents [14–16]. However, honeybee venom, or apitoxin, is also composed of other peptides such as adolapin, apamin, and MCD peptide, and enzymes such as phospholipase A2 and hyaluronidase [16]. Although melittin is the most bioactive component of apitoxin, its bioactivity is enhanced by other components of bee venom [17]. In this sense, it has been demonstrated that melittin and phospholipase A2 have synergistic activity. Melittin exposes membrane phospholipids through pore formation to the catalytic site of phospholipase A2 [14]. Although the antimicrobial properties of melittin have been studied in depth, only limited studies have evaluated the antimicrobial ability of apitoxin, and very few strains were included [12,18]. It is therefore necessary to determine whether the apitoxins obtained in different geographic locations and tested in different studies show similar inhibition values. It is also important that these types of studies include a large collection of wild strains to increase the significance of the data obtained.

Therefore, the aim of this study is to evaluate the antimicrobial activity and determine the minimum inhibitory concentration (MIC) of five apitoxins obtained from apiaries located in different parts of Ecuador on a large collection of wild-strain foodborne pathogens. For this purpose, 50 *Salmonella* strains belonging to different serotypes and subspecies and 8 *L. monocytogenes* strains were included in this study. These pathogens were selected because *Salmonella* spp. and *L. monocytogenes* are two of the main foodborne pathogens in the European Union, with 91,662 and 2480 confirmed cases of human infections in 2017, respectively. Moreover, by including these two pathogens, we tested both Gram-positive and Gram-negative bacteria.

2. Results

The amount of apitoxin collected each time was between 29 and 40 mg. The amount collected from apitoxin 1 was 34.33 ± 2.98 mg, from apitoxin 2 was 36.55 ± 1.46 mg, from apitoxin 3 was 37.25 ± 4.95 mg, from apitoxin 4 was 39.66 ± 0.67 mg, and from apitoxin 5 was 39.66 ± 0.78 mg.

There were significant differences ($p < 0.05$) in the amounts between apitoxin 1 and apitoxins 4 and 5. There were no significant differences ($p < 0.05$) in the concentration of melittin in the apitoxins tested in this study, with values around 129 μg/mL. The five tested apitoxins showed antimicrobial activity against all *S. enterica* and *L. monocytogenes* strains included in this study. In *S. enterica*, the MIC values ranged between 256 and 1024 μg/mL (Table 1), but most of the strains showed an MIC value of 512 μg/mL. The lowest inhibitory concentration in Apitoxins 1 and 4 was 256 μg/mL, and four and three strains, respectively, showed an MIC value of 1024 μg/mL. On the other hand, apitoxin 5 showed the most strains in which growth was inhibited at 256 μg/mL.

Table 1. Minimum inhibitory concentrations (MICs) of five apitoxins tested in 50 *Salmonella* strains isolated from poultry.

			MIC (μg/mL)				
Strain	**Source**	**Code**	**Apitoxin 1**	**Apitoxin 2**	**Apitoxin 3**	**Apitoxin 4**	**Apitoxin 5**
S. Anatum	PF	A1	512	256	256	512	512
S. Anatum	PF	A6	512	512	512	512	512
S. Anatum	PF	A15	512	512	512	512	512
S. enterica subspecies *arizonae*	PF	AZ1	512	256	256	512	512
S. enterica subspecies *arizonae*	PF	AZ6	512	256	256	512	256
S. enterica subspecies *arizonae*	PF	AZ12	256	256	256	512	512
S. enterica subspecies *arizonae*	PF	AZ16	1024	512	512	512	512
S. enterica subspecies *arizonae*	PF	AZ20	512	256	512	512	512
S. enterica subspecies *arizonae*	PF	AZ21	512	512	256	256	256
S. Bardo	PF	B2	512	512	512	512	512
S. Bardo	PF	B3	512	512	512	512	512
S. Bredeney	PF	BR1	1024	512	512	512	512
S. Dabou	PF	DA1	512	512	512	512	256
S. Drac	PF	DC4	1024	512	512	512	512
S. Enteritidis	CK	ET1	512	512	256	256	256
S. Enteritidis	PF	ET2	512	512	256	256	512
S. Infantis	PF	I1	256	256	256	512	256
S. Infantis	PF	I2	256	256	256	512	256
S. Infantis	PF	I3	256	256	256	512	256
S. Infantis	PF	I4	256	256	256	1024	256
S. Infantis	PF	I7	512	512	512	256	256
S. Infantis	PF	I12	512	512	512	512	512
S. Infantis	PF	I11	512	512	512	512	512
S. Infantis	PF	I18	512	512	512	256	512
S. Isangi	PF	IG1	512	512	512	512	512
S. Isangi	PF	IG9	512	512	512	512	512
S. Montevideo	PF	M1	512	512	512	512	512
S. Mbandaka	PF	MB1	512	512	512	512	256
S. Ndolo	PF	ND1	512	512	512	512	512
S. Ndolo	PF	ND2	512	512	512	512	512
S. Ndolo	PF	ND5	512	512	512	256	256
S. Newport	PF	N1	512	512	512	512	512
S. Newport	PF	N6	512	512	512	512	512
S. Rissen	PF	R1	512	512	512	512	256
S. enterica subspecies *salamae*	PF	SA1	512	512	512	1024	512

Table 1. *Cont.*

			MIC (μg/mL)				
Strain	**Source**	**Code**	**Apitoxin 1**	**Apitoxin 2**	**Apitoxin 3**	**Apitoxin 4**	**Apitoxin 5**
S. enterica subspecies *salamae*	PF	SA2	1024	512	512	1024	512
S. enterica subspecies *salamae*	PF	SA3	1024	1024	1024	1024	1024
S. Seftenberg	PF	S1	512	512	512	512	512
S. Stanleyville	PF	ST1	512	512	512	512	512
S. Thompson	PF	TM1	1024	512	512	512	512
S. Typhimurium	CK	T2	512	256	512	512	256
S. Typhimurium	CK	T3	512	512	256	512	512
S. Typhimurium	PF	T6	512	512	512	512	512
S. Typhimurium	PF	T10	512	512	256	512	256
S. Typhimurium	PF	T12	512	512	512	512	256
S. Typhimurium	PF	T13	512	256	256	512	256
S. Typhimurium	PF	T18	512	512	512	512	512
S. Typhimurium	PF	T21	256	512	256	256	256
S. Typhimurium	PF	T24	512	512	512	512	512
S. Typhimurium	CC	CECT 4395	512	512	512	512	512
			n (%)	n (%)	n (%)	n (%)	n (%)
MIC (μg/mL)		256	6 (12%)	11 (22%)	15 (30%)	7 (14%)	17 (34%)
		512	38 (76%)	38 (76%)	34 (68%)	39 (78%)	33 (64%)
		1024	6 (12%)	1 (2%)	1 (2%)	4 (8%)	1 (2%)

CC, culture collection; CK, chicken meat; PF, poultry farm.

The value of MIC_{90} and MIC_{50} for *S. enterica* was 512 μg/mL for four of the apitoxins tested. However, the MIC_{90} of apitoxin 1 was 1024 μg/mL, which could indicate lower antimicrobial activity. There were no significant differences ($p > 0.05$) in MIC values between the five apitoxins tested. It is also remarkable that four *S.* Infantis strains showed an MIC of 256 μg/mL in four of the apitoxins tested. This indicates a higher susceptibility of those strains to apitoxin in comparison with the other strains of *S.* Infantis. In this sense, there were significant differences ($p < 0.05$) in resistance results between *S.* Infantis and the other strains of *S. enterica* subspecies *enterica*. On the other hand, the strain *S. enterica* subspecies *salamae* SA3 showed higher values, with an MIC of 1024 μg/mL in the five apitoxins tested. In fact, *S. enterica* subspecies *salamae* was significantly more resistant ($p < 0.05$) than *S. enterica* subspecies *arizonae* or *S. enterica* subspecies *enterica*.

In the case of *L. monocytogenes*, the MIC values observed were lower than those found in *S. enterica* strains, ranging between 16 and 32 μg/mL (Table 2). There were differences in the MIC_{50} of the apitoxins used in this study. For apitoxins 1, 2, and 4, the MIC_{50} was 16 μg/mL, and for apitoxins 3 and 5, it was 32 μg/mL. All apitoxins had an MIC_{90} of 32 μg/mL. It is also remarkable that all tested *L. monocytogenes* strains showed an MIC of 32 μg/mL with apitoxin 5. The MIC value of *L. monocytogenes* was significantly lower ($p < 0.05$) than that of *Salmonella* spp.

Table 2. Minimum inhibitory concentrations of five apitoxins tested in eight *L. monocytogenes* strains isolated from foodstuff.

			MIC (μg/mL)				
Strain	**Source**	**Code**	**Apitoxin 1**	**Apitoxin 2**	**Apitoxin 3**	**Apitoxin 4**	**Apitoxin 5**
L. monocytogenes	RM	LHICA 1	16	16	32	16	32
L. monocytogenes	RM	LHICA 2	16	16	32	16	32
L. monocytogenes	CH	LHICA 3	32	16	32	32	32
L. monocytogenes	CH	LHICA 4	32	32	16	32	32
L. monocytogenes	CH	LHICA 5	16	16	32	32	32
L. monocytogenes	FP	LHICA 6	16	16	16	32	32
L. monocytogenes	FP	LHICA 7	32	32	32	16	32
L. monocytogenes	CC	CECT 934	32	16	32	16	32
			n (%)	n (%)	n (%)	n (%)	n (%)
MIC (μg/mL)		16	4	6	2	4	0
		32	4	2	6	4	8

CC, culture collection; CH, cheese; FP, fish product; RM, rabbit meat.

3. Discussion

The discovery and evaluation of new and natural antimicrobial substances is one of the main strategies to decrease the use of antibiotics and avoid the increase in multidrug-resistant strains [6]. In the last several decades, compounds isolated from natural products have shown promising activity against resistant bacteria [19]. In this sense, venoms have been shown to be composed of various substances, such as antimicrobial peptides, with high inhibitory activity [15,20,21]. In this study the antimicrobial activity of bee venom was tested. Different works have observed that the main components of apitoxin, such as pure melittin and phospholipase A, have high antimicrobial activity against different bacterial pathogens [15]. However, a very limited number of studies have tested the inhibitory capacity of pure apitoxin in bacteria, and the information available on foodborne pathogens is currently insufficient [22]. The results of this study show significant differences in MIC values between *Salmonella* (256–1024 μg/mL) and *L. monocytogenes* (16–32 μg/mL) strains. A previous study that evaluated the activity of commercially available apitoxin against oral bacteria such as *Enterococcus faecalis* and *Streptococcus salivarius* found MIC values between 20 and 40 μg/mL [18], very similar to the results observed in *L. monocytogenes* in this study. Additionally, a recent study found an MIC of 7.2 μg/mL in strains of Gram-positive *Staphylococcus aureus* bacteria [12]. In the same way, Picoli et al. [23] observed that *S. aureus* had lower MIC values (6–7 μg/mL) than Gram-negative *Escherichia coli* (40–42.5 μg/mL) and *Pseudomonas aeruginosa* (65–70 μg/mL) bacteria for the AMP of bee venom melittin. These differences can be related to structural differences between Gram-positive and Gram-negative bacteria. In this sense, it has been suggested that melittin can penetrate the peptidoglycan layer of Gram-positive bacteria more easily than the membrane of Gram-negative bacteria, which is protected by a layer of lipopolysaccharides [13,14]. In the same way, the phospholipase A2 present in apitoxin causes phospholipid membrane degradation, resulting in cell death [13,14]. However, the outer membrane in Gram-negative bacteria reduces the efficacy of phospholipase A2 by reducing the interaction of this enzyme with the cytoplasmic membrane [16]. Therefore, the combination of apitoxin with other substances that disrupt the outer membrane of Gram-negative bacteria could increase the antimicrobial activity of apitoxin. One of the main advantages of the present study in comparison with the studies previously described is the number of strains included. *Salmonella* spp. are composed of more than 2600 serotypes and six subspecies, which differ in their pathogenicity [24]. The results of this study show that the observed MIC values were very stable through the *Salmonella enterica* species, but there were some significant differences between some subspecies and serotypes of *Salmonella enterica* subspecies *enterica*. In addition, the differences observed between the five apitoxins were not due to different concentrations of melittin, as no significant differences were observed between them.

4. Materials and Methods

4.1. Apitoxin Collection

Bee venom, or apitoxin, was collected from 5 *Apis mellifera* apiaries in Ecuador: El Inca (apitoxin 1), Apiary Caranqui (apitoxin 2), Apiary Clatura (apitoxin 3), Apiary Cotacachi (apitoxin 4), and Apiary ECAA (apitoxin 5) (Figure 1). The collections were made between 11:00 and 13:00 from 9 January to 28 May 2016, with an interval of 21 days between collections, until 5 collections per apiary and hive were completed. Apitoxin was collected by using an electric stimulus, as previously described [25]. Briefly, when bees land on a woven copper wire located inside the beehive, an electric stimulus is applied, causing the release of bee venom without killing the bees. The bee venom is collected on glass slides, where the apitoxin crystallizes. The glass slides are transported to the laboratory, where the crystallized bee venom is detached with a scraper, collected in microtubes, and weighed. This crude apitoxin was used in subsequent analyses.

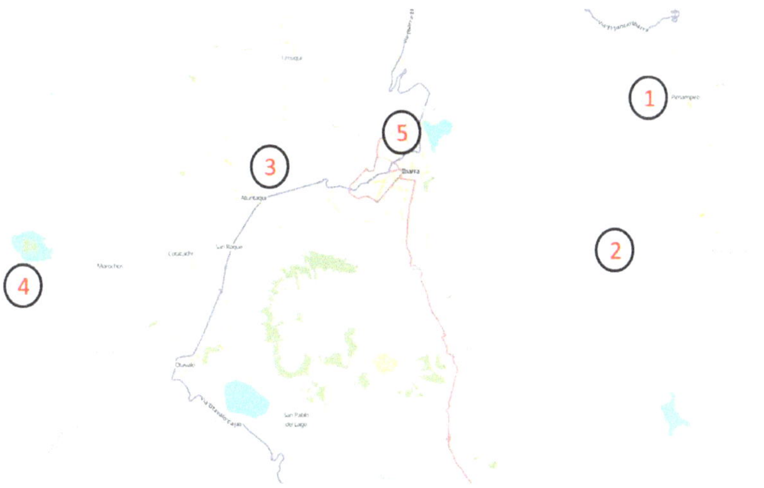

Figure 1. Geographic location of five apiaries in the province of Imbabura (Ecuador). 1: Apiary El Inca; 2: Apiary Caranqui; 3: Apiary Clatura; 4: Apiary Cotacachi; and 5: Apiary ECAA.

4.2. Mellitin Determination of Apitoxin by HPLC-UV

The melittin content of the 5 apitoxin samples was determined according the method developed by Rybak-Chmielewska and Szczêsna [26] with some modifications. A melittin standard with 96.5% purity was obtained from Sigma-Aldrich (Germany). Briefly, 5 mg of apitoxin was mixed with 5 mL of ultrapure water and sonicated for 5 min, and the liquid was filtered through a 0.45 µm polytetrafluoroethylene syringe filter and collected in an amber glass vial. A volume of 5 µL of 85% phosphoric acid was added to the vial. The samples were analyzed by HPLC in a Jasco LC-Net II/ADC (Jasco, Spain) coupled with a UV-2070 detector (Jasco). A Machery-Nagel C-18 column with a length of 250 mm, internal diameter of 4 mm, and particle size of 5 µm was used. Gradient chromatography was performed by the linear method with 5–80% of eluent (acetonitrile in 20% phosphoric acid) for 45 min with flow velocity of the moving phase at 1 mL·min^{-1}. Melittin was identified at 220 nm wavelength. The data were collected through the use of Chrom NAV software (Jasco).

4.3. *Salmonella and L. monocytogenes Strains*

A total of 50 *S. enterica* and 8 *L. monocytogenes* strains, including culture collection strains *Salmonella* CECT 4395 and *L. monocytogenes* CECT 934, were used in this study. The rest of the *Salmonella* strains were isolated in our laboratory from poultry farms within the framework of the national *Salmonella* control plan and from chicken meat according ISO 6579:2017 [27]. All *Salmonella* strains were serotyped using the Kauffman–White typing scheme for the detection of somatic (O) and flagellar (H) antigens with standard antisera (Bio-Rad Laboratories, Irvine, CA, USA). The rest of the *L. monocytogenes* strains were isolated in our laboratory from food products (rabbit meat, cheese, fish products) by routine analysis for the food industry according to ISO 11290-1:2017 [28]. *Salmonella* strains were kept at −20 °C in Tryptic Soy Broth (TSB; Oxoid, Basingstoke, UK) supplemented with 20% glycerol, and *L. monocytogenes* strains were kept in Brain Heart Infusion (PanReac AppliChem, Barcelona, Spain) supplemented with 20% glycerol until use.

4.4. *Determination of Minium Inhibitory and Biocidal Concentrations*

The MIC of the 5 apitoxins included in this study was determined according Clinical and Laboratory Standards Institute (CLSI) guidelines by using the broth microdilution method. Briefly, an initial stock of 4096 μg/mL of each apitoxin was prepared. Dilutions of apitoxin in Mueller–Hinton agar from 2048 μg/mL to 2 μg/mL were made. *Salmonella* and *L. monocytogenes* strains were grown in nutrient agar (PanReac, AppliChem, Spain) for 24 h at 37 °C. Isolated colonies were used to obtain a saline suspension of 0.5 McFarland equivalent to 10^8 colony-forming units (CFU)/mL. This suspension was diluted to 1:20 to obtain a final concentration of 10^6 CFU/mL. The broth volume in a 96-well microtiter plate was 0.1 mL, and 0.01 mL of the diluted bacterial suspension was inoculated to a final bacterial concentration of 10^4 CFU/mL. The 96-well microtiter plates were incubated for 24 h at 37 °C, and the MIC value of each strain with each apitoxin was determined. The MIC was defined as the lowest concentration of antimicrobial agent that completely inhibited the visual growth of the organism in the wells.

4.5. *Stastitical Analysis*

GraphPad Prism 8 (GraphPad, San Diego, CA, USA) was used in this research for statistical analysis. Chi-squared tests were performed to evaluate the differences between the 5 apitoxins tested and between genera, subspecies, and serotypes. Analysis of variance (one-way ANOVA) and Tukey's honestly significant difference test ($p < 0.05$) were used to determine the differences between the amounts of apitoxin collected from the apiaries.

5. Conclusions

This study increases the information available on the antimicrobial capacity of apitoxin against foodborne pathogens. The results demonstrate that apitoxin is a potential alternative agent to inhibit the growth of common foodborne pathogens in the food chain at low concentrations, especially in *L. monocytogenes* strains. Therefore, apitoxin can potentially be used alone as an alternative to common antimicrobials or even in combination with them to enhance the antimicrobial activity of both substances. Future studies should be focused on developing new models to apply this substance at different steps of the food chain in order to translate in vitro results to real-life applications.

Author Contributions: Conceptualization, A.L. and C.M.F.; methodology, A.L. and V.A.; validation, A.L.; formal analysis, A.L.; investigation, A.L. and P.R.; resources, V.A. and A.C.; writing—original draft preparation, A.L. and P.R.; writing—review and editing, A.L., C.M.F., and J.M.M.; supervision, C.M.F. and B.V.; funding acquisition, V.A. and A.C. All authors have read and agreed to the published version of the manuscript.

Funding: This research received no external funding.

Conflicts of Interest: The authors declare no conflicts of interest.

References

1. Durand, G.A.; Raoult, D.; Dubourg, G. Antibiotic discovery: History, methods and perspectives. *Int. J. Antimicrob. Agents* **2019**, *53*, 371–382. [CrossRef] [PubMed]
2. Flemming, A. *Penicillin, Nobel Lecturer*; The Nobel Foundation: Stockholm, Sweden, 1945; pp. 1–11.
3. Roca, I.; Akova, M.; Baquero, F.; Carlet, J.; Cavaleri, M.; Coenen, S.; Cohen, J.; Findlay, D.; Gyssens, I.; Heure, O.E.; et al. The global threat of antimicrobial resistance: Science for intervention. *New Microbes New Infect.* **2015**, *6*, 22–29. [CrossRef]
4. European Commission. *A European One Health Action Plan against Antimicrobial Resistance*; European Commission: Brussels, Belgium, 2017; pp. 1–24.
5. Hudson, J.A.; Frewer, L.J.; Jones, G.; Brereton, P.A.; Whittingham, M.J.; Stewart, G. The agri-food chain and antimicrobial resistance: A review. *Trends Food Sci. Technol.* **2017**, *69*, 131–147. [CrossRef]
6. Ghosh, C.; Sarkar, P.; Issa, R.; Haldar, J. Alternatives to Conventional Antibiotics in the Era of Antimicrobial Resistance. *Trends Microbiol.* **2019**, *27*, 323–338. [CrossRef]
7. Palmieri, G.; Balestrieri, M.; Proroga, Y.T.R.; Falcigno, L.; Facchiano, A.; Riccio, A.; Capuano, F.; Marrone, R.; Neglia, G.; Anastasio, A. New antimicrobial peptides against foodborne pathogens: From in silico design to experimental evidence. *Food Chem.* **2016**, *211*, 546–554. [CrossRef]
8. Palmieri, G.; Balestrieri, M.; Capuano, F.; Proroga, Y.T.R.; Pomilio, F.; Centorame, P.; Riccio, A.; Marrone, R.; Anastasio, A. Bactericidal and antibiofilm activity of bactenecin-derivative peptides against the food-pathogen *Listeria monocytogenes*: New perspectives for food processing industry. *Int. J. Food Microbiol.* **2018**, *279*, 33–42. [CrossRef] [PubMed]
9. Laverty, G.; Gorman, S.P.; Gilmore, B.F. The potential of antimicrobial peptides as biocides. *Int. J. Mol. Sci.* **2011**, *12*, 6566–6596. [CrossRef] [PubMed]
10. Colagiorgi, A.; Festa, R.; Di Ciccio, P.A.; Gogliettino, M.; Balestrieri, M.; Palmieri, G.; Anastasio, A.; Ianieri, A. Rapid biofilm eradication of the antimicrobial peptide 1018-K6 against *Staphylococcus aureus*: A new potential tool to fight bacterial biofilms. *Food Control* **2020**, *107*, 106815. [CrossRef]
11. Wang, L.; Zhao, X.; Xia, X.; Zhu, C.; Qin, W.; Xu, Y.; Hang, B.; Sun, Y.; Chen, S.; Zhang, H.; et al. Antimicrobial Peptide JH-3 Effectively Kills *Salmonella enterica* Serovar Typhimurium Strain CVCC541 and Reduces Its Pathogenicity in Mice. *Probiotics Antimicrob. Proteins* **2019**, *11*, 1379–1390. [CrossRef]
12. Marques Pereira, A.F.; Albano, M.; Bérgamo Alves, F.C.; Murbach Teles Andrade, B.F.; Furlanetto, A.; Mores Rall, V.L.; Delazari dos Santos, L.; de Oliveira Orsi, R.; Fernandes Júnior, A. Influence of apitoxin and melittin from *Apis mellifera* bee on *Staphylococcus aureus* strains. *Microb. Pathog.* **2020**, *141*, 104011. [CrossRef]
13. Memariani, H.; Memariani, M.; Moravvej, H.; Shahidi-Dadras, M. Melittin: A venom-derived peptide with promising anti-viral properties. *Eur. J. Clin. Microbiol. Infect. Dis.* **2020**, *39*, 5–17. [CrossRef] [PubMed]
14. Memariani, H.; Memariani, M.; Shahidi-Dadras, M.; Nasiri, S.; Akhavan, M.M.; Moravvej, H. Melittin: From honeybees to superbugs. *Appl. Microbiol. Biotechnol.* **2019**, *103*, 3265–3276. [CrossRef]
15. Pascoal, A.; Estevinho, M.M.; Choupina, A.B.; Sousa-Pimenta, M.; Estevinho, L.M. An overview of the bioactive compounds, therapeutic properties and toxic effects of apitoxin. *Food Chem. Toxicol.* **2019**, *134*, 110864. [CrossRef] [PubMed]
16. Wehbe, R.; Frangieh, J.; Rima, M.; Obeid, D.E.; Sabatier, J.-M.; Fajloun, Z. Bee venom: Overview of main compounds and bioactivities for therapeutic interests. *Molecules* **2019**, *24*, 2997. [CrossRef]
17. Koumanov, K.; Momchilova, A.; Wolf, C. Bimodal regulatory effect of melittin and phospholipase A 2-activating protein on human type II secretory phospholipase A 2. *Cell Biol. Int.* **2003**, *27*, 871–877. [CrossRef]
18. Leandro, L.F.; Mendes, C.A.; Casemiro, L.A.; Vinholis, A.H.C.; Cunha, W.R.; De Almeida, R.; Martins, C.H.G. Antimicrobial activity of apitoxin, melittin and phospholipase A2 of honey bee (*Apis mellifera*) venom against oral pathogens. *Anais Acad. Bras. Cienc.* **2015**, *87*, 147–155. [CrossRef] [PubMed]
19. Gyawali, R.; Ibrahim, S.A. Natural products as antimicrobial agents. *Food Control* **2014**, *46*, 412–429. [CrossRef]
20. El-Aziz, T.M.A.; Soares, A.G.; Stockand, J.D. Snake venoms in drug discovery: Valuable therapeutic tools for life saving. *Toxins* **2019**, *11*, 564. [CrossRef]
21. Santos, P.P.; Pereira, G.R.; Barros, E.; Ramos, H.J.O.; Oliveira, L.L.; Serrão, J.E. Antibacterial activity of the venom of the Ponerine and *Pachycondyla striata* (*Formicidae*: *Ponerinae*). *Int. J. Trop Insect. Sci.* **2020**, *40*, 393–402. [CrossRef]

22. Arteaga, V.; Lamas, A.; Regal, P.; Vázquez, B.; Miranda, J.M.; Cepeda, A.; Franco, C.M. Antimicrobial activity of apitoxin from *Apis mellifera* in *Salmonella enterica* strains isolated from poultry and its effects on motility, biofilm formation and gene expression. *Microb. Pathog.* **2019**, *137*, 103771. [CrossRef]
23. Picoli, T.; Peter, C.M.; Zani, J.L.; Waller, S.B.; Lopes, M.G.; Boesche, K.N.; Vargas, G.D.; Hübner, S.D.O.; Fischer, G. Melittin and its potential in the destruction and inhibition of the biofilm formation by *Staphylococcus aureus, Escherichia coli* and *Pseudomonas aeruginosa* isolated from bovine milk. *Microb. Pathog.* **2017**, *112*, 57–62. [CrossRef] [PubMed]
24. Lamas, A.; Miranda, J.M.; Regal, P.; Vázquez, B.; Franco, C.M.; Cepeda, A. A comprehensive review of non-enterica subspecies of *Salmonella enterica*. *Microbiol. Res.* **2018**, *206*, 60–73. [CrossRef] [PubMed]
25. Guatemal Sánchez, A.; Jáuregui Sierra, D.; Arteaga Cadena, V.; Aguirre, S.M.A. Determinación de las condiciones óptimas de un equipo extractor de apitoxina en abejas (*Apis mellifera*). *Rev. Electron. Vet.* **2017**, *18*, 1–11.
26. Rybak-Chmielewska, H.; Szczêsna, T. HPLC study of chemical composition of honeybee (*Apis mellifera* L.) venom. *J. Apic. Sci.* **2004**, *48*, 103–109.
27. International Standarization Organization. *Microbiology of the food chain—Horizontal method for the detection, enumeration and serotyping of Salmonella—Part 1: Detection of Salmonella spp. (ISO 6579-1:2017)*; International Standarization Organization: Geneva, Switzerland, 2017.
28. International Organization for Standardization ISO 11290-1:2017. *Microbiology of the Food Chain—Horizontal Method for the Detection and Enumeration of Listeria Monocytogenes and of Listeria spp.—Part 1: Detection Method*; International Standarization Organization: Geneva, Switzerland, 2017; pp. 1–36.

antibiotics

MDPI

Article

A Trypsin Inhibitor from *Moringa oleifera* Flowers Modulates the Immune Response In Vitro of *Trypanosoma cruzi*-Infected Human Cells

Isabella Coimbra Vila Nova [1], Leyllane Rafael Moreira [2], Diego José Lira Torres [2], Kamila Kássia dos Santos Oliveira [2], Leydianne Leite de Siqueira Patriota [1], Luana Cassandra Breitenbach Barroso Coelho [1], Patrícia Maria Guedes Paiva [1], Thiago Henrique Napoleão [1], Virgínia Maria Barros de Lorena [2] and Emmanuel Viana Pontual [3,*]

1 Departamento de Bioquímica, Centro de Biociências, Universidade Federal de Pernambuco, Recife 50670-901, Pernambuco, Brazil; isabella.coimbra@ufpe.br (I.C.V.N.); leydianne.patriota@ufpe.br (L.L.d.S.P.); luana.coelho@ufpe.br (L.C.B.B.C.); patricia.paiva@ufpe.br (P.M.G.P.); thiago.napoleao@ufpe.br (T.H.N.)
2 Departamento de Imunologia, Centro de Pesquisas Aggeu Magalhães, Fundação Oswaldo Cruz, Recife 50670-901, Brazil; leyllanemoreira@gmail.com (L.R.M.); diegolira18ufpe@gmail.com (D.J.L.T.); kamilakassia@outlook.com (K.K.d.S.O.); lorena@cpqam.fiocruz.br (V.M.B.d.L.)
3 Departamento de Morfologia e Fisiologia Animal, Universidade Federal Rural de Pernambuco, Recife 52171-900, Brazil
* Correspondence: emmanuel.pontual@ufrpe.br; Tel.: +55-81-3320-6395

Received: 15 July 2020; Accepted: 12 August 2020; Published: 14 August 2020

check for updates

Abstract: *Trypanosoma cruzi* causes the lethal Chagas disease, which is endemic in Latin America. Flowers of *Moringa oleifera* (Moringaceae) express a trypsin inhibitor (MoFTI) whose toxicity to *T. cruzi* trypomastigotes was previously reported. Here, we studied the effects of MoFTI on the viability of human peripheral blood mononuclear cells (PBMCs) as well as on the production of cytokines and nitric oxide (NO) by *T. cruzi*-infected PBMCs. Incubation with MoFTI (trypsin inhibitory activity: 62 U/mg) led to lysis of trypomastigotes (LC_{50} of 43.5 μg/mL) but did not affect the viability of PBMCs when tested at concentrations up to 500 μg/mL. A selectivity index > 11.48 was determined. When *T. cruzi*-infected PBMCs were treated with MoFTI (43.5 or 87.0 μg/mL), the release of the pro-inflammatory cytokine TNF-α and INF-γ, as well as of NO, was stimulated. The release of the anti-inflammatory cytokine IL-10 also increased. In conclusion, the toxicity to *T. cruzi* and the production of IL-10 by infected PBMCs treated with MoFTI suggest that this molecule may be able to control parasitemia while regulating the inflammation, preventing the progress of Chagas disease. The data reported here stimulate future investigations concerning the in vivo effects of MoFTI on immune response in Chagas disease.

Keywords: cytokines; cytotoxicity; immunomodulatory agent; *Moringa oleifera*; protease inhibitor; trypanocidal agent

1. Introduction

The Chagas disease, also known as American trypanosomiasis, is an endemic and lethal disease common in Latin America. It is caused by the protozoan *Trypanosoma cruzi* and the transmission occurs when vertebrates come in contact with the feces of infected triatomine insects, popularly known as "kissing bugs". The current situation of the Chagas disease is a concern, since it is estimated that about 6 to 7 million people worldwide are infected with *T. cruzi* [1].

The life cycle of *T. cruzi* is complex, comprising several evolutionary forms. The vector harbors epimastigotes and metacyclic trypomastigotes in its gut, while the blood trypomastigotes and the intracellular amastigotes are found in vertebrate hosts. Trypomastigotes and amastigotes represent the main targets for therapies [2,3].

The World Health Organization [1] recommends treating Chagas disease using benznidazole and nifurtimox. Both medicines are effective if given early in the acute phase, but their effectiveness is reduced in the more advanced stages; in addition, there is an increase in the frequency of adverse effects with increasing patient age [4,5]. The most common side effects are rashes, fever, generalized edema, lymphadenopathy, myalgia, arthralgia, gastrointestinal disorders, neutropenia, thrombocytopenic purpura, and peripheral polyneuropathy [6].

This scenario stimulates the search for new trypanocidal agents that are more effective in the chronic phase of Chagas disease and less toxic to hosts [7]. The processes of infection of host cells by *T. cruzi* and the survival of the parasite depend on the activity of important proteases; thus, an imbalance in the activity of these enzymes can cause damages to the parasite [8–10]. In this sense, *T. cruzi* enzymes represent interesting targets for studies of new therapeutic approaches [11].

It is well reported that some compounds (called protease inhibitors) are able to interact with protease molecules at specific sites, leading to the reduction or blockage of their activities [12]. The deregulation of proteases is a triggering factor for the onset of various pathologies and recent research has pointed out protease inhibitors as promising pharmacological agents [13,14]. Antioxidant, anti-inflammatory, immunomodulatory, antiviral, antimicrobial and antiparasitic activities of these molecules have been demonstrated [15–18]. Sangenito et al. [19] reported that inhibitors of HIV aspartyl peptidase affected the integrity of cellular structures of *T. cruzi* trypomastigotes, leading to metabolic disorders.

Moringa oleifera Lamarck (Moringaceae) is a pantropical tree (Figure 1A) that has aroused interest due to its medicinal properties and use as a source of oil and biogas, for example [20–22]. Its flowers (Figure 1B) contain a 18-kDa protein called MoFTI with trypsin inhibitory activity and toxicity to *T. cruzi* trypomastigotes with a lethal concentration that led to lysis of 50% of parasites (LC_{50}) of 41.20 µg/mL [11]. MoFTI was more toxic to the parasites than to murine macrophages and Vero cells, with selectivity indexes (SI) of 9.8 and >12, respectively [11]. The statement that MoFTI is a trypanocidal agent stimulated the investigations described in the present manuscript. Here, we tested the hypothesis that this inhibitor may interfere with the immune response of human peripheral blood mononuclear cells (PBMCs) infected with *T. cruzi*. In this sense, the assessment of MoFTI effects the viability and production of cytokines and nitric oxide (NO) by infected PBMCs are reported.

Figure 1. The *Moringa oleifera* tree (**A**) and its flowers (**B**).

2. Results and Discussion

The affinity chromatography of *M. oleifera* flower extract in the Trypsin–Agarose column resulted in isolation of MoFTI (Figure 2A), which was able to inhibit the hydrolyze of N-α-benzoyl-DL-arginine-p-nitroanilide (BApNA) by trypsin in a dose-dependent way (Figure 2B). The specific trypsin inhibitory activity of MoFTI was 62 U/mg, agreeing with previous reports [11,22]. Pontual et al. [22] reported that MoFTI showed an inhibition constant (Ki) of 2.4 μM on bovine trypsin.

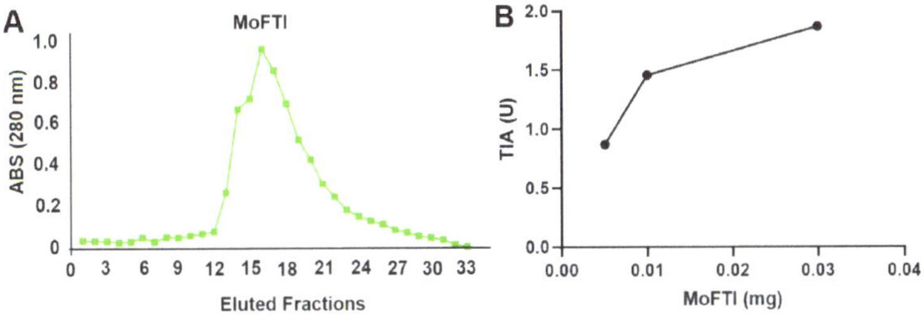

Figure 2. Isolation of MoFTI. (**A**) Affinity chromatography of *M. oleifera* flower extract in Trypsin–Agarose column. The elution step with 0.1 M KCl-HCl pH 2.0 can be seen and fractions of 1.0 mL were collected. (**B**) Trypsin inhibitor activity (TIA) of MoFTI on bovine trypsin.

After verifying that the protease inhibitor domain of MoFTI was active, we checked whether its antiparasitic property was also active. It was found that incubation with this inhibitor led to lysis of *T. cruzi* trypomastigotes, since the number of parasite cells counted in treatments with MoFTI was lower than in the negative control. After 24 h, the LC_{50} value was 43.5 (26.2–60.9) μg/mL. Similarly, Pontual et al. [11] reported an LC_{50} value of 41.20 μg/mL for MoFTI on *T. cruzi* trypomastigotes. In this same work, the authors showed that MoFTI was low in toxicity to murine peritoneal macrophages (50% cytotoxic concentration, CC_{50}, of 407.01 μg/mL) and did not interfere with the viability of Vero cells at concentrations up to 500 μg/mL.

In the present work, we investigated the effect of MoFTI on viability of human PBMCs. After 24, 72 and 120 h, it was revealed that MoFTI did not significantly ($p > 0.05$) affect the ability of these cells to metabolize the 3-[4,5-dimethylthiazol-2yl]-diphenyl tetrazolium bromide (MTT), in comparison with negative control (Figure 3). Therefore, it was assumed that, under the conditions used here, MoFTI was not toxic to human PBMCs. As can be seen in Figure 3A, the significantly ($p < 0.05$) greater number of viable cells in treatment with the inhibitor at 15.62 μg/mL, compared to the negative control group, suggest that MoFTI induced cell proliferation after exposure to 24 h. Benznidazole was also not able to interfere with PBMCs viability in comparison with the negative control.

It is well known that molecules with potential for use in therapy of infectious diseases need to be toxic to parasites without affecting the viability of host cells, or at least causing much more damage to parasites than to hosts [23]. The ratio between the CC_{50} value for PBMCs (>500 μg/mL) and LC_{50} for trypomastigotes showed that MoFTI was selective (SI > 11.48) for the parasite regarding these human cells. To the best of our knowledge, this is the first report of MoFTI effects on human cells and this datum suggests that this protein can be an interesting starting material for the production of a new drug for Chagas disease therapy. In fact, it can be expected that a protease inhibitor will end up interacting unwantedly with important enzymes of host cells; however, this is not the only report of a non-cytotoxic protease inhibitor to human cells. For example, the trypsin inhibitor from *Tecoma stans* leaves (TesTI) was also not toxic to human PBMCs [24].

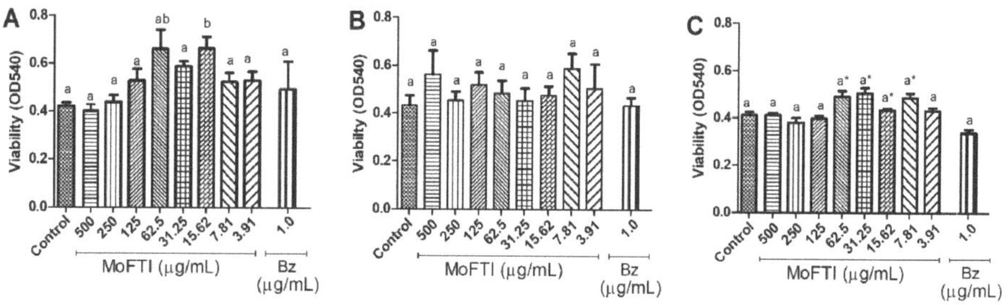

Figure 3. Effects of *M. oleifera* flower trypsin inhibitor (MoFTI) and benznidazole (Bz) on the viability of human peripheral blood mononuclear cells (PBMCs) after exposure by 24 h (**A**), 72 h (**B**), and 120 h (**C**). Different letters (a,b) indicate significant differences between the negative control and the other treatments by analysis of variance (ANOVA) followed by the Kruskal–Wallis test ($p < 0.05$). The asterisk (*) indicates significant differences between Bz and the other treatments. Control: negative control (untreated cells).

Previous reports have shown that plant compounds can act as immunomodulatory agents, and this can be interesting from a therapeutic point of view since, when these agents modulate the production of cytokines and other immune mediators, they can increase body defense against pathogens or pathological conditions, even when there is no direct toxicity to the causative agent [25]; because of this, we evaluated the effect of MoFTI on the release of cytokines by PBMCs uninfected or infected by *T. cruzi*. Interestingly, no alterations of cytokine production were observed regarding to negative control group when uninfected PBMCs were exposed to MoFTI (Figures 4 and 5). After 48-h incubation, MoFTI was not able to affect the release of interferon (IFN) γ by infected cells (Figure 4A). On the other hand, the release of tumor necrosis factor (TNF) α and interleukin (IL) 10 (Figure 4B,C, respectively) by *T. cruzi*-infected PBMCs was stimulated by MoFTI at 87.0 μg/mL (2 × LC_{50}) in comparison with the control group.

TNF-α is a cytotoxic factor associated with Th1 response against microorganisms, while IL-10 is a Th2 anti-inflammatory interleukin that inhibits the release of pro-inflammatory cytokines. Clinical studies showed that individuals with the chronic cardiac Chagas disease produce pro-inflammatory cytokines, such as TNF-α, to control *T. cruzi* infection. Simultaneously, anti-inflammatory cytokines, especially IL-10, are released to prevent damages to the host tissues and to slow the progress of cardiac complications [26,27]. Individuals at the chronic phase of the indeterminate form of Chagas disease produce higher levels of IL-10 and this is the reason why they do not progress to the clinical stage of Chagas cardiomyopathy [28]. In this sense, the trypanocidal activity of MoFTI, along with the simultaneous stimulation of TNF-α and IL-10 release, suggests that this protein may be able to control parasitemia while regulating the inflammation, preventing the progress of Chagas disease.

Still, after 48-h exposure, increased release of IL-6 was detected for PBMCs infected with *T. cruzi* and treated with MoFTI (2 × LC_{50}) at a similar level to that shown by the untreated infected cells (Figure 4D). This datum suggests that the release of IL-6 occurred because of *T. cruzi* infection and not due to MoFTI activity. IL-6 is a pleiotropic cytokine that influences antigen-specific immune responses and inflammatory reactions [29,30]. The profile of the other cytokines investigated here did not change in the untreated infected cells. The release of IL-4, IL-2 and NO by infected cells was not significantly affected by MoFTI after 48-h exposure (Figure 4E–G, respectively).

Unlike the results obtained for 48 h, the infected PBMCs had the release of INF-γ strongly increased in response to treatment for 120 h with both concentrations of MoFTI (Figure 5A) regarding the control. However, no differences were found for the release of TNF, IL-10, IL-6, IL-4 and IL-2 between the treatment groups (Figure 5B–F). When the infected PBMCs were treated with MoFTI at both concentrations, a significant increase in NO release was recorded (Figure 5G), but this does not seem to depend on the infection, because a similar result was detected for uninfected PBMCs treated with the inhibitor. The NO release may have occurred due to direct action of MoFTI or in response to the production of INF-γ [31]. Benznidazole did not affect cytokine release by both uninfected and infected cells.

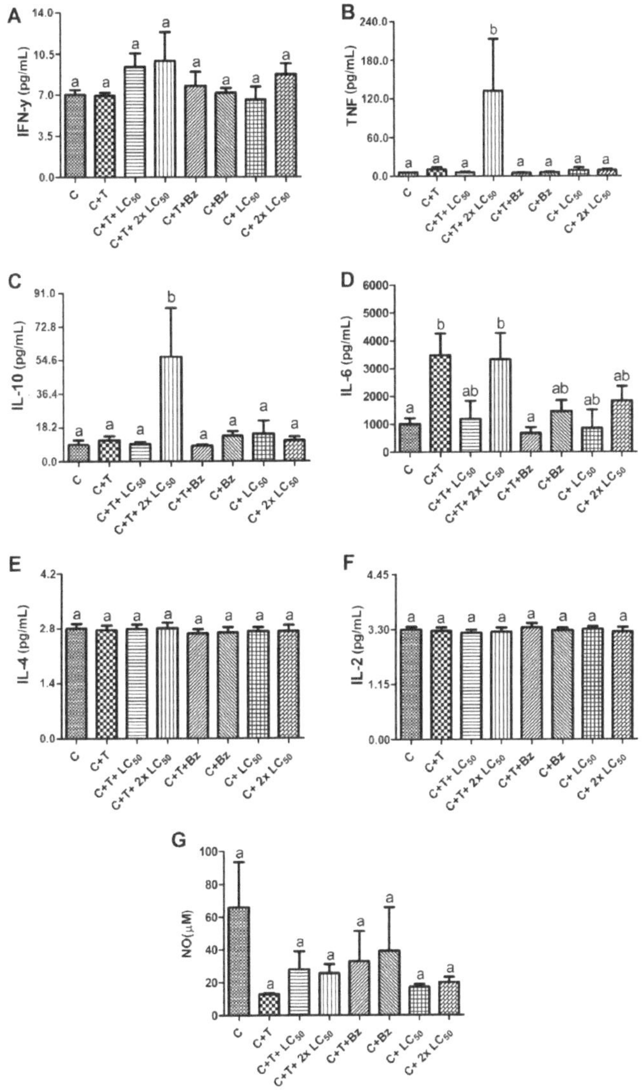

Figure 4. Effect of *M. oleifera* flower trypsin inhibitor (MoFTI) on the production of cytokines (**A–F**) and nitric oxide (**G**) by human peripheral blood mononuclear cells (PBMCs) infected or not with *T. cruzi* trypomastigotes after 48-h exposure. The treatments were: negative control (C = untreated and uninfected PBMCs); untreated PBMCs infected with trypomastigotes (C + T); infected PBMCs treated with MoFTI at 43.5 μg/mL (C + T + LC_{50}) and 87.0 μg/mL (C + T + 2 × LC_{50}); infected PBMCs treated with benznidazole (C + T + Bz); uninfected PBMCs treated with benznidazole (C + Bz); uninfected PBMCs treated with MoFTI at 43.5 μg/mL (C + LC_{50}) and 87.0 μg/mL (C+ 2 × LC_{50}). Different letters (a,b) indicate significant differences between treatments by analysis of variance (ANOVA) followed by the Tukey post-test ($p < 0.05$).

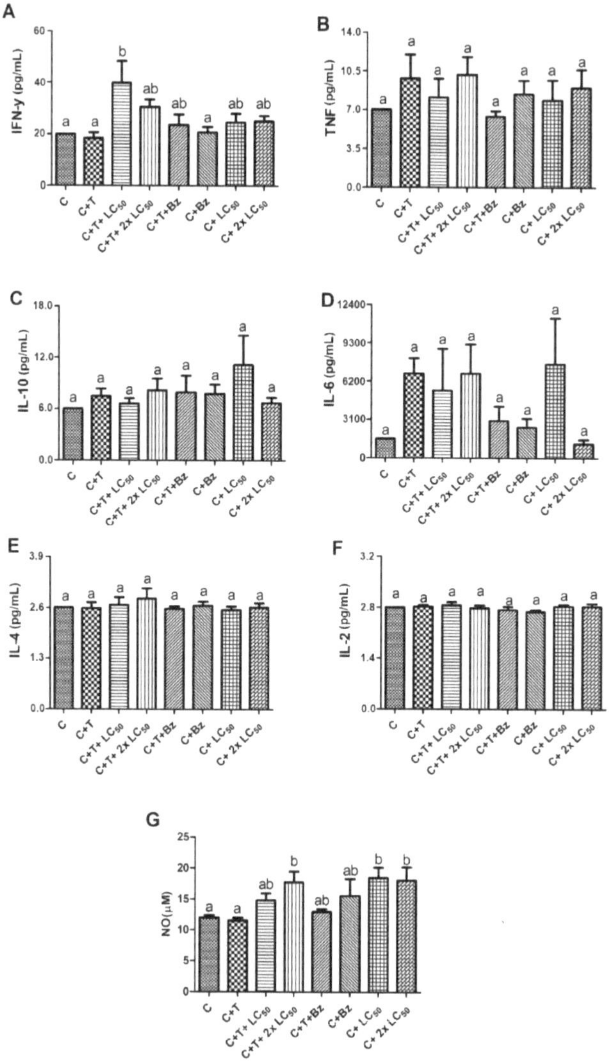

Figure 5. Effect of *M. oleifera* flower trypsin inhibitor (MoFTI) on the production of cytokines (**A**–**F**) and nitric oxide (**G**) by human peripheral blood mononuclear cells (PBMCs) infected or not with *T. cruzi* trypomastigotes after 120-h exposure. The treatments were: negative control (C = untreated and uninfected PBMCs); untreated PBMCs infected with trypomastigotes (C + T); infected PBMCs treated with MoFTI at 43.5 μg/mL (C + T + LC_{50}) and 87.0 μg/mL (C + T + 2 × LC_{50}); infected PBMCs treated with benznidazole (C + T + Bz); uninfected PBMCs treated with benznidazole (C + Bz); uninfected PBMCs treated with MoFTI at 43.5 μg/mL (C + LC_{50}) and 87.0 μg/mL (C+ 2 × LC_{50}). Different letters (a,b) indicate significant differences between treatments by analysis of variance (ANOVA) followed by the Tukey post-test ($p < 0.05$).

NO plays an important role in the defense of macrophages against *T. cruzi* by damaging parasite biochemistry and causing, for example, the inhibition of metalloproteins that mediate crucial metabolic processes, including cruzipain. This enzyme participates in parasite nutrition and the infection of host cells [32].

3. Materials and Methods

3.1. Isolation of MoFTI

The collection of *M. oleifera* flowers occurred in Recife city (8°02′57.9″ S, 34°56′47.8″ W), Pernambuco, Brazil, and a voucher specimen (number 73345) is deposited at the herbarium *Dárdano de Andrade Lima* (*Instituto Agronômico de Pernambuco*, Recife, Brazil). MoFTI was purified as described by Pontual et al. [22]. The procedure started from the maceration (10 min at 27 °C) of fresh flowers (50 g) with distilled water (100 mL) in a blender. The mixture was filtered through gauze and centrifuged (9000× *g*, 15 min, 4 °C) to remove suspended material. The crude preparation was dialyzed (6 h) against 0.1 M Tris-HCl, pH 8.0, containing 0.02 M $CaCl_2$ and loaded (4 mL; 66.8 mg of proteins) onto a Trypsin–Agarose (Sigma-Aldrich, St. Louis, MO, USA) column (4.5 × 1.0 cm). MoFTI was eluted with 0.1 M KCl-HCl, pH 2.0, and the presence of proteins in the collected fractions was accompanied by the measurement of absorbance at 280 nm. After dialysis (16 h) against distilled water, MoFTI was lyophilized to dryness and resuspended to a concentration of 1000 µg/mL in distilled water, for assessment of trypsin inhibitory activity, or in Rockwell Park Memorial Institute 1640 (RPMI 1640) complete medium (Sigma-Aldrich), for the assays with *T. cruzi* or PBMCs. Protein concentration was determined according to Lowry et al. [33] using bovine serum albumin (31.25–500 µg/mL) as standard.

3.2. Trypsin Inhibitory Activity

The ability of MoFTI to inhibit trypsin was assayed according to Pontual et al. [22]. In a microtiter plate, bovine trypsin (5 µL, 0.1 mg/mL in 0.1 M Tris-HCl, pH 8.0, containing 0.02 M $CaCl_2$) was added to 5 µL of 8.0 mM BApNA and MoFTI (0.005 to 0.03 mg/mL). Next, each well received Tris buffer to complete a volume of 200 µL. After incubation (30 min at 37 °C), the absorbance at 405 nm was measured using a microplate reader (Multiskan, Thermo Fisher Scientific, Waltham, MA, USA). One unit of trypsin inhibitor activity corresponded to the amount of MoFTI able to reduce the absorbance by 0.01 in accordance with trypsin activity in absence of the inhibitor.

3.3. Obtaining of T. cruzi Trypomastigotes

Cryopreserved trypomastigotes (10^7 trypomastigotes/mL) were thawed in a water bath at 37 °C. After centrifugation (400× *g*, 10 min, 22° C), the supernatant was discarded, and the pellet resuspended with 5 mL of complete Roswell Park Memorial Institute (RPMI) 1640 medium. The suspension was distributed in culture bottles containing Vero cells and incubated (37 °C, 5% CO_2) for 24 h. Next, the supernatant was removed to withdraw parasites that did not infect cells. RPMI 1640 complete medium (5 mL) was added and the cultures were incubated for 7 days. During this time, the multiplication of intracellular parasites was daily observed using an inverted microscope. After cell disruption, the trypomastigotes were collected for the subsequent assays.

3.4. Trypanocidal Activity of MoFTI

Culture-derived trypomastigotes (Y strain) were maintained by weekly passages in Vero cells cultured in complete RPMI 1640 medium. Trypomastigotes (10^6 parasites/mL, 100 µL) were placed in 96-well plates and treated with MoFTI (100 µL, 6.25–100 µg/mL) in RPMI medium supplemented with 10% fetal bovine serum (FBS). In negative control, parasites were incubated with complete RPMI medium in the absence of MoFTI. Benznidazole (1.0 µg/mL) was used as positive control. The assays were incubated for 24 h at 37 °C. Next, the number of trypomastigotes was determined by counting cells using a Neubauer chamber in a E100 LED microscope (Nikon, Melville, NY, USA). The percentage of lysis (%) was estimated regarding the number of trypomastigotes in the negative control (100%). Three independent experiments were performed in quadruplicate and the concentration (µg/mL) of MoFTI that leads to lysis of 50% of trypomastigote cells (LC_{50}) for 24 h was calculated using the software MedCalc version 17.9.7 (MedCalc Software bvba, Ostend, Belgium).

3.5. Isolation of Human PBMCs

The blood collection was performed following the protocol approved (process: 07511612.2.0000.5190) by the Research Ethics Committee of the Instituto Aggeu Magalhães/Fundação Oswaldo Cruz (IAM/FIOCRUZ). The blood was collected from ten healthy individuals (6 women and 4 men) through a vacuum system (Vacutainer; BD Biosciences, Franklin Lakes, NJ, USA) in tubes containing sodium heparin, and homogenized (15 mL) with an aliquot of filtered and sterile phosphate buffered saline (PBS), pH 7.2 (15 mL). Next, 15 mL of this mixture was added to 50 mL Falcon polypropylene tubes containing 15 mL of Ficoll-Hypaque (GE Healthcare Life Sciences, Uppsala, Sweden). After centrifugation (900× *g*, 30 min, 20 °C), PBMCs appeared as a ring between Ficoll and plasma and were collected using a sterile transfer pipette and placed in 15 mL Falcon polypropylene tubes. PBMCs were resuspended using 14 mL of incomplete RPMI 1640 medium containing 1% penicillin/streptomycin and centrifuged (400× *g*, 10 min). This procedure was repeated twice. Following, the cells were resuspended in 2 mL of RPMI 1640 complete medium supplemented with FBS (10%) and containing 1% penicillin/streptomycin. The cells (10 µL) were subsequently stained with Trypan Blue (Sigma-Aldrich) dye (90 µL) and counted in a Neubauer chamber. The number of cells were recorded and adjusted to the desired concentration of 10^6 cells/mL.

3.6. Effect of MoFTI on PBMCs Viability

The effect of MoFTI on viability of PBMCs was assayed measuring the activity of mitochondrial succinate dehydrogenase [34]. Cell suspension (10^6 cells) was placed in 96-well culture plates with complete RPMI 1640 medium and exposed to MoFTI (3.9 to 500.0 µg/mL). In negative control, PBMCs were incubated with RPMI 1640 medium in absence of MoFTI. Benznidazole (1.0 µg/mL) was also tested. The plates were incubated at 37 °C and 5% CO_2 for 24, 72 and 120 h. Next, the culture medium was removed and RPMI medium containing 5.0 mg/mL of MTT (Sigma-Aldrich) was added. After incubation (37 °C, 5% CO_2) for 3 h, the culture medium containing MTT was removed and 100 µL of dimethyl sulfoxide was added. The presence of formazan crystals derived from MTT reduction was immediately recorded by measuring the absorbance at 540 nm. The experiment was conducted in triplicates and the selectivity index (SI) was calculated from the ratio between cytotoxicity to PBMCs (CC_{50}) and the LC_{50} of MoFTI.

3.7. Treatment of *T. cruzi*-Infected PBMCs with MoFTI

PBMCs (1 mL, 2 × 10^6 cells) in complete RPMI medium were placed in 48-well polystyrene culture plates, which were incubated (37 °C, 24 h) to fix adherent cells (mainly monocytes). Next, 0.5 mL of the RPMI medium was removed from each well, and trypomastigotes (0.5 mL, 10^6 cells) were added. The plates were incubated (37 °C, 5% CO_2) for 2 h to allow the infection of PBMCs. The mixture (RPMI medium containing infected adherent and non-adherent cells) was treated with MoFTI at 43.5 µg/mL (LC_{50}) and 87.0 µg/mL (2 × LC_{50}). In negative control, the mixture was maintained without MoFTI, while benznidazole (1.0 µg/mL) was used in positive control. To compare the cell response in the presence or absence of parasites, uninfected cells were incubated in the absence or presence of MoFTI (LC_{50} and 2 × LC_{50}) or benznidazole (1.0 µg/mL), as described above. The plates were incubated (37 °C, 5% CO_2) for 48 or 120 h, and 700 µL of the supernatant from each well was removed and immediately stored at −20 °C for later use to measure cytokine and nitric oxide (NO) levels.

3.8. Effect of MoFTI on Cytokine Release by *T. cruzi*-Infected PBMCs

The release of the cytokines IL-2, IL-4, IL-6, IL-10, IFN-γ and TNF-α by PBMCs was quantified using the Cytometric Bead Array (CBA) system, following the instructions of the manufacturer (BD Biosciences). The data were acquired on the FACScalibur flow cytometer. The acquisitions and analyses were performed using the CellQuestPro software (BD Biosciences) and 5000 events were acquired within the lymphocyte population. The analyses were performed using the BD CBA software.

3.9. *Effect of MoFTI on NO Production by T. cruzi-Infected PBMCs*

Aliquots (50 μL) of the supernatants removed from the plates referred to in Section 3.7 were added to the Griess reagent (50 μL) in 96 well microplates, according to Resende et al. [35]. After incubation (28 °C, 15 min), NO production was estimated by measuring the absorbance at 540 nm and a standard curve of nitrite (3.12–400.0 μM).

3.10. *Statistical Analysis*

The results regarding cytokine and NO levels were statistically evaluated using the GraphPad Prism 5.0 software (GraphPad Software Inc, San Diego, CA, USA). The data that passed the Kolmogorov–Smirnov normality test were analyzed using analysis of variance (ANOVA) followed by the Tukey post-test, while the data that did not pass the normality test were analyzed by the Kruskal–Wallis test followed by the Dunnet post-test. Significance level at 5% was considered.

4. Conclusions

The data reported herein point to the trypanocidal trypsin inhibitor of *M. oleifera* flowers as an immunomodulatory agent on *T. cruzi*-infected human PBMCs by stimulating the release of pro-inflammatory (TNF-α and INF-γ) and anti-inflammatory (IL-10) cytokines, as well as NO. In addition, MoFTI was more toxic to the parasite cells than to the human immune cells. Our findings stimulate future investigations of MoFTI in vivo effects on immune response in Chagas disease. It should be remembered that MoFTI is a proteinaceous inhibitor. In this sense, some points must be addressed to develop drug formulations using MoFTI, such as its stability and immunogenicity. Finally, strategies for producing MoFTI at a large scale must be designed so that it can be inserted in the pharmaceutical industry.

Author Contributions: Conceptualization, I.C.V.N., T.H.N., V.M.B.d.L. and E.V.P.; methodology, I.C.V.N., L.R.M., D.J.L.T., K.K.d.S.O., L.L.d.S.P., T.H.N., V.M.B.d.L. and E.V.P.; software, I.C.V.N., L.R.M., D.J.L.T., K.K.d.S.O. and L.L.d.S.P.; validation, I.C.V.N., L.R.M., D.J.L.T., K.K.d.S.O., L.L.d.S.P., T.H.N., V.M.B.d.L. and E.V.P.; formal analysis, I.C.V.N., T.H.N., V.M.B.d.L. and E.V.P.; investigation, I.C.V.N., L.R.M., D.J.L.T., K.K.d.S.O., L.L.d.S.P., T.H.N., V.M.B.d.L. and E.V.P.; resources, T.H.N., V.M.B.d.L. and E.V.P.; data curation, I.C.V.N., L.R.M., D.J.L.T., K.K.d.S.O. and L.L.d.S.P.; writing—original draft preparation, I.C.V.N.; writing—review and editing, L.C.B.B.C., P.M.G.P., T.H.N., V.M.B.d.L. and E.V.P.; visualization, L.C.B.B.C., P.M.G.P., T.H.N., V.M.B.d.L. and E.V.P.; supervision, V.M.B.d.L. and E.V.P.; project administration, V.M.B.d.L. and E.V.P.; funding acquisition, L.C.B.B.C., P.M.G.P., T.H.N., V.M.B.d.L. and E.V.P. All authors have read and agreed to the published version of the manuscript.

Funding: This research was funded by Conselho Nacional de Desenvolvimento Científico e Tecnológico (CNPq; 408789/2016-6), Coordenação de Aperfeiçoamento de Pessoal de Nível Superior (CAPES, Finance Code 001) and Fundação de Amparo à Ciência e Tecnologia do Estado de Pernambuco (FACEPE; APQ-0108-2.08/14; APQ-0661-2.08/15) for financial support.

Conflicts of Interest: The authors declare no conflict of interest. The funders had no role in the design of the study; in the collection, analyses, or interpretation of data; in the writing of the manuscript, or in the decision to publish the results.

References

1. World Health Organization. Chagas Disease. Available online: http://www.who.int/en/news-room/factsheets/detail/chagas-disease-(american-trypanosomiasis) (accessed on 20 June 2020).
2. Souza, W. Basic cell biology of *Trypanosoma cruzi*. *Curr. Pharm. Des.* **2002**, *8*, 269–285. [CrossRef] [PubMed]
3. Echeverria, L.E.; Morillo, C.A. American trypanosomiasis (Chagas disease). *Infect. Dis. Clin. North Am.* **2019**, *33*, 119–134. [CrossRef] [PubMed]
4. Ferreira, A.M.; Damasceno, R.F.; Monteiro-Junior, R.S.; Oliveira, I.A.C.D.; Prates, T.E.C.; Nunes, M.C.P.; Haikal, D.S.A. Reações adversas ao benzonidazol no tratamento da Doença de Chagas: Revisão sistemática de ensaios clínicos randomizados e controlados. *Cad. Saúde Colet.* **2019**, *27*, 354–362. [CrossRef]
5. Francisco, A.F.; Jayawardhana, S.; Olmo, F.; Lewis, M.D.; Wilkinson, S.R.; Taylor, M.C.; Kelly, J.M. Challenges in Chagas Disease drug development. *Molecules* **2020**, *25*, 2799. [CrossRef]

6. Molina-Morant, D.; Fernández, M.L.; Bosch-Nicolau, P.; Sulleiro, E.; Bangher, M.; Salvador, F.; Sanchez-Montalva, A.; Ribeiro, A.L.P.; de Paula, A.M.B.; Eloi, S.; et al. Efficacy and safety assessment of different dosage of benznidazol for the treatment of Chagas disease in chronic phase in adults (MULTIBENZ study): Study protocol for a multicenter randomized Phase II non-inferiority clinical trial. *Trials* **2020**, *21*, 1–10. [CrossRef]
7. Juárez-Saldivar, A.; Schroeder, M.; Salentin, S.; Haupt, V.J.; Saavedra, E.; Vázquez, C.; Reyes-Espinosa, F.; Herrera-Mayorga, V.; Villalobos-Rocha, J.C.; García-Pérez, C.A.; et al. Computational drug repositioning for Chagas Disease using protein-ligand interaction profiling. *Int. J. Mol. Sci.* **2020**, *21*, 4270. [CrossRef]
8. Cazzulo, J.J.; Couso, R.; Raimondi, A.; Wernstedt, C.; Hellman, U. Further characterization and partial amino acid sequence of a cysteine proteinase from *Trypanosoma cruzi*. *Mol. Biochem. Parasitol.* **1989**, *33*, 33–42. [CrossRef]
9. Cazzulo, J. Proteinases of *Trypanosoma cruzi*: Potential targets for the chemotherapy of Chagas disease. *Curr. Top. Med. Chem.* **2002**, *2*, 1261–1271. [CrossRef]
10. Ferreira, R.A.A.; Pauli, I.; Sampaio, T.S.; Souza, M.L.; Ferreira, L.L.G.; Magalhães, L.G.; Rezende, C.O., Jr.; Ferreira, R.S.; Krogh, R.; Dias, L.C.; et al. Structure-based and molecular modeling studies for the discovery of cyclic imides as reversible cruzain inhibitors with potent anti-*Trypanosoma cruzi* activity. *Front. Chem.* **2019**, *7*, 798. [CrossRef]
11. Pontual, E.V.; Pires-Neto, D.F.; Fraige, K.; Higino, T.M.M.; Carvalho, B.E.A.; Alves, N.M.P.; Lima, T.A.; Zingali, R.B.; Coelho, L.C.B.B.; Bolzani, V.S.; et al. A trypsin inhibitor from *Moringa oleifera* flower extract is cytotoxic to *Trypanosoma cruzi* with high selectivity over mammalian cells. *Nat. Prod. Res.* **2018**, *32*, 2940–2944. [CrossRef]
12. Marathe, K.R.; Patil, R.H.; Vishwakarma, K.S.; Chaudhari, A.B.; Maheshwari, V.L. Protease inhibitors and their applications: An overview. In *Studies Natural Products Chemistry*; Atta-ur-Rahman, Ed.; Elsevier: Amsterdam, The Netherlands, 2019; Volume 62, pp. 211–242.
13. Agbowuro, A.A.; Huston, W.M.; Gamble, A.B.; Tyndall, J.D. Proteases and protease inhibitors in infectious diseases. *Med. Res. Rev.* **2018**, *38*, 1295–1331. [CrossRef] [PubMed]
14. Cotabarren, J.; Lufrano, D.; Parisi, M.G.; Obregón, W.D. Biotechnological, biomedical, and agronomical applications of plant protease inhibitors with high stability: A systematic review. *Trends Plant. Sci.* **2020**, *262*, 110398. [CrossRef]
15. Shamsi, T.N.; Parveen, R.; Afreen, S.; Azam, M.; Sen, P.; Sharma, Y.; Haque, Q.M.R.; Fatma, T.; Manzoor, N.; Fatima, S. Trypsin inhibitors from *Cajanus cajan* and *Phaseolus limensis* possess antioxidant, anti-inflammatory, and antibacterial activity. *J. Diet. Suppl.* **2018**, *15*, 939–950. [CrossRef] [PubMed]
16. Blisnick, A.; Šimo, L.; Grillon, C.; Fasani, F.; Brûlé, S.; Le Bonniec, B.; Prina, E.; Marsot, M.; Relmy, A.; Blaise-Boisseau, S.; et al. The immunomodulatory effect of IrSPI, a tick salivary gland serine protease inhibitor involved in *Ixodes ricinus* tick feeding. *Vaccines* **2019**, *7*, 148. [CrossRef] [PubMed]
17. Mishra, U.N.; Reddy, M.V.; Prasad, D.T. Plant serine protease inhibitor (SPI): A potent player with bactericidal, fungicidal, nematicidal and antiviral properties. *Int. J. Cardiovasc. Sci.* **2020**, *8*, 2985–2993. [CrossRef]
18. Pramanik, A.; Paik, D.; Pramanik, P.K.; Chakraborti, T. Serine protease inhibitors rich *Coccinia grandis* (L.) Voigt leaf extract induces protective immune responses in murine visceral leishmaniasis. *Biomed. Pharmacother.* **2019**, *111*, 224–235. [CrossRef]
19. Sangenito, L.S.; Menna-Barreto, R.F.; Oliveira, A.C.; d'Avila-Levy, C.M.; Branquinha, M.H.; Santos, A.L. Primary evidence of the mechanisms of action of HIV aspartyl peptidase inhibitors on *Trypanosoma cruzi* trypomastigote forms. *Int. J. Antimicrob. Agents.* **2018**, *52*, 185–194. [CrossRef]
20. Makkar, H.A.; Becker, K. Nutritional value and antinutritional components of whole and ethanol extracted *Moringa oleifera*. *Anim. Feed Sci. Technol.* **1996**, *63*, 211–228. [CrossRef]
21. Karadi, R.V.; Gadge, N.B.; Alagawadi, K.R.; Savadi, R.V. Effect of *Moringa oleifera* Lam. root-wood on ethylene glycol induced urolithiasis in rats. *J. Ethnopharmacol.* **2006**, *105*, 306–311. [CrossRef]
22. Pontual, E.V.; Santos, N.D.L.; Moura, M.C.; Coelho, L.C.B.B.; Navarro, D.M.A.F.; Napoleão, T.H.; Paiva, P.M.G. Trypsin inhibitor from *Moringa oleifera* flowers interferes with survival and development of *Aedes aegypti* larvae and kills bacteria inhabitant of larvae midgut. *Parasitol. Res.* **2014**, *113*, 727–733. [CrossRef]
23. Bermudez, J.; Davies, C.; Simonazzi, A.; Real, J.P.; Palma, S. Current drug therapy and pharmaceutical challenges for Chagas disease. *Acta Trop.* **2016**, *156*, 1–16. [CrossRef] [PubMed]

24. Patriota, L.L.S.; Procópio, T.F.; Souza, M.F.D.; Oliveira, A.P.S.; Carvalho, L.V.N.; Pitta, M.G.R.; Rego, M.J.B.M.; Paiva, P.M.G.; Pontual, E.V.; Napoleão, T.H. A trypsin inhibitor from *Tecoma stans* leaves inhibits growth and promotes ATP depletion and lipid peroxidation in *Candida albicans* and *Candida krusei*. *Front. Microbiol.* **2016**, *7*, 611. [CrossRef]
25. Jantan, I.; Ahmad, W.; Bukhari, S.N.A. Plant-derived immunomodulators: An insight on their preclinical evaluation and clinical trials. *Front. Plant. Sci.* **2015**, *6*, 655. [CrossRef] [PubMed]
26. Dutra, W.O.; Menezes, C.A.; Magalhães, L.M.; Gollob, K.J. Immunoregulatory networks in human Chagas disease. *Parasite Immunol.* **2014**, *36*, 377–387. [CrossRef] [PubMed]
27. Vallejo, A.; Monge-Maillo, B.; Gutiérrez, C.; Norman, F.F.; López-Vélez, R.; Pérez-Molina, J.A. Changes in the immune response after treatment with benznidazole versus no treatment in patients with chronic indeterminate Chagas disease. *Acta Trop.* **2016**, *164*, 117–124. [CrossRef] [PubMed]
28. Álvarez, J.M.; Fonseca, R.; Silva, H.B.; Marinho, C.R.; Bortoluci, K.R.; Sardinha, L.R.; D'império-Lima, M.R. Chagas disease: Still many unsolved issues. *Mediators Inflamm.* **2014**, *2014*, 9. [CrossRef] [PubMed]
29. Kishimoto, T. Interleukin-6: Discovery of a pleiotropic cytokine. *Arthritis Res. Ther.* **2006**, *8*, S2. [CrossRef] [PubMed]
30. Jones, S.A.; Jenkins, B.J. Recent insights into targeting the IL-6 cytokine family in inflammatory diseases and cancer. *Nat. Rev. Immunol.* **2018**, *18*, 773–789. [CrossRef]
31. Roffê, E.; Rothfuchs, A.G.; Santiago, H.C.; Marino, A.P.M.P.; Ribeiro-Gomes, F.L.; Eckhaus, M.; Antonelli, L.R.V.; Murphy, P.M. IL-10 limits parasite burden and protects against fatal myocarditis in a mouse model of *Trypanosoma cruzi* infection. *J. Immunol.* **2012**, *188*, 649–660. [CrossRef]
32. Cardoso, M.S.; Reis-Cunha, J.L.; Bartholomeu, D.C. Evasion of the immune response by *Trypanosoma cruzi* during acute infection. *Front. Immunol.* **2016**, *6*, 659. [CrossRef]
33. Lowry, O.H.; Rosebrough, N.J.; Farr, A.L.; Randall, R.J. Protein measurement with the folin phenol reagent. *J. Biol. Chem.* **1951**, *193*, 265–275. [PubMed]
34. Mosmann, T. Rapid colorimetric assay for cellular growth and survival: Application to proliferation and cytotoxicity assays. *J. Immunol. Methods.* **1983**, *65*, 55–63. [CrossRef]
35. Resende, L.A.; Roatt, B.M.; Aguiar-Soares, R.D.O.; Viana, K.F.; Mendonca, L.Z.; Lanna, M.F.; Silveira-Lemos, D.; Corrêa-Oliveira, R.; Martins-Filho, O.A.; Fujiwara, R.T.; et al. Cytokine and nitric oxide patterns in dogs immunized with LBSap vaccine, before and after experimental challenge with *Leishmania chagasi* plus saliva of *Lutzomyia longipalpis*. *Vet. Parasitol.* **2013**, *198*, 371–381. [CrossRef] [PubMed]

antibiotics

MDPI

Article

Anthelminthic Activity of Assassin Bug Venom against the Blood Fluke *Schistosoma mansoni*

Miray Tonk [1,2], Andreas Vilcinskas [1,2,3], Christoph G. Grevelding [4] and Simone Haeberlein [4,*]

1 Institute for Insect Biotechnology, Justus Liebig University of Giessen, Heinrich-Buff-Ring 26-32, 35392 Giessen, Germany; miray.tonk@agrar.uni-giessen.de (M.T.); andreas.vilcinskas@agrar.uni-giessen.de (A.V.)

2 LOEWE Centre for Translational Biodiversity Genomics (LOEWE-TBG), Senckenberganlage 25, 60325 Frankfurt, Germany

3 Fraunhofer Institute for Molecular Biology and Applied Ecology, Department of Bioresources, Winchester Strasse 2, 35394 Giessen, Germany

4 Institute of Parasitology, BFS, Justus Liebig University of Giessen, Schubertstr. 81, 35392 Giessen, Germany; christoph.grevelding@vetmed.uni-giessen.de

* Correspondence: simone.haeberlein@vetmed.uni-giessen.de

Received: 31 July 2020; Accepted: 28 September 2020; Published: 1 October 2020

check for updates

Abstract: Helminths such as the blood fluke *Schistosoma mansoni* represent a major global health challenge due to limited availability of drugs. Most anthelminthic drug candidates are derived from plants, whereas insect-derived compounds have received little attention. This includes venom from assassin bugs, which contains numerous bioactive compounds. Here, we investigated whether venom from the European predatory assassin bug *Rhynocoris iracundus* has antischistosomal activity. Venom concentrations of 10–50 μg/mL inhibited the motility and pairing of *S. mansoni* adult worms in vitro and their capacity to produce eggs. We used EdU-proliferation assays to measure the effect of venom against parasite stem cells, which are essential for survival and reproduction. We found that venom depleted proliferating stem cells in different tissues of the male parasite, including neoblasts in the parenchyma and gonadal stem cells. Certain insect venoms are known to lyse eukaryotic cells, thus limiting their therapeutic potential. We therefore carried out hemolytic activity assays using porcine red blood cells, revealing that the venom had no significant effect at a concentration of 43 μg/mL. The observed anthelminthic activity and absence of hemolytic side effects suggest that the components of *R. iracundus* venom should be investigated in more detail as potential antischistosomal leads.

Keywords: assassin bug; *Rhynocoris iracundus*; *Schistosoma mansoni*; venom; in vitro culture; natural compound; stem cells; cell proliferation

1. Introduction

Helminths (parasitic worms) infect more than 3.5 billion people worldwide, causing significant morbidity and economic losses [1,2]. Novel anthelminthic compounds are urgently needed to achieve better control of this important group of parasites given the limited availability of effective vaccines and drugs [3–5]. Among helminths, blood flukes (schistosomes) such as *Schistosoma mansoni* cause schistosomiasis, a neglected tropical disease that globally affects more than 200 million people and causes 200,000 deaths each year [6,7]. Male and female schistosomes mate in the blood vessels of their host and produce hundreds of eggs per day, which, if trapped in the liver, can trigger chronic diseases including liver fibrosis [6,8]. The treatment of schistosomiasis currently relies on a limited drug repertoire, with praziquantel as the current gold standard [9]. The continual use of this drug since its approval in the 1980s likely promotes emergence of resistant helminth populations, as evidenced by animal studies and human drug administration programs [10–12]. The discovery of alternative antischistosomal drugs is therefore a high priority in neglected tropical disease research [13].

Natural products represent a treasure trove for the discovery of new drugs, particularly novel anti-infectives. Plant-derived natural products have been extensively studied for their antischistosomal activity, whereas animal-derived compounds have received comparatively little attention [14], despite being the focus of drug discovery for various other therapeutic applications [15,16]. Only a few studies have reported on the antischistosomal activity of bee, scorpion, frog and snake venoms [17–21]. Venoms are injected by animals into the body of their victims using stings, spines or bites [22–24]. These complex fluids include proteolytic enzymes, biogenic amines, neurotoxic peptides, neurotransmitters, and compounds that bind to and disrupt the function of multiple molecular targets in the victim [25]. Assassin bugs (Reduviidae) are a family of predaceous hemipteran insects comprising ~6800 species [26]. They are known for their potent venom, which is injected via a straw-like proboscis to paralyze and liquefy other invertebrates as prey. Assassin bugs can also use their venom defensively against (mainly vertebrate) predators [25,27]. The composition and function of assassin bug venom is poorly understood, but more than 200 compounds have recently been identified in two reduviid species: *Platymeris biguttatus* L. and *Psytalla horrida* (both Hemiptera, Reduviidae) [28]. This is an important step toward the repurposing of venom toxins for biomedical applications. Here, we investigated the potential anthelminthic properties of venom from the European predatory assassin bug *Rhynocoris iracundus* against adult *S. mansoni*. We assessed the effects of the venom on parasite motility, reproduction, and cell proliferation in vitro for a cultivation period of 3 days.

2. Results

2.1. *Assassin Bug Venom Reduces Motility, Pairing, Attachment and Egg Production in S. mansoni*

Venom was collected from *R. iracundus* by physical stimulation (Figure 1). The venom was tested for its anthelminthic activity against pairs of adult *S. mansoni* using an in vitro culture system over a period of 72 h. To assess the vitality of the worms, we determined their motility and the percentage of worms fit enough to (a) maintain the pairing state and (b) attach via their suckers to the base of the culture plate. As a positive control, worm couples were treated with different concentrations of praziquantel which caused death to all worms at 5 μM (Supplementary Figure S1). While pairs of worms in the control group remained motile and attached, those treated with 25 or 50 μg/mL of venom showed an overall loss of vitality (Figure 2, Videos S1–S4). Both males and females treated with the high dose of venom also became stunted (Figure 2C). At a venom concentration of 25 μg/mL, the motility of worms was significantly inhibited after 72 h, with male worms often being more affected (motility score 1) than females (scores 1 or 2). At the higher venom concentration (50 μg/mL), a significant loss of motility was observed already after 24 h (Figure 3A). Some (25 μg/mL) or all (50 μg/mL) worms were unable to attach to the base of the culture plate or maintain their pairing status (Figure 3B,C). Finally, a dose-dependent reduction in egg production was observed, while the shape of eggs appeared normal (Figure 3D–F). The lowest tested concentration of venom (10 μg/mL) had a slight impact on motility in some worms, but significantly reduced pairing stability and egg production (Figure 3C,D). Taken together, these results confirmed that *R. iracundus* venom affects *S. mansoni* motility, pairing stability, attachment and fecundity, starting at concentrations as low as 10 μg/mL.

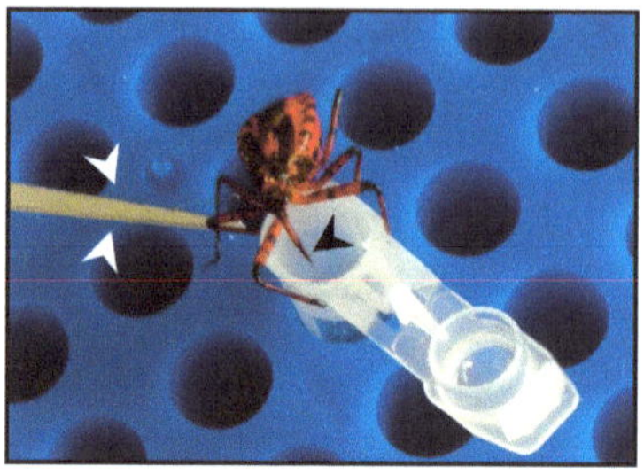

Figure 1. The European predatory assassin bug *Rhynocoris iracundus*. Stimulation of *R. iracundus* on the hind legs using entomological forceps (white arrow heads) encourages the insect to use its proboscis (black arrow head) to inject venom through laboratory film (Parafilm) stretched over a collection tube containing phosphate-buffered saline (PBS).

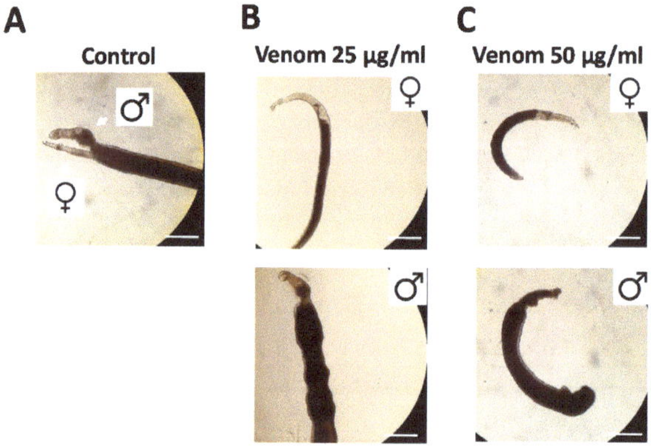

Figure 2. *Rhynocoris iracundus* venom affects the vitality of *Schistosoma mansoni*. Worm pairs were treated with different concentrations of venom (25 or 50 µg/mL). Representative images show worms after 72 h. (**A**) Untreated control worms remained paired and attached via their suckers to the base of the culture plate. The addition of venom at (**B**) 25 µg/mL or (**C**) 50 µg/mL induced the separation of pairs and detachment from the plate. Scale bars = 250 µm.

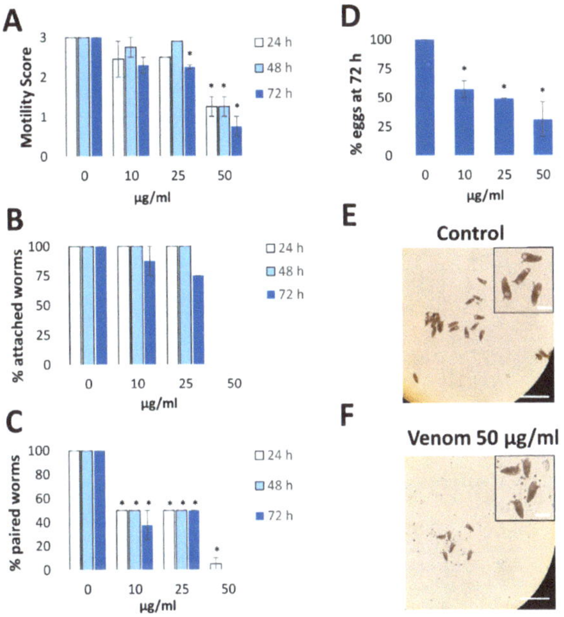

Figure 3. Effect of *Rhynocoris iracundus* venom on *Schistosoma mansoni* motility, pairing and egg production. Worm pairs were treated with different concentrations of venom (10–50 µg/mL) for a period of 72 h. We measured (**A**) motility, (**B**) the percentage of worms attached to the base of the plate, and (**C**) pairing stability every 24 h. (**D**) The number of eggs produced within 72 h relative to the untreated control. The shape of the eggs appeared normal after venom treatment (inserts). Graphs show a summary of two experiments with 5–8 worm pairs (mean ± SEM). Significant differences vs. the control are indicated (* $p < 0.05$, Wilcoxon rank sum test). (**E**,**F**) Representative images showing the number of eggs produced by untreated control worms and venom-treated worms (50 µg/mL). Scale bars = 250 µm, for inserts = 60 µm.

2.2. *Proliferating Stem Cells Are Depleted by Assassin Bug Venom*

Antischistosomal effects may be associated with a decrease in the number of proliferating stem cells [29], which are considered essential for parasite development and survival [30]. We therefore investigated whether *R. iracundus* venom had a similar effect. Because stem cells are the only proliferating cells in adult schistosomes [31], we made use of the thymidine analog EdU (5-ethynyl-2-deoxyuridine) in order to visualize proliferating stem cells in whole-mount worms. EdU-positive stem cells were observed throughout the parenchyma of male and female worms (Figure 4A). These are known as neoblasts and have been shown to provide a constant stream of new cells for the development of the tegument, gastrodermis and potentially other tissues [31]. EdU-positive stem cells were also abundant in the gonads: spermatogonia in testes and oogonia in the ovary (Figure 4A), which give rise to germ cells. The analysis of venom-treated female worms by confocal laser scanning microscopy (CLSM) revealed no obvious change in the number of EdU-positive stem cells compared to untreated controls. However, the number of proliferating stem cells in males treated with 50 µg/mL venom fell to near zero in both the parenchyma and gonads (Figure 4B). To quantify this effect, we performed 3D image analysis to determine the numbers of EdU-positive stem cells and of Hoechst-positive total cells in the testes (Figure 5A–D) and the parenchyma (Figure 5E–H). This revealed a significant reduction in the frequency of stem cells and of the density of stem cells per defined tissue volume with 50 µg/mL venom. This was observed for both, spermatogonial stem cells (Figure 5C,D) and parenchymatic neoblasts (Figure 5G,H).

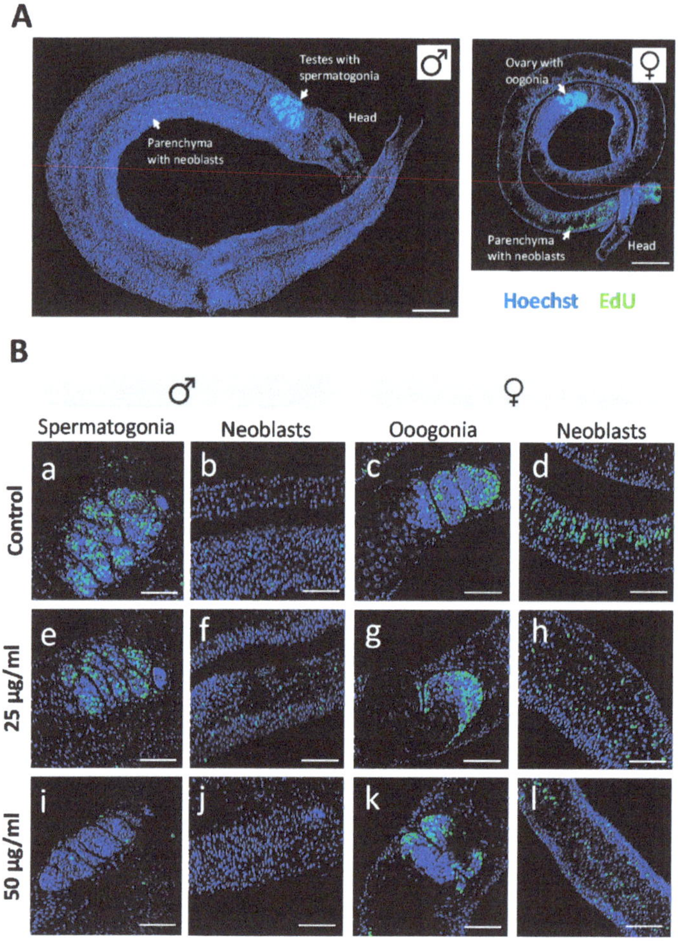

Figure 4. Effect of *Rhynocoris iracundus* venom on the proliferation of *Schistosoma mansoni* stem cells. (**A**) Overview of the location of parenchymal stem cells (neoblasts) and gonadal stem cells (spermatogonia and oogonia) in male and female worms. Stem cells are labeled with EdU (green), and nuclei are counterstained with Hoechst 33342 (blue). Scale bars = 100 µm. (**B**) Worm pairs were treated for 72 h with 25 or 50 µg/mL of venom or cultured without venom as a control. EdU was added during the final 24-h period. The abundance of EdU-positive proliferating stem cells was comparable in worms of the control group (a–d) and those treated with 25 µg/mL of venom (e–h) whereas 50 µg/mL of venom reduced the number of proliferating stem cells in males (i, j) but not in females (k, l). Scale bars = 50 µm. Representative images of four worms per treatment group are shown.

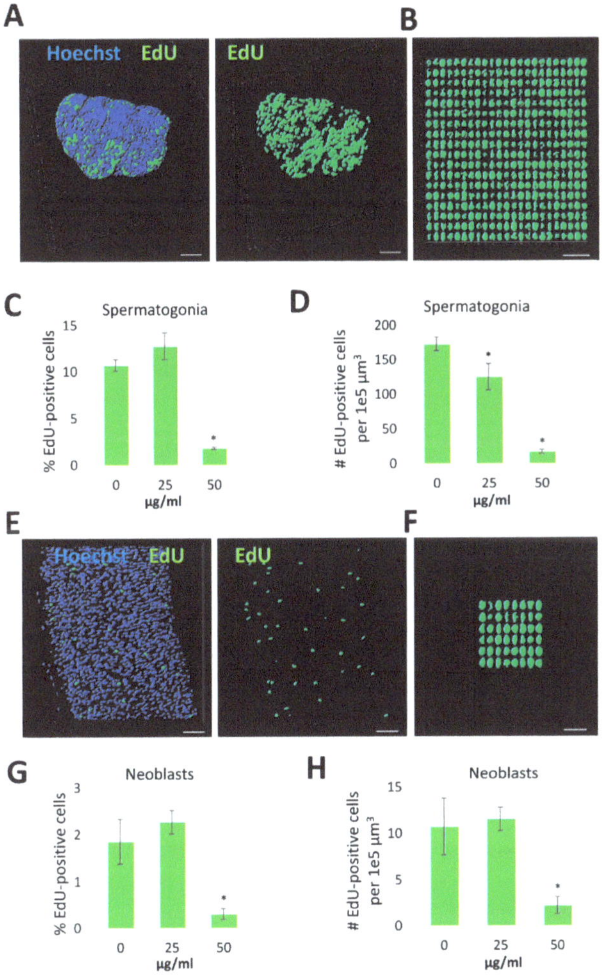

Figure 5. Reduction in stem cell frequency and density in male *Schistosoma mansoni* treated with *Rhynocoris iracundus* venom. Worm pairs were treated for 72 h with 25 or 50 µg/mL of venom or cultured without venom as a control. Proliferating stem cells were labeled with EdU and nuclei of all cells with Hoechst 33342. Cell numbers were quantified in z-stacks using the software package "IMARIS for cell biologists" (Bitplane). The percentage of EdU-positive cells related to the total cell number (C, G) and the number of EdU-positive cells per 1e5 μm^3 tissue were calculated (D, H). (**A**) Representative images of testes from one worm which was digitally separated from the surrounding tissue using IMARIS. Nuclei are depicted in blue, stem cells in green. Scale bar = 40 µm. (**B**) All EdU-positive stem cells (spermatogonia) from the testes shown in (A) were aligned and quantified. Scale bar = 25 µm. The frequency (**C**) and density (**D**) of spermatogonial stem cells in testes were calculated. (**E**) Representative images of parenchyma from one worm after processing with IMARIS. Nuclei are depicted in blue, stem cells in green. Scale bar = 30 µm. (**F**) All EdU-positive stem cells (neoblasts) from the parenchymatic area shown in (E) were aligned and quantified. Scale bar = 15 µm. The frequency (**G**) and density (**H**) of neoblasts were calculated. Four worms per treatment group were analyzed. Statistical differences compared to the untreated group are indicated with * $p < 0.05$ (Wilcoxon rank sum test).

We used carmine red staining to gain a deeper insight into the cellular composition of the testicular lobes and to assess effects on cell differentiation. Control males typically featured pronounced testicular lobes filled with a large number of large spermatogonia, various stages of maturing cells, and mature spermatozoa (Figure 6A). In contrast, the testicular lobes were shrunken after venom treatment, included atypical cell-free areas, and lacked most of the large spermatogonial stem cells. The few remaining spermatogonia showed evidence of intracellular degradation (Figure 6B). Given the abundance of spermatozoa in the lobes and seminal vesicle (Figure 6B) and the reduction of stem cell frequency and density, these results argue for the selective depletion of proliferating stem cells by assassin bug venom.

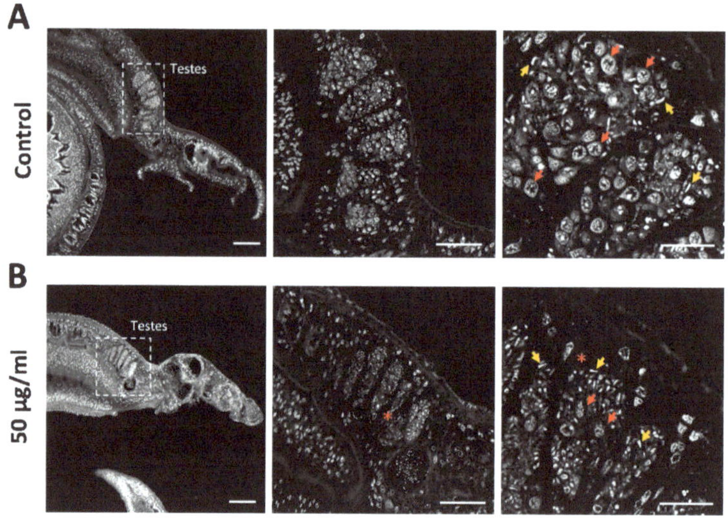

Figure 6. *Rhynocoris iracundus* venom reduces the number of spermatogonia in the testes of *Schistosoma mansoni*. Pairs of worms were treated with 50 µg/mL of venom for 72 h and males were stained with carmine red to reveal morphological details. (**A**) Control males feature typically pronounced testicular lobes filled with large spermatogonia (red arrows show examples) and different stages of maturing cells. Mature spermatozoa appear as small white comma-shaped cells (yellow arrows). (**B**) Testicular lobes in venom-treated males appear shrunken, include atypical cell-free areas (marked with *), and lack most of the spermatogonia, whereas mature spermatozoa are still present. The remaining spermatogonia show evidence of intracellular degradation. Scale bars = 100 µm (left), 50 µm (center), 20 µm (right).

2.3. Hemolytic Analysis of Assassin Bug Venom

Certain insect venoms are known for their ability to lyse eukaryotic cells, which limits their suitability as therapeutic leads [32,33]. To assess the hemolytic activity of the crude venom, we carried out hemolytic assays using porcine red blood cells, with 10% Triton X-100 as a positive control (100% lysis). The crude venom at a concentration of 43 µg/mL caused only 6.3% hemolysis, which can be regarded as non-significant (Figure 7).

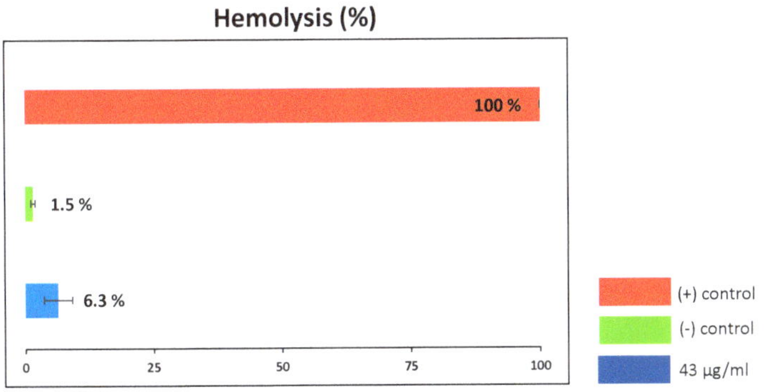

Figure 7. Hemolytic activity of *Rhynocoris iracundus* venom against porcine red blood cells. Relative proportion of cells lysed by *R. iracundus* venom (43 µg/mL) compared to 10% Triton X- 100 as a positive (+) control (100% lysis) and PBS as a negative (–) control.

3. Discussion

The aim of the study was to test whether venom from *R. iracundus* has anthelminthic activity and might therefore be of interest in drug discovery research. Our data reveal that venom reduced the vitality and egg production of *S. mansoni* adult worms, which was paralleled by the depletion of proliferating stem cells in male worms.

3.1. Antischistosomal Effects of Assassin Bug Venom

Reduced motor activity and detachment are important antischistosomal phenotypes. In vivo, both phenotypes would very likely result in the detachment of worms from the endothelial walls of mesenteric veins and thereby the displacement and degradation of the parasite by its host. Venom clearly reduced motility and caused detachment of worms during in vitro culture. Furthermore, diminished egg production was observed, which would reduce the pathological effect of helminths in vivo because fewer eggs accumulate in the liver [6]. It is unclear whether the impairment of egg production is a direct or indirect effect of the venom. A direct effect would require venom components to interfere with pathways involved in oogenesis, as an example. However, we would argue for a rather indirect effect: when separated from their male partners, female worms arrest egg production within a few days [34]. This seems more likely because exposure to 10 µg/mL of the venom triggered the separation of mating pairs and fewer eggs were laid, but the overall fitness of most females (in terms of motility and substrate attachment) was unaffected.

The antischistosomal effects of *R. iracundus* venom are difficult to compare with other insect-derived compounds due to the sparse literature published in this field. Bee venom and bee propolis (a complex beehive product) have previously been tested in vivo in mouse models of schistosomiasis. Both products reduced the pathogen burden [21], possibly reflecting their known immunomodulatory capacity within the host [35]. However, the potential direct effects of these compounds on worm vitality were not assessed in vitro, which leaves the question unanswered whether bee-derived compounds have a direct influence on the parasite. Recently, we demonstrated a direct schistosomicidal effect for the alkaloid harmonine [29], which is produced by the harlequin ladybird *Harmonia axyridis* (Coleoptera, Coccinellidae) as a bioweapon [36]. In *S. mansoni*, harmonine not only affected motility, pairing, substrate attachment and egg laying, but also caused damage to the tegument [29], which is the physiologically active surface layer of schistosomes [37]. *R. iracundus* venom triggered mild antischistosomal effects at 10 µg/mL and severe effects at 50 µg/mL, whereas harmonine was more active, triggering mild effects at 5 µg/mL and severe effects at 10 µg/mL. It has to be taken into account that harmonine is a defined compound, whereas assassin bug venom is a complex mixture of ~220 different enzymes, toxins and other compounds [38]. In future studies, it will be important to identify the active antischistosomal components of the crude venom, and such components are likely to be active against *S. mansoni* at a much lower concentration than the crude venom.

3.2. *Antiproliferative Effect of Assassin Bug Venom*

The importance of stem cells for growth and development has been demonstrated in various helminths, including *S. mansoni* [30,39]. Compounds affecting stem cell proliferation and hence the viability of schistosomes are therefore attractive drug candidates. The venom of *R. iracundus* caused a strong depletion of proliferating stem cells in male but not in female worms. Together with the more severe reduction of motility, males appeared more sensitive to assassin bug venom compared to females. This may reflect the fact that paired females are mostly shielded from the environment, here the culture medium containing venom, by the male's body. However, we find this unlikely because one early effect of the venom is to cause pair separation, which would expose females to the venom after ~24 h. A more plausible explanation for these phenotypes is based on sex-dependent differences in the efficiency of uptake and/or mode of action of the venom. Interestingly, lady-beetle-derived harmonine also impaired stem cell proliferation, but it affected both sexes. Enzyme activity assays suggested this may involve the inhibition of a schistosome acetylcholine esterase [29]. It is unclear whether the depletion of EdU-positive cells by harmonine reflects cell cycle arrest or cell death among the stem cell population. Our experiments with *R. iracundus* venom suggest that EdU-positive cell depletion is not based on an arrest in cell differentiation because differentiated spermatozoa were still present. Schistosome stem cells appear more sensitive towards venom than other cells, indicating that the mechanism of action targets proliferating rather than quiescent cells.

The available literature indicates a double-edged effect of animal-derived venom components on the proliferation of various cell types. Either cell proliferation was promoted, as reported for cobra, scorpion and lizard venom components tested against embryonic stem cells and mesenchymal stem cells [40,41], or venom components inhibited proliferation, as demonstrated for bufalin (a steroid hormone) and bombesin (a peptide hormone) isolated from toad venom and tested against stem cells [42,43]. Hormones may also be responsible for the anti-proliferative effect of *R. iracundus* venom. In addition, venom necrotoxins and cytotoxins might be involved, both of which typically kill cells [44]. Redulysins have been found in the venoms of other assassin bugs and were defined as putative pore-forming proteins with a cytolytic motif [45,46]. Therefore, we assume that *R. iracundus* redulysins may play a role for the observed cytotoxic effects against schistosome stem cells, with support from other compounds.

3.3. *Venom as Source for Antischistosomal Compounds*

Results of the hemolytic assay indicated that the crude venom is not hemolytic, and from this perspective appears suitable for biotechnological applications and for the development of therapeutic leads. The absence of hemolytic activity is particularly important in the context of antischistosomal drugs, which must be bioavailable and efficacious in the blood where schistosomes live. Once active components in *R. iracundus* venom have been identified in future studies, cytotoxicity testing against different cell lines would be crucial. Together with the characterization of EC50 values against *S. mansoni*, this will allow for judging whether the selectivity is suitable to pursue venom components, e.g., to preclinical animal studies.

4. Materials and Methods

4.1. *Ethical Statement*

Syrian hamsters (*Mesocricetus auratus*) were used as model hosts in accordance with the European Convention for the Protection of Vertebrate Animals used for Experimental and Other Scientific Purposes (ETS No 123; revised Appendix A). The experiments were approved by the Regional Council (Regierungspraesidium) Giessen (V54-19 c 20/15 h 02 GI 18/10 Nr. A 14/2017).

4.2. Production of Adult Worms

Freshwater snails of the genus *Biomphalaria glabrata* were used as the intermediate host for a Liberian strain (Bayer AG, Monheim) of *S. mansoni* [47,48]. Syrian hamsters from Janvier (France) were infected at 8 weeks of age by the paddling method [48]. In brief, hamsters were exposed to shallow water containing 1700-2000 cercariae for 45 min during which cercariae penetrated the host's skin. Adult worm couples were collected by hepatoportal perfusion of hamsters 46 days post-infection [49]. Worms were cultured in M199 medium (Sigma-Aldrich, Germany) supplemented with 10% newborn calf serum (Sigma), 1% 1 M HEPES and 1% ABAM solution (10,000 units/mL penicillin, 10 mg/mL streptomycin and 25 mg/mL amphotericin B) at 37 °C in a 5% CO_2 atmosphere.

4.3. Assassin Bug Collection and Rearing

The adult *R. iracundus* specimens were collected from North Rhine-Westphalia, Germany, with permission granted (Permission No. 425-104.1713) from the nature conservation authority (Obere Naturschutzbehörde) as part of the County Government of Rhineland-Palatinate. The insects used in this study were reared on a diet of mealworm larvae (*Tenebrio molitor* L.) in a ventilated box under constant conditions (24 ± 1 °C, 55–75% relative humidity).

4.4. Venom Collection

In order to stimulate the production of venom used by *R. iracundus* for defense purposes, hind legs were gently pressed with entomological forceps to mimic a predatory attack (Figure 1). This induced the insects to display a defense posture and to penetrate laboratory film (Parafilm) stretched across the opening of a pre-cooled 200-μL Eppendorf tube containing 100 μL phosphate-buffered saline (PBS). Following this procedure, the tubes were centrifuged briefly. Four specimens of *R. iracundus* were used and venom was collected every 2–3 days. The protein content was determined using the Pierce bicinchoninic acid (BCA) assay kit (Thermo Fisher Scientific, Germany). Venom then was stored at –20 °C.

4.5. Evaluation of the Physiological Effects of Venom

The anthelminthic activity of *R. iracundus* venom against adult pairs of *S. mansoni* was assessed in vitro. The worms were cultured in 96-well plates in supplemented M199 medium (one worm pair per well) mixed with different concentrations of the venom (10, 25 or 50 μg/mL) or the same volume of PBS as a negative control. The worms were incubated at 37 °C in a 5% CO_2 atmosphere for 72 h, and the medium plus venom was refreshed every 24 h. Venom-induced effects on worm motility, pairing stability and attachment to the culture plate were assessed every 24 h using an inverted microscope (Labovert, Germany). Worm motility was scored as recommended by WHO-TDR [50], with the scores 3 (normal motility), 2 (reduced motility), 1 (minimal and sporadic movements) and 0 (no movement within 30 s was considered dead). Egg numbers per well were counted after the 72-h culture period.

4.6. Proliferation Assay and CLSM

To assess the potential effect of venom on cell proliferation, EdU was added to a final concentration of 10 μM for the last 24 h of the in vitro culture period. The worms were then fixed with 4% paraformaldehyde, stained with the Click-iT Plus EdU Alexa Fluor 488 imaging kit (Thermo Fisher Scientific) and counterstained with Hoechst 33342 as previously described [29,51]. Morphological effects on testicular cells were assessed by fixing worms in AFA (66.5% ethanol, 1.1% paraformaldehyde, 2% glacial acetic acid) and staining with CertistainH carmine red (Merck, Germany) as previously described [52,53]. A TSC SP5 inverse confocal laser scanning microscope (Leica, Germany) was used for imaging. AlexaFluor488 and carmine red were excited using an argon-ion laser at 488 nm, and Hoechst at 405 nm. Optical section thickness and background signals were defined by setting the pinhole size to

1 Airy unit in the Leica LAS AF software. Z-stacks were acquired by CLSM with a step-size of 0.3 µm for quantification of EdU-positive stem cells and Hoechst-positive total cell numbers. For each worm, testes and two selected parenchymatic tissue areas were manually selected using the software package "IMARIS for cell biologists" (Bitplane, Switzerland). Cells were quantified applying the automatic surface creation of the software. To minimize background noise or counting of artifacts, a threshold was set prior to cell quantification that excluded objects <3 µm.

4.7. Hemolytic Activity Assay

Porcine blood was obtained from a local butcher and was mechanically treated to remove coagulants. Red blood cells were harvested by centrifugation (1500× *g*, 3 min, room temperature) and washed three times with PBS. A cell suspension was prepared with a dilution factor of 1:10 in PBS. Crude venom (final concentration 43 µg/mL) was mixed with the red blood cells (4.8×10^7 cells/mL) in a 96-well plate and incubated for 1 h at 37 °C. Venom-induced hemolysis was then measured in relation to 10% Triton X-100 as a positive control (set at 100%) and PBS as a negative control [54].

4.8. Statistical Analysis

Homogeneity of variance was checked with Levene's test (https://www.statskingdom.com/230var_levenes.htmL). Statistical significance was tested using the nonparametric Wilcoxon rank sum test (https://ccb-compute2.cs.uni-saarland.de/wtest/) [55]. $p < 0.05$ was considered statistically significant.

5. Conclusions

We have demonstrated antischistosomal effects of venom from the European predatory assassin bug *R. iracundus*. The effects included impairment of motility, pairing stability, attachment and egg production. Thus, assassin bug venom not only affects prey invertebrates but also helminths. Furthermore, the venom also caused the ablation of proliferating stem cells in male schistosomes. These phenotypes are reminiscent of the effects induced by paralytic and cytolytic assassin bug venoms used to subdue invertebrate prey [25,27]. The observed anthelminthic effects, together with the absence of hemolytic activity, warrant further studies to identify the antischistosomal components of *R. iracundus* venom and assess their suitability for therapeutic applications in the field of parasitology. The transcriptomic and proteomic data recently obtained for this venom will greatly facilitate future research in this direction [38] and provide insight into a new and underexploited resource for the development of anthelminthic drugs.

Supplementary Materials: The following are available online at http://www.mdpi.com/2079-6382/9/10/664/s1. Figure S1: Effect of praziquantel on *Schistosoma mansoni* motility as a positive control for the in vitro culture assay. Worm pairs were treated with different concentrations of praziquantel (0.1–5 µM) for a period of 72 h. Motility was measured every 24 h and compared to DMSO-treated control worms. The graph shows a summary of two experiments with 10 worm pairs per experiment (mean ± SEM). Significant differences vs the control are indicated (*$p < 0.05$, Wilcoxon rank sum test), Video S1: *Schistosoma mansoni* pair in the control group. The male worm was attached via its suckers to the base of the culture plate and showed normal motility. The female resides within the ventral grove of the male partner. Normal motility involves whole body movements (displayed by the male in the video from 10 sec onwards), Video S2: *Schistosoma mansoni* pair after treatment with 25 µg/mL *Rhynocoris iracundus* venom for 72 h, showing reduced motility (motility score 2). The male worm was detached with its sucker from the base of the culture plate and did not show whole-body movements, Video S3: *Schistosoma mansoni* male treated with 50 µg/mL *Rhynocoris iracundus* venom for 72 h, showing severe loss of motility (little movement detected, confined to the posterior end, motility score 1), Video S4: *Schistosoma mansoni* female treated with 50 µg/mL *Rhynocoris iracundus* venom for 72 h, showing severe loss of motility (little movement detected, confined to the anterior and posterior ends, motility score 1).

Author Contributions: Conceptualization, M.T. and S.H.; Methodology, M.T. and S.H.; Investigation, S.H.; Resources, A.V. and C.G.G.; Writing—Original Draft Preparation, M.T. and S.H.; Writing—Review & Editing, M.T., S.H., A.V. and C.G.G.; Visualization, M.T. and S.H.; Supervision, S.H.; Funding Acquisition, M.T., S.H., A.V. and C.G.G. All authors have read and agreed to the published version of the manuscript.

Funding: SH and CGG would like to acknowledge funding by the LOEWE Centre for Novel Drug Targets against Poverty-Related and Neglected Tropical Infectious Diseases (DRUID), which is part of the excellence initiative of the Hessen State Ministry of Higher Education, Research and the Arts (HMWK). MT and AV would like to acknowledge generous funding by the HMWK via the LOEWE Centre for Translational Biodiversity Genomics (LOEWE-TBG) and the LOEWE Center for Insect Biotechnology and Bioresources.

Acknowledgments: The authors thank Zeinab Waad Zadiq for experimental support, and Christina Scheld, Bianca Kulik and Georgette Stovall for excellent technical assistance in the maintenance of the *S. mansoni* life cycle. We are grateful to Nicolai Rügen for insect collection and technical support for venom collection, Paul Bauer for maintaining *R. iracundus*, and Irina Häcker for providing the porcine blood. The authors thank Richard Twyman for editing the manuscript.

Conflicts of Interest: The authors declare no conflict of interest. The funders had no role in the design of the study; in the collection, analysis, or interpretation of data; in the writing of the manuscript, or in the decision to publish the results.

References

1. Hotez, P.J.; Bundy, D.A.P.; Beegle, K.; Brooker, S.; Drake, L.; de Silva, N.; Montresor, A.; Engels, D.; Jukes, M.; Chitsulo, L.; et al. Helminth infections: Soil-transmitted helminth infections and schistosomiasis. In *Disease Control Priorities in Developing Countries*; Jamison, D.T., Breman, J.G., Measham, A.R., Alleyne, G., Claeson, M., Evans, D.B., Jha, P., Mills, A., Musgrove, P., Eds.; Oxford University Press: Washington, DC, USA; New York, NY, USA, 2006.
2. Feigin, V. Global, regional, and national incidence, prevalence, and years lived with disability for 310 diseases and injuries, 1990–2015: A systematic analysis for the Global Burden of Disease Study 2015. *Lancet* **2016**, *388*, 1545–1602.
3. Newman, D.J.; Cragg, G.M. Natural products as sources of new drugs over the 30 years from 1981 to 2010. *J. Nat. Prod.* **2012**, *75*, 311–335. [CrossRef] [PubMed]
4. Neves, B.J.; Andrade, C.H.; Cravo, P.V. Natural products as leads in schistosome drug discovery. *Molecules* **2015**, *20*, 1872–1903. [CrossRef] [PubMed]
5. Moser, W.; Schindler, C.; Keiser, J. Drug Combinations against Soil-Transmitted Helminth Infections. *Adv. Parasitol.* **2019**, *103*, 91–115.
6. Colley, D.G.; Bustinduy, A.L.; Secor, W.E.; King, C.H. Human schistosomiasis. *Lancet* **2014**, *383*, 2253–2264. [CrossRef]
7. Hotez, P.J.; Alvarado, M.; Basanez, M.G.; Bolliger, I.; Bourne, R.; Boussinesq, M.; Brooker, S.J.; Brown, A.S.; Buckle, G.; Budke, C.M.; et al. The global burden of disease study 2010: Interpretation and implications for the neglected tropical diseases. *PLoS Negl. Trop. Dis.* **2014**, *8*, e2865. [CrossRef] [PubMed]
8. Cheever, A.W.; Macedonia, J.G.; Mosimann, J.E.; Cheever, E.A. Kinetics of egg production and egg excretion by *Schistosoma mansoni* and *S. japonicum* in mice infected with a single pair of worms. *Am. J. Trop. Med. Hyg.* **1994**, *50*, 281–295. [PubMed]
9. Doenhoff, M.J.; Cioli, D.; Utzinger, J. Praziquantel: Mechanisms of action, resistance and new derivatives for schistosomiasis. *Curr. Opin. Infect. Dis.* **2008**, *21*, 659–667. [CrossRef]
10. Fallon, P.G.; Doenhoff, M.J. Drug-resistant schistosomiasis: Resistance to praziquantel and oxamniquine induced in *Schistosoma mansoni* in mice is drug specific. *Am. J. Trop. Med. Hyg.* **1994**, *51*, 83–88. [CrossRef]
11. Botros, S.S.; Bennett, J.L. Praziquantel resistance. *Expert Opin. Drug Discov.* **2007**, *2*, S35–S40. [CrossRef]
12. Mwangi, I.N.; Sanchez, M.C.; Mkoji, G.M.; Agola, L.E.; Runo, S.M.; Cupit, P.M.; Cunningham, C. Praziquantel sensitivity of Kenyan *Schistosoma mansoni* isolates and the generation of a laboratory strain with reduced susceptibility to the drug. *Int. J. Parasitol. Drugs Drug Resist.* **2014**, *4*, 296–300. [CrossRef] [PubMed]
13. Cioli, D.; Pica-Mattoccia, L.; Basso, A.; Guidi, A. Schistosomiasis control: Praziquantel forever? *Mol. Biochem. Parasitol.* **2014**, *195*, 23–29. [CrossRef] [PubMed]
14. De Moraes, J. Natural products with antischistosomal activity. *Future Med. Chem.* **2015**, *7*, 801–820. [CrossRef] [PubMed]
15. Herzig, V.; Cristofori-Armstrong, B.; Israel, M.R.; Nixon, S.A.; Vetter, I.; King, G.F. Animal toxins-Nature's evolutionary-refined toolkit for basic research and drug discovery. *Biochem. Pharmacol.* **2020**, 114096. [CrossRef]
16. Mohamed Abd El-Aziz, T.; Garcia Soares, A.; Stockand, J.D. Snake venoms in drug discovery: Valuable therapeutic tools for life saving. *Toxins* **2019**, *11*, 564. [CrossRef]

17. El-Asmar, M.F.; Swelam, N.; Abdel Aal, T.M.; Ghoneim, K.; Hodhod, S.S. Factor(s) in the venom of scorpions toxic to *Schistosoma mansoni* (intestinal belharzia) cercariae. *Toxicon* **1980**, *18*, 711–715. [CrossRef]
18. Stábeli, R.G.; Amui, S.F.; Sant'Ana, C.D.; Pires, M.G.; Nomizo, A.; Monteiro, M.C.; Romão, P.R.; Guerra-Sá, R.; Vieira, C.A.; Giglio, J.R.; et al. Bothrops moojeni myotoxin-II, a Lys49-phospholipase A2 homologue: An example of function versatility of snake venom proteins. *Comp. Biochem. Physiol. C Toxicol. Pharmacol.* **2006**, *142*, 371–381.
19. De Moraes, J.; Nascimento, C.; Miura, L.M.; Leite, J.R.; Nakano, E.; Kawano, T. Evaluation of the *in vitro* activity of dermaseptin 01, a cationic antimicrobial peptide, against *Schistosoma mansoni*. *Chem. Biodivers.* **2011**, *8*, 548–558. [CrossRef]
20. Hassan, E.A.; Abdel-Rahman, M.A.; Ibrahim, M.M.; Soliman, M.F. In vitro antischistosomal activity of venom from the Egyptian snake *Cerastes cerastes*. *Rev. Soc. Bras. Med. Trop.* **2016**, *49*, 752–757. [CrossRef]
21. Mohamed, A.H.; Hassab El-Nabi, S.E.; Bayomi, A.E.; Abdelaal, A.A. Effect of bee venom or proplis on molecular and parasitological aspects of *Schistosoma mansoni* infected mice. *J. Parasit. Dis.* **2016**, *40*, 390–400. [CrossRef]
22. Bailey, P.C. The feeding behaviour of a sit-and wait-predator, *Ranatra dispar* (Heteroptera: Nepidae): Optimal foraging and feeding dynamics. *Oecologia* **1986**, *68*, 291–297. [CrossRef] [PubMed]
23. Sano-Martins, I.S.; González, C.; Anjos, I.V.; Díaz, J.; Gonçalves, L.R.C. Effectiveness of *Lonomia* antivenom in recovery from the coagulopathy induced by *Lonomia orientoandensis* and *Lonomia casanarensis* caterpillars in rats. *PLoS Negl. Trop. Dis.* **2018**, *12*, e0006721. [CrossRef] [PubMed]
24. Arif, F.; Williams, M. *Hymenoptera Stings (Bee, Vespids and Ants)*; StatPearls Publishing LLC: Treasure Island, FL, USA, 2020.
25. Walker, A.A.; Weirauch, C.; Fry, B.G.; King, G.F. Venoms of heteropteran insects: A treasure trove of diverse pharmacological toolkits. *Toxins* **2016**, *8*, 43. [CrossRef] [PubMed]
26. Hwang, W.S.; Weirauch, C. Evolutionary history of assassin bugs (insecta: Hemiptera: Reduviidae): Insights from divergence dating and ancestral state reconstruction. *PLoS ONE* **2012**, *7*, e45523. [CrossRef]
27. Edwards, J.S. The action and composition of the saliva of an assassin bug *Platymeris rhadamanthus* Gaerst (Hemiptera, Reduviidae). *J. Exp. Biol.* **1961**, *38*, 61–77.
28. Fischer, G.; Conceicao, F.R.; Leite, F.P.; Dummer, L.A.; Vargas, G.D.; Hubner Sde, O.; Dellagostin, O.A.; Paulino, N.; Paulino, A.S.; Vidor, T. Immunomodulation produced by a green propolis extract on humoral and cellular responses of mice immunized with SuHV-1. *Vaccine* **2007**, *25*, 1250–1256. [CrossRef]
29. Kellershohn, J.; Thomas, L.; Hahnel, S.R.; Grunweller, A.; Hartmann, R.K.; Hardt, M.; Vilcinskas, A.; Grevelding, C.G.; Haeberlein, S. Insects in anthelminthics research: Lady beetle-derived harmonine affects survival, reproduction and stem cell proliferation of *Schistosoma mansoni*. *PLoS Negl. Trop. Dis.* **2019**, *13*, e0007240. [CrossRef]
30. Wendt, G.R.; Collins, J.J., 3rd. Schistosomiasis as a disease of stem cells. *Curr. Opin. Genet. Dev.* **2016**, *40*, 95–102. [CrossRef]
31. Collins, J.J., 3rd; Wang, B.; Lambrus, B.G.; Tharp, M.E.; Iyer, H.; Newmark, P.A., 3rd; Wang, B.; Lambrus, B.G.; Tharp, M.E.; Iyer, H.; et al. Adult somatic stem cells in the human parasite *Schistosoma mansoni*. *Nature* **2013**, *494*, 476–479.
32. Monincová, L.; Budesínský, M.; Slaninová, J.; Hovorka, O.; Cvacka, J.; Voburka, Z.; Fucík, V.; Borovicková, L.; Bednárová, L.; Straka, J.; et al. Novel antimicrobial peptides from the venom of the eusocial bee *Halictus sexcinctus* (Hymenoptera: Halictidae) and their analogs. *Amino Acids* **2010**, *39*, 763–775.
33. Mortari, M.R.; do Couto, L.L.; dos Anjos, L.C.; Mourão, C.B.; Camargos, T.S.; Vargas, J.A.; Oliveira, F.N.; Gati Cdel, C.; Schwartz, C.A.; Schwartz, E.F. Pharmacological characterization of *Synoeca cyanea* venom: An aggressive social wasp widely distributed in the Neotropical region. *Toxicon* **2012**, *59*, 163–170. [CrossRef] [PubMed]
34. Erasmus, D.A. A comparative study of the reproductive system of mature, immature and "unisexual" female *Schistosoma mansoni*. *Parasitology* **1973**, *67*, 165–183. [CrossRef] [PubMed]
35. Cornara, L.; Biagi, M.; Xiao, J.; Burlando, B. Therapeutic properties of bioactive compounds from different honeybee products. *Front. Pharmacol.* **2017**, *8*, 412. [CrossRef]
36. Vilcinskas, A.; Stoecker, K.; Schmidtberg, H.; Rohrich, C.R.; Vogel, H. Invasive harlequin ladybird carries biological weapons against native competitors. *Science* **2013**, *340*, 862–863. [CrossRef]

37. Van Hellemond, J.J.; Retra, K.; Brouwers, J.F.; van Balkom, B.W.; Yazdanbakhsh, M.; Shoemaker, C.B.; Tielens, A.G. Functions of the tegument of schistosomes: Clues from the proteome and lipidome. *Int. J. Parasitol.* **2006**, *36*, 691–699. [CrossRef]
38. Tonk, M.; Institute for Insect Biotechnology, Justus Liebig University of Giessen, Giessen, Germany. Personal communication, 2020.
39. Koziol, U.; Rauschendorfer, T.; Zanon Rodriguez, L.; Krohne, G.; Brehm, K. The unique stem cell system of the immortal larva of the human parasite *Echinococcus multilocularis*. *Evodevo* **2014**, *5*, 10. [CrossRef] [PubMed]
40. Zhou, H.; Li, D.; Shi, C.; Xin, T.; Yang, J.; Zhou, Y.; Hu, S.; Tian, F.; Wang, J.; Chen, Y. Effects of Exendin-4 on bone marrow mesenchymal stem cell proliferation, migration and apoptosis *in vitro*. *Sci. Rep.* **2015**, *5*, 12898. [CrossRef]
41. Miao, Z.; Lu, Z.; Luo, S.; Lei, D.; He, Y.; Wu, H.; Zhao, J.; Zheng, L. Murine and Chinese cobra venom-derived nerve growth factor stimulate chondrogenic differentiation of BMSCs *in vitro*: A comparative study. *Mol. Med. Rep.* **2018**, *18*, 3341–3349. [CrossRef]
42. Assimakopoulos, S.F.; Tsamandas, A.C.; Georgiou, C.D.; Vagianos, C.E.; Scopa, C.D. Bombesin and neurotensin exert antiproliferative effects on oval cells and augment the regenerative response of the cholestatic rat liver. *Peptides* **2010**, *31*, 2294–2303. [CrossRef]
43. Liu, J.; Zhang, Y.; Sun, S.; Zhang, G.; Jiang, K.; Sun, P.; Zhang, Y.; Yao, B.; Sui, R.; Chen, Y.; et al. Bufalin induces apoptosis and improves the sensitivity of human glioma stem-like sells to temozolamide. *Oncol. Res.* **2019**, *27*, 475–486. [CrossRef]
44. White, J. Bites and stings from venomous animals: A global overview. *Ther. Drug Monit.* **2000**, *22*, 65–68. [CrossRef] [PubMed]
45. Walker, A.A.; Madio, B.; Jin, J.; Undheim, E.A.; Fry, B.G.; King, G.F. Melt with this kiss: Paralyzing and liquefying venom of the assassin bug *Pristhesancus plagipennis* (hemiptera: Reduviidae). *Mol. Cell. Proteom.* **2017**, *16*, 552–566. [CrossRef] [PubMed]
46. Walker, A.A.; Robinson, S.D.; Undheim, E.A.B.; Jin, J.; Han, X.; Fry, B.G.; Vetter, I.; King, G.F. Missiles of mass disruption: Composition and glandular origin of venom used as a projectile defensive weapon by the assassin bug *Platymeris rhadamanthus*. *Toxins* **2019**, *11*, 673. [CrossRef] [PubMed]
47. Gönnert, R. Schistosomiasis-Studien. II. Über die Eibildung bei Schistosoma mansoni und das Schicksal der Eier im Wirtsorganismus. *Z. Trop. Parasitol.* **1955**, *6*, 33–52.
48. Dettman, C.D.; Higgins-Opitz, S.B.; Saikoolal, A. Enhanced efficacy of the paddling method for schistosome infection of rodents by a four-step pre-soaking procedure. *Parasitol. Res.* **1989**, *76*, 183–184. [CrossRef]
49. Grevelding, C.G. The female-specific W1 sequence of the Puerto Rican strain of *Schistosoma mansoni* occurs in both genders of a Liberian strain. *Mol. Biochem. Parasitol.* **1995**, *71*, 269–272. [CrossRef]
50. Ramirez, B.; Bickle, Q.; Yousif, F.; Fakorede, F.; Mouries, M.A.; Nwaka, S. Schistosomes: Challenges in compound screening. *Expert Opin. Drug Discov.* **2007**, *2*, S53–S61. [CrossRef]
51. Hahnel, S.; Quack, T.; Parker-Manuel, S.J.; Lu, Z.; Vanderstraete, M.; Morel, M.; Dissous, C.; Cailliau, K.; Grevelding, C.G. Gonad RNA-specific qRT-PCR analyses identify genes with potential functions in schistosome reproduction such as SmFz1 and SmFGFRs. *Front. Genet.* **2014**, *5*, 170. [CrossRef]
52. Neves, R.H.; de Lamare Biolchini, C.; Machado-Silva, J.R.; Carvalho, J.J.; Branquinho, T.B.; Lenzi, H.L.; Hulstijn, M.; Gomes, D.C. A new description of the reproductive system of *Schistosoma mansoni* (Trematoda: Schistosomatidae) analyzed by confocal laser scanning microscopy. *Parasitol. Res.* **2005**, *95*, 43–49. [CrossRef]
53. Beckmann, S.; Grevelding, C.G. Imatinib has a fatal impact on morphology, pairing stability and survival of adult *Schistosoma mansoni in vitro*. *Int. J. Parasitol.* **2010**, *40*, 521–526. [CrossRef]
54. Tonk, M.; Pierrot, C.; Cabezas-Cruz, A.; Rahnamaeian, M.; Khalife, J.; Vilcinskas, A. The *Drosophila melanogaster* antimicrobial peptides Mtk-1 and Mtk-2 are active against the malarial parasite *Plasmodium falciparum*. *Parasitol. Res.* **2019**, *118*, 1993–1998. [CrossRef] [PubMed]
55. Marx, A.; Backes, C.; Meese, E.; Lenhof, H.P.; Keller, A. EDISON-WMW: Exact dynamic programing solution of the Wilcoxon-Mann-Whitney test. *Genom. Proteom. Bioinform.* **2016**, *14*, 55–61. [CrossRef] [PubMed]

antibiotics

MDPI

Article

Assessment of the Sensitivity of Some Plant Pathogenic Fungi to 6-Demethylmevinolin, a Putative Natural Sensitizer Able to Help Overcoming the Fungicide Resistance of Plant Pathogens

Larisa Shcherbakova *, Maksim Kartashov, Natalia Statsyuk *, Tatyana Pasechnik and Vitaly Dzhavakhiya

All-Russian Research Institute of Phytopathology, Institute str., vl. 5, Bolshie Vyazemy, Moscow 143050, Russia; maki505@mail.ru (M.K.); beefarmer@yandex.ru (T.P.); vitaly@vniif.ru (V.D.)

* Correspondence: larisa@vniif.ru (L.S.); nataafg@gmail.com (N.S.)

Received: 28 October 2020; Accepted: 23 November 2020; Published: 25 November 2020

check for updates

Abstract: Agricultural fungicides contaminate the environment and promote the spread of fungicide-resistant strains of pathogenic fungi. The enhancement of pathogen sensitivity to these pesticides using chemosensitizers allows the reducing of fungicide dosages without a decrease in their efficiency. Using Petri plate and microplate bioassays, 6-demethylmevinolin (6-DMM), a putative sensitizer of a microbial origin, was shown to affect both colony growth and conidial germination of *Alternaria solani*, *A. alternata*, *Parastagonospora nodorum*, *Rhizoctonia solani*, and four *Fusarium* species (*F. avenaceum*, *F. culmorum*, *F. oxysporum*, *F. graminearum*) forming a wheat root rot complex together with *B. sorokiniana*. Non- or marginally toxic 6-DMM concentrations suitable for sensitizing effect were determined by the probit analysis. The range of determined concentrations confirmed a possibility of using 6-DMM as a putative sensitizer for the whole complex of root rot agents, other cereal pathogens (*A. alternata*, *P. nodorum*), and some potato (*R. solani*, *A. solani*) and tomato (*A. solani*) pathogens. Despite the different sensitivities of the eight tested pathogens, 6-DMM lacked specificity to fungi and possessed a mild antimycotic activity that is typical of other known pathogen-sensitizing agents. The pilot evaluation of the 6-DMM sensitizing first confirmed a principal possibility of using it for the sensitization of *B. sorokiniana* and *R. solani* to triazole- and strobilurin-based fungicides, respectively.

Keywords: chemosensitization; antifungal compounds; plant pathogenic fungi; fungicide resistance; 6-demethylmevinolin; environmental pollution

1. Introduction

To meet an increasing demand for crop products, high-yielding cultivars of agricultural plants are now grown all over the world. One of the main conditions for fully realizing a potential productivity of such cultivars is a successful control of crop pathogens, primarily fungi, which may cause diseases resulting in up to 70–80% of yield losses [1]. In many countries, including Russia, causative agents of foot/root rots (*Fusarium* spp., *Bipolaris* spp., *Rhizoctonia* spp.), leaf and/or glume blotches (*Bipolaris* spp., *Alternaria* spp., *Parastagonospora* spp.) head blight, kernel smudge (*Fusarium* spp., *Alternaria* spp.), and wilt (*F. oxysporum*) belong to the most widespread and detrimental pathogens of such economically important crops as cereals, potato and tomato [2–4]. As in the case of other plant pathogenic fungi, a common practice to control these agents, and thus efficiently prevent significant yield losses caused by these pathogens, is the use of chemical fungicides. However, like medical antibiotics and antimycotics, the effectiveness of agricultural fungicides is under threat because of the evolution of fungicide resistance, which is often developed soon after a new fungicide is introduced due to improper and/or extensive fungicide application practices [5,6]. A wide use of modern fungicides resulted in a significant increase in the frequency of high fungicide resistance and multiple or cross-resistance of various plant pathogenic fungi. Fungicides with a single-site mode of action, such as Quinone outside Inhibitors or DeMethylation Inhibitors (QoI- and DMI-fungicides, respectively) commonly used against above-mentioned pathogens can reduce or lose their protective efficacy relatively quickly [7–12]. Moreover, in some cases, the time required for the emergence of resistant strains since the start of a commercial application of such fungicides do not exceed several years [13]. The attempts to control resistant forms by increasing the dosages or frequency of fungicide applications increase the total costs for plant protection and only complicate the problem resulting in an enhanced accumulation of such forms and stimulating their spreading across populations [14]. Apart from this problem, the excess of fungicides spread into the environment causing significant contamination of terrestrial and water ecosystems and multiple negative effects on soil microbiota, insects, vertebrates, as well as in poisoning of food and feed [15–17]. Thus, the solving of a fungicide resistance problem and associated problems of the excess fungicide applications and environment pollution become one of the dominant trends in the current plant protection science.

Different strategies are proposed for agricultural practices to prevent or minimize negative side effects of extensive fungicidal treatments accompanied with increasing resistance of fungi and environmental risks [18,19]. A promising new approach intended to reduce fungicidal dosages without any mitigation of the antifungal effect is the enhancement of a pathogen sensitivity to fungicides (chemosensitization). This can be accomplished by the co-application of a fungicide with certain natural or chemical eco-friendly compounds (sensitizers) at concentrations, which should meet the following requirements. First, both fungicide and sensitizer, applied alone, should provide no or insignificant fungitoxic effect. Second, both compounds applied together should effectively suppress a target pathogen, preferably in a synergistic or in some cases in an additive manner [20,21]. The synergy between ineffective doses of the components occurs since a sensitizer and a fungicide attack different pathways of fungal metabolism or distant stages of the same metabolic pathway.

The chemosensitization approach was initially developed in medicine to overcome the resistance of pathogenic fungal hospital strains to antimycotic drugs. To date, a number of secondary plant metabolites, as well as their synthetic analogues, which can be used as chemosensitizing agents against human-infecting fungi, have been revealed. Some of them possess an antifungal activity, but significantly less than that of commercial antimycotics [20]. In contrast, only a few chemosensitization studies were recently performed for agricultural purposes [21–23], with a confirmed effect on plants in some cases [22,23]. The cited researchers demonstrated that not only plants but also bacteria [22] or filamentous fungi [23–25] may serve as the sources of metabolites significantly enhancing the sensitivity of plant pathogens to agricultural fungicides including such widely-used and rather persistent ones as triazoles [26].

The first research step to search for chemosensitizing compounds is the in vitro testing of the antifungal effect of putative sensitizers to determine their working concentration range and to make sure their toxicity is significantly lower than that of a fungicide used against the target pathogen. Recently, we found non-fungitoxic metabolites of plant origin, which enhanced the sensitivity of five cereal pathogens to one of DMI-fungicides [27]. Earlier, we also screened a range of microbial metabolites for sensitizing activity and briefly reported that a secondary metabolite of *Penicillium citrinum*, 6-demethylmevinolin (6-DMM) was the most promising as a putative sensitizer among them. 6-DMM was much less fungitoxic compared to tebuconazole, and it enhanced the sensitivity of a wheat and barley pathogen *B. sorokiniana* to one of tebuconazole-based fungicides [25]. We supposed that 6-DMM might be applied to improve the protective effect of triazoles against other soil or foliar pathogenic fungi attacking plants under field conditions, in the case that 6-DMM does not possess narrow antifungal and sensitizing activities towards a single pathogen. Therefore, the current study is the first step towards confirming this assumption and it aims to evaluate the effect of 6-DMM on the in vitro growth and germination of some other triazole-controlled pathogenic fungi in order to determine non- or marginally toxic 6-DMM concentrations for these species and to compare their sensitivity to this putative sensitizer. Here, we focused mainly on a 6-DMM activity towards *Fusarium* species (*F. avenaceum*, *F. culmorum*, *F. graminearum*, *F. oxysporum*), which most often form the common pathogenic complex of wheat or barley root rots together with *B. sorokiniana* [28–30]. Additionally, the effect on other important crop pathogens, such as *Alternaria alternata*, *A. solani*, causing early bight of tomato and potato, *Parastagonospora nodorum*, a causative agent of wheat glume/leaf blotch and *Rhizoctonia solani*, potato stem canker and black scurf agent, was tested. We also demonstrated 6-DMM sensitized *B. sorokiniana* to tebuconazole, and *R. solani* to azoxystrobin formulated as Folicur®, EC 250 and Quadris®, SC 250, respectively.

2. Results

2.1. The Effect of 6-DMM on the Fungal Growth and Spore Germination

The growth-inhibitory effect of 6-DMM towards several plant pathogens was evaluated after culturing the fungi on PDA containing 6-DMM at seven (*F. graminearum*), eight (*F. avenaceum*, *F. culmorum*), nine (*F. oxysporum*), or six (*A. alternata*, *A. solani*, *P. nodorum*, *R. solani*) concentrations that ranged, depending on the pathogen, from 5 to 800 μg/mL. Additionally, the conidia of some fungi were exposed to five (*Alternaria* spp.) or at least six (*Fusarium* spp.) 6-DMM concentrations varied in a wide range from 10 ng/mL to 15 μg/mL. As a result, nominally fungitoxic, sub-fungitoxic, and strongly or totally inhibiting concentration ranges causing 2–15%, 30–60% or 80–100% growth suppression, respectively, were selected for each pathogen. According to the obtained data, 6-DMM was found to affect both colony growth and conidial germination of all tested fungi. As in case of *B. sorokiniana* [25], this microbial metabolite was much less toxic for them than the tebuconazole-based fungicide (Figure 1). In our experiments, minimum inhibitory concentration (MIC) values of the fungicide exceeded those of 6-DMM for *F. avenaceum* and *F. culmorum* at least five-fold, and were either an order (*F. graminearum*, *F. oxysporum*, *A. alternata*) or several orders (*A. solani*, *P. nodorum*, *R. solani*) of magnitude higher.

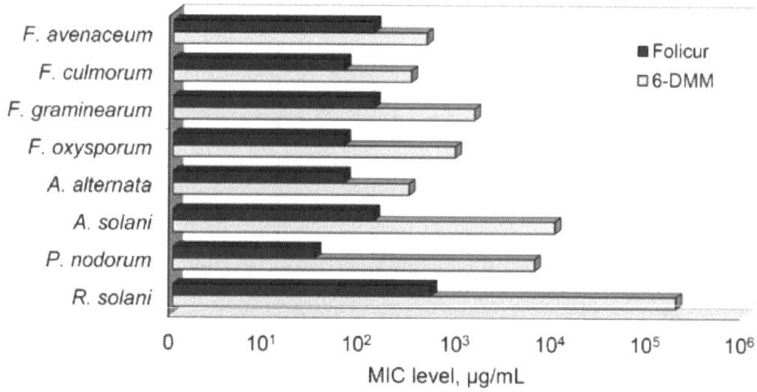

Figure 1. Minimum inhibitory concentrations (MIC) of 6-demethylmevinolin (6-DMM) and a tebuconazole-based fungicide (Folicur® EC 250) preventing visible growth of fungal colonies.

In general, 6-DMM concentrations causing 50% inhibition of *Fusarium* spp., *A. alternata*, and *P. nodorum* growth or their spore germination were significantly higher compared to ED_{50} for *A. solani* and *R. solani* (Table 1, Figure 2A,C).

Table 1. Inhibitory concentrations of 6-dimethylmevinolin (6-DMM) for eight plant pathogenic fungi calculated by a probit analysis.

Pathogen	Inhibitory Concentrations of 6-DMM *						
	Colony Growth				Spore Germination		
	ED_{10}, μg/mL	ED_{50}, μg/mL	ED_{95}, mg/mL	R^2	ED_{50}, μg/mL	ED_{95}, mg/mL	R^2
B. sorokiniana **	6.0 [a]	9.5 [a]	0.08 [a]	0.977	1.30 [a]	4.8 [a]	0.939
F. culmorum	1.0 [b]	38.0 [b]	0.17 [a]	0.923	0.35 [b]	3.2 [b]	0.917
F. avenaceum	1.3 [b]	57.5 [c]	0.26 [b]	0.940	0.40 [b]	4.1 [a]	0.986
F. oxysporum	1.1 [b]	63.1 [c]	0.55 [c]	0.932	0.80 [c]	4.7 [a]	0.956
F. graminearum	1.6 [c]	141.3 [d]	0.74 [d]	0.973	1.25 [d]	7.1 [c]	0.913
A. alternata	4.8 [a]	30.9 [b]	0.17 [a]	0.979	3.60 [e]	5.2 [d]	0.964
A. solani	85.8 [d]	401.2 [e]	13.16 [d]	0.891	3.85 [e]	5.4 [d]	0.991
P. nodorum	7.1 [a]	36.3 [b]	7.59 [e]	0.946	nd ***	nd	nd
R. solani	50.0 [d]	398.0 [e]	50.12 [f]	0.883	nd	nd	nd

* Different uppercase letters within the column indicate statistically significant differences ($p \leq 0.05$). ** For *B. sorokiniana*, ED_{10} is the minimum 6-DMM concentration determined experimentally, which provided effective sensitization of this pathogen to tebuconazole [25], while ED_{50} and ED_{95} were calculated in this work, using previously obtained data. ED means effective dose (see 4.5). *** Not determined.

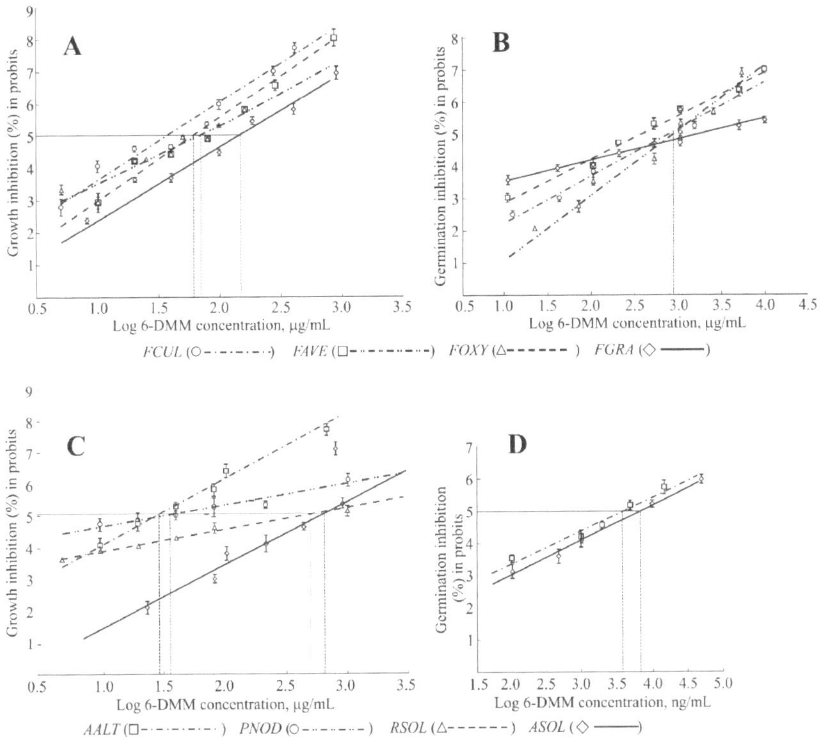

Figure 2. Dose-response curves showing inhibitory effect of 6-DMM on the growth (**A**,**C**) and germination (**B**,**D**) of various plant pathogenic fungi. FCUL, *Fusarium culmorum*; FAVE, *F. avenaceum*; FOXY, *F. oxysporum*; FGRA, *F. graminearum*; AALT, *Alternaria alternata*; PNOD, *Parastagonospora nodorum*; ASOL, *A. solani*; RSOL, *Rhizoctonia solani*. Vertical dotted lines cross the *X*-axis at the points corresponding to $LogED_{50}$ for each of pathogens. Bars on the graphs (**A**–**D**) indicate SD and SE, respectively. The markers correspond to logarithms of the tested 6-DMM concentrations.

Interestingly, ED_{10} values were almost equal for four *Fusarium* fungi, insignificantly differed in *A. alternata* or *P. nodorum* and did not exceed the minimum 6-DMM concentration previously found to enhance *B. sorokiniana* sensitivity to tebuconazole [25]. These data point to a common range of 6-DMM nominal toxic concentrations, which probably could be used in the studies on sensitization of all these cereal pathogens.

When culturing on PDA supplemented with 6-DMM, *A. solani* manifested a much lower sensitivity to the putative sensitizer compared to *A. alternata*, for which the inhibitory effect of both nominally toxic and the growth suppressing 6-DMM concentrations was dozens of times higher as compared to *A. solani*. Besides, *A. solani* was completely insensitive to doses sensitizing *B. sorokiniana*. In contrast, no difference in the sensitivity to 6-DMM was found between these causative agents of Alternaria diseases in the spore germination tests (Table 1). Moreover, the dose-response patterns obtained for germinating conidia and colony growth of the pathogens were similar in both cases (Figure 2C,D).

With respect to the response to growth suppression, potato-damaging *R. solani* took almost the same position as the most insensitive *A. solani* (Table 1), at the same time showing the dose-response character similar to *P. nodorum* (Figure 2C).

2.2. *The Sensitizing Effect of 6-DMM on* B. sorokiniana *and* R. solani

For in vitro testing of the 6-DMM sensitizing activity, *B. sorokiniana* and *R. solani* were chosen, for which the maximum and minimum growth inhibition ED_{95} and MIC levels were determined (Table 1, Figure 1).

The pilot Petri plate bioassays involving co-applications of 6-DMM and Folicur® or Quadris® fungicides at different concentration combinations showed the inhibition of fungal growth to be significantly ($p < 0.05$) enhanced when both tested pathogens were cultured on fungicide-containing PDA amended with 6-DMM (Figures 3–6). For *B. sorokiniana*, the synergistic interaction of 6-DMM and Folicur® was observed in 15 of 32 concentration combinations tested by the checkerboard assay (Figure 3). In other cases, the effect was rather additive ($E_r \geq E_e$, $p > 0.05$; data not shown). The most pronounced sensitizing effect was observed for 6-DMM concentrations equal to 4, 6, or 8 μg/mL applied together with Folicur® at a ratio of 1.5:1, 2:1, or 4:1 (Figure 4). In these cases, the growth-inhibiting effect in relation to the pathogen demonstrated a 16-fold increase. For example, the MIC value of the Folicur® used alone was 64 μg/mL, while, in the presence of the sensitizer, the complete inhibition of the fungal colony growth was observed at 4 μg/mL. The fractional inhibitory concentration indices (FICIs) varied from 0.27 to 0.36 (0.32 on average) confirming the synergetic interactions in these tebuconazole/6-DMM combinations and suggesting a significant increase in the *B. sorokiniana* sensitivity to the fungicide. A statistically significant excess of E_r over E_e evidencing the effect of synergism was revealed even for those combinations, where the 6-DMM dose reached 10 μg/mL (54.7% growth inhibition), i.e., exceeded ED_{50} (Figure 5).

In the experiment with *R. solani*, the synergistic enhancement of the growth-inhibiting effect was also registered for the 6-DMM and Quadris® (azoxystrobin) combinations (Figure 6). The E_r values significantly exceeded E_e values in 10 of 16 combinations tested, while another six variants of the combined use demonstrated either an additive effect, or a less than 10% increase in the inhibiting action of one of the components.

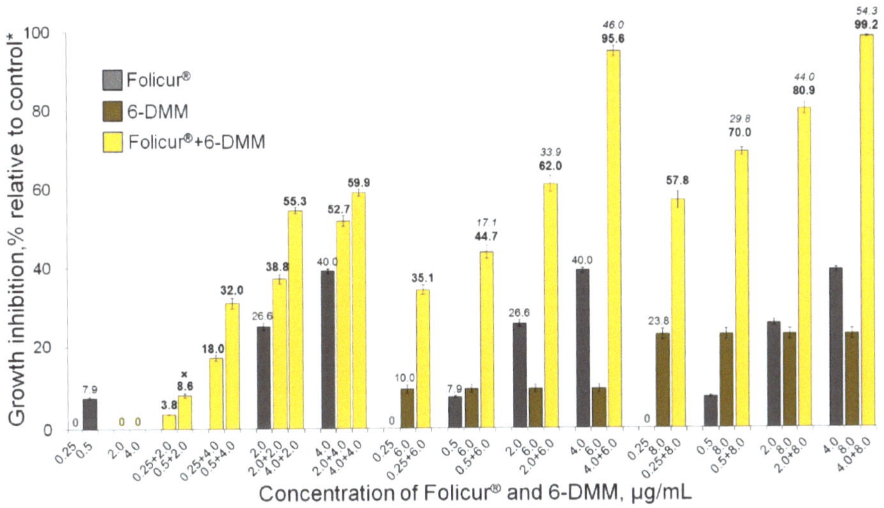

Figure 3. Enhancing the inhibitory effect of Folicur® EC 250 on the in vitro growth of *Bipolaris sorokiniana* due the fungicide combination with 6-DMM. The numbers in bold or regular above the columns indicate E_r values, while italic numbers show E_e values related to the same fungicide/6-DMM concentration combination. E_r shows the inhibition of the fungal growth (%) when the fungicide and 6-DMM were co-applied; E_e is the inhibition calculated for an estimated additive effect of the fungicide and 6-DMM (%). In the case when $E_r > E_e$ at $p \leq 0.05$, a synergistic interaction between the fungicide and the sensitizer is confirmed (see Materials and Methods, Section 4.4 and [31]). The case of an additive effect is marked with "×". Each histogram column represents the mean of three experiments (two diameter measuring of each colony, three colonies per each individual or combined application in each of three independent assays). Y-bars indicate standard error (SE) of the mean. The difference between treatments is significant at $p \leq 0.05$ (*t*-test for independent variables). * Control colonies were cultured on potato dextrose agar free of 6-DMM and the fungicide.

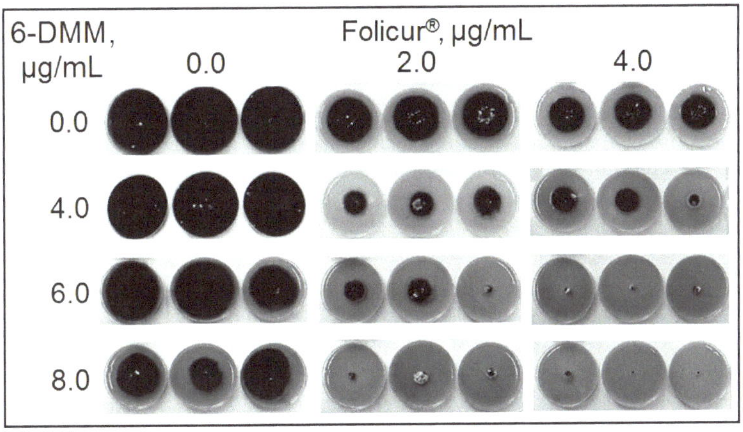

Figure 4. *Bipolaris sorokiniana* colonies grown for 9 days on potato dextrose agar supplemented with 6-DMM, Folicur® EC 250, or their combinations. The shown Petri plate cultures represent a typical picture for one of three performed checkerboard assays.

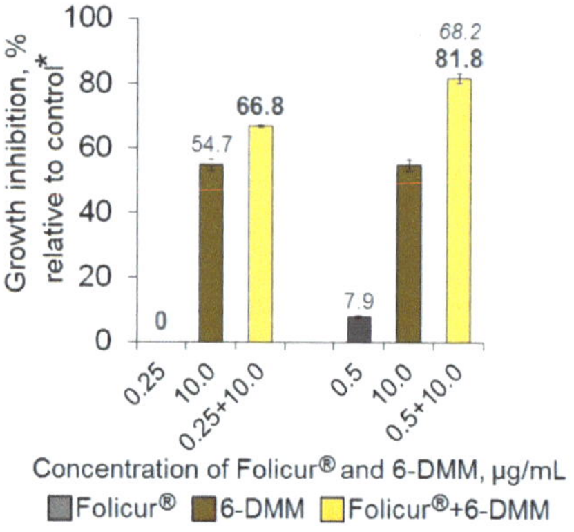

Figure 5. Synergistic augmentation of the in vitro suppression of the *Bipolaris sorokiniana* growth by ineffective Folicur® doses after their combining with 6-DMM (10.0 μg/mL). Each bar represents the mean of three experiments. The italic number above the column indicates E_e, other numbers indicate E_r values; $E_r > E_e$ at $p < 0.05$ (see Materials and Methods, Section 4.4). * Control colonies were cultured on potato dextrose agar free of 6-DMM and the fungicide.

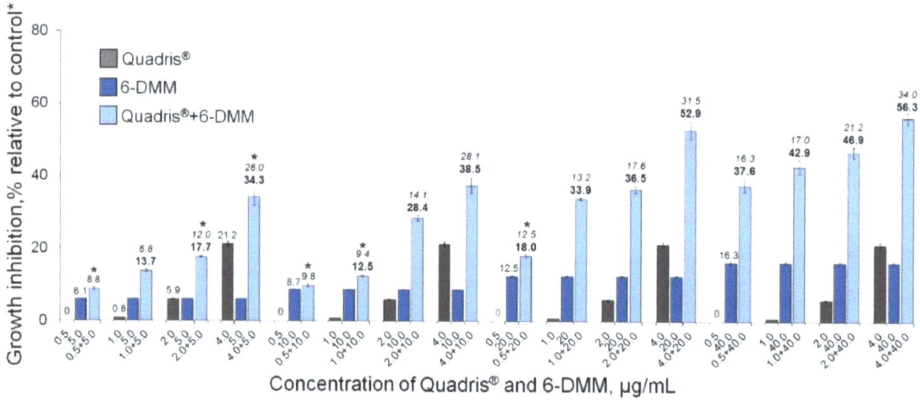

Figure 6. The inhibitory effect of Quadris® SC 250 on the in vitro growth of *Rhizoctonia solani* augmented due the fungicide combination with 6-DMM. Each bar represents the mean of three experiments. Concentration combinations providing an additive effect are marked with asterisks; for other combinations, $E_r > E_e$ at $p \leq 0.05$ (see Materials and Methods 4.4). The difference between treatments is significant at $p \leq 0.05$ (*t*-test for independent variables). * Control colonies were cultured on potato dextrose agar free of 6-DMM and the fungicide.

3. Discussion

The antifungal effect of 6-DMM was evaluated using different plant pathogenic fungi, such as wheat pathogens *A. alternata*, *F. avenaceum*, *F. culmorum*, *F. graminearum*, and *P. nodorum*; tomato/potato pathogen *A. solani* causing early bight of these two crops, *F. oxysporum* infecting wheat and tomato, and *R. solani*, an important potato pathogen also able to infect a wide range of other crops. All these fungi are controlled with various triazoles, and tebuconazole is one of the most often applied among them [32]. The strains tested in our work demonstrated stable in vitro growth, active spore production, and the sensitivity to tebuconazole. Tebuconazole has been reported to show a good effect against a wide range of *Fusarium* species and other fungi during plant treatments, and its in vitro inhibitory effect has been well documented [33–36]. On the other hand, a widespread application of tebuconazole-containing formulations may promote selection and accumulation of tebuconazole-resistant strains in fungal populations [34,37,38] that decreases or even nullifies the sensitivity to tebuconazole resulting in in vitro variability of its fungicidal activity [39].

Fusarium species included in this study form a wheat root rot pathogenic complex able to cause significant yield losses in cereals across the world [29]. Since Fusarium root rot belongs to seed-transmitted diseases, seed treatment with fungicides still remains the main way to control it in spite of reports about the appearance of fungicide-resistant strains of target pathogens. In this context, the development of approaches providing the reduction in fungicide dosages without decreasing its antifungal effect is a very important practical task. The results of this study allowed us to determine the range of non- or marginally toxic 6-DMM concentrations for the studied *Fusarium* species. All four species were less sensitive to 6-DMM than *B. sorokiniana* at high growth-inhibiting concentrations and differed by the sensitivity to this microbial compound. However, high reliability of the approximation in the probit analysis confirmed with the R^2 level determination (Table 1), similar dose-response pattern for 6-DMM in all four species (Figure 2A), and lower ED_{10} values (as compared to the dosage providing sensitizing effect for *B. sorokiniana*; Table 1) may evidence that 6-DIMM is a putative sensitizer to triazole fungicides for *Fusarium* pathogens and that it will be possible to select a working concentration of this compound providing an increased sensitivity of all components of the wheat root rot complex towards triazole fungicides.

A. solani, a foliar pathogen of tomato and potato, is considered to be poorly controlled by chemicals [40], while *R. solani* is a great problem in potato tuber storage. Early bight of tomato caused by *A. solani* is one of the most devastating diseases of this crop, while *A. alternata* is a soilborne fungus that often accompanies both Fusarium root rot agents, and *A. solani* on potato and tomato. *R. solani* on potato is often controlled by azoxistrobin and other strobilurins. These QoI-fungicides belong to the antifungal compounds with a high risk of resistance development in pathogens, and strobilurin-resistant *R. solani* strains have been reported for many pathogenic populations [8,11,41,42]. Our pilot findings on enhancing the sensitivity of these agents to Quadris® open an avenue to further investigations on the chemosensitization of such resistant strains.

We found that strains pathogenic to *Solanum* plants were much less sensitive to the growth-suppressing action of 6-DMM as compared to strains of cereal pathogens (Table 1, Figure 2A,C). Additionally, a drastic distinction in the 6-DMM effect on the growth of *A. solani* and *A. alternata* was observed. In contrast, the germination of *A. solani* conidia was inhibited by 6-DMM as effectively as in *A. alternata* and other pathogens.

Like many other known chemosensitizers, which generally have a mild antimicrobial activity and lack specificity to fungi [20], 6-DMM was revealed to possess such properties. Summing up all the presented results, one can conclude that they confirm a low fungitoxicity of 6-DMM, pointing to a wide spectrum of its antifungal activity, and suggesting that 6-DMM may have promise as a natural remedy for combination with chemicals to reduce their content in fungicidal formulations and their xenobiotic impact in agriculture, and probably, to enhance the protective effect of chemical fungicides by sensitization of various plant pathogens.

Fusarium, *Alternaria*, and *Rhizoctonia* fungi produce a range of highly toxic secondary metabolites including trichothecene and polyketide mycotoxins [43,44]. These compounds cause mycotoxicoses in animals and are carcinogens and/or allergens for humans, while their producers are among nosocomial infections that are especially dangerous for immunodeficient patients. Earlier we showed that 6-DMM suppressed the biosynthesis of a polyketide aflatoxin B1 in a toxigenic *Aspergillus flavus* [45]. In this regard, there is a suggestion that the antifungal effect of 6-DMM may be accompanied by the inhibition of the biosynthesis of toxins in other toxigenic fungi. In this case, this putative sensitizer will have an additional beneficial effect.

The pilot studies on the assessment of the sensitizing effect of 6-DMM presented in this study first confirmed a principal possibility of using 6-DMM for the sensitization of *B. sorokiniana* and *R. solani* to the fungicides based on triazole (Folicur®) and strobilurin (Quadris®), respectively, which are widely used for the crop protection against these dangerous plant pathogenic fungi. A number of concentration combinations resulting in a synergistic enhancement of the in vitro fungicidal effect were revealed for both pathogens. Being less responsive to 6-DMM alone (Table 1), *R. solani* was also less responsive to sensitization by this agent compared to *B. sorokiniana*, and the higher doses of the sensitizer were needed to synergistically augment the Quadris® effect against *R. solani*. Nevertheless, the results obtained for the fungicide/sensitizer co-applications at the most effective concentration combinations towards *B. sorokiniana* (4.0 + 8.0 μg/mL of Folicur® + 6-DMM) and *R. solani* (4.0 + 40.0 μg/mL of Quadris® + 6-DMM) suggest some prospects for 6-DMM combined with antifungal compounds of a different chemical nature and mode of action to improve their inhibitory effect. These findings also confirm the ability of 6-DMM to enhance the sensitivity of fungi belonging to different taxa, infecting different crops, and considerably differed in their pathogenesis processes.

The study on a sensitization of other plant pathogenic fungi with 6-DMM is in progress. The further investigations of this microbial product could contribute to overcoming the resistance of plant pathogens to agricultural fungicides and the reduction in the fungicide contamination of agricultural areas and products, as well as the total environmental pollution with toxic xenobiotics.

4. Materials and Methods

4.1. Fungal Strains and Their Culturing

Pathogen strains *A. alternata* MRD1-12, *Bipolaris sorokiniana* Ir-01-38, *F. avenaceum* Br-04-60, *F. culmorum* OR-02-37, *F. graminearum* FG-30, *F. oxysporum* KF-1713-4, and *R. solani* 100,063 isolated from various agricultural plants were provided by the State Collection of Plant Pathogenic Microorganisms of the All-Russian Research Institute of Phytopathology (ARRIP). The strains *A. solani* MO-VNIIF-9-2018 and *P. nodorum* B-9/47 were provided by the ARRIP work collections.

Plant pathogens' stock cultures maintained on potato dextrose agar (PDA) slants were resumed by culturing for 5–15 days (depending on the growth rate of a certain fungus) in Petri plates on the same medium to obtain spore-producing actively growing colonies. The aerial mycelia of these colonies at the log growth stage were used in the further experiments for inoculations of the fresh control or 6-DMM-containing PDA by placing a piece of the mycelium into the center of a 90-mm Petri plate. Conidia of *Fusarium* and *Alternaria* fungi were collected from a colony surface by flooding the mycelium with sterilized water and gently rubbing with a glass rod. Spore suspensions were filtered through sterile cotton wool to remove mycelial debris, and spore concentrations were counted using a hemocytometer. To simulate spore production in *Alternaria solani*, its cultures were flooded with sterilized ice water, irradiated with UV-A (315–360 nm) [46]. The Petri plates with treated colonies were allowed to dry for 3 days at room temperature under a diffused light [47] in sterilized laminar boxes followed by incubation at 24 °C for 7 days.

4.2. Preparation of 6-DMM and Fungicide Samples for Microbiological Experiments

Since 6-DMM lactone produced by the microbiological synthesis and extracted from *P. citrinum* culture liquid [48] was insoluble in aquatic media, we transformed it into a water-soluble sodium salt prior to addition to the nutrient media. To do this, a preparation of 6-DMM was dissolved in hot ethanol, the ethanol solution was amended with an equimolar amount of NaOH, incubated for 30 min at gently mixing and diluted with distilled water so that the final 6-DMM content was not lower than 10% in order to prevent a pellet formation during storage that occurred in less concentrated solutions. A 6-DMM concentration in the resulting sample was determined by HPLC as described earlier [48]. Just prior to microbiological tests, portions of this sample were diluted with ethanol to prepare 1% 6-DMM stock solution, whose aliquots were added aseptically into sterilized warm melted PDA up to necessary concentrations before PDA inoculation with the pathogens.

Commercial samples of Folicur®, EC 250 (a.i. tebuconazole) as well as Quadris® SC 250 (a.i. azoxystrobin) used in sensitization tests with *B. sorokiniana* and *R. solani*, respectively, were dissolved in distilled water and sterilized by filtration through a 0.22-μm Millipore membrane. The minimal volumes of the stock filtrate were added to PDA as described above.

4.3. In Vitro Assessment of a 6-DMM Effect on the Fungal Growth and Spore Germination

The influence of 6-DMM on the fungal growth was studied by a Petri plate bioassay involving the cultivation of fungi on PDA supplemented with different 6-DMM concentrations. The tested fungi were grown at 25–27 °C in the dark until the control colonies grown on 6-DMM-free PDA completely covered the agar surface. A growth inhibitory effect was determined by diminishing the average colony diameter measured on 6-DMM-containing PDA as compared to the control that was calculated after measuring the minimum and maximum diameters of colonies (six replications per treatment for each pathogen) in two perpendicular directions.

To assess the effect of 6-DMM on the germination of *Fusarium* spp. and *Alternaria* spp., a microplate test [23,49] was applied. Briefly, fungal conidia were incubated for 5–6 h at room temperature in distilled water (control) or in aquatic solutions supplemented with the 6-DMM sodium salt. The number of germinated conidia (among 500 ones of each pathogen per treatment) was counted in the control and 6-DMM-containing spore suspensions using an inverted microscope followed by calculation of the percentage of conidia germination inhibition as compared to the control.

4.4. In Vitro Assessment of a 6-DMM Sensitizing Effect

The sensitization experiments were designed under the principle of a double-dilution test and a checkerboard assay [20,50,51]. Marginally fungicidal or sub-fungicidal concentrations of the fungicides and non-fungitoxic or marginally toxic doses of the putative sensitizer were selected in preliminary experiments including the pathogen cultivation in the presence of either 6-DMM or each fungicide. Then, the fungi were grown as described above on PDA supplemented with marginally fungicidal or sub-fungicidal concentrations of Folicur® (*B. sorokiniana*) or Quadris® (*R. solani*) causing 0–10% or 11–40% colony growth inhibition, respectively, and on PDA with 6-DMM in the concentration range of 2–10 μg/mL (*B. sorokiniana*) or 5–40 μg/mL (*R. solani*), which provided no or marginally toxic effect. In parallel, the fungi were grown on PDA containing both fungicide and 6-DMM taken at the same concentrations as alone. Control cultures were grown under the same conditions on PDA without any supplements. Except for the cases where both the sensitizer and the fungicides showed no detectable suppression of a fungal growth, the synergy in 6-DMM/fungicide combinations was determined by the Limpel formula [31]:

$$E_e = \frac{(X+Y)-XY}{100} < E_r\ (p \leq 0.05),$$

where E_e is the expected additive inhibiting effect of the use of both components (%), X and Y represent the level of a spore germination inhibition by each of the components alone (%), and E_r is the inhibition obtained by a joint use of 6-DMM and a fungicide. In addition, fractional inhibitory concentration indices (FICIs) [20,50] were calculated; according to generally accepted protocols, FICIs ≤ 0.5 were interpreted as a confirmation of synergistic interactions [50,52] between 6-DMM and a fungicide.

4.5. Data Analysis and Statistical Treatment

Average 6-DMM inhibitory concentrations capable of 10, 50 or 95% suppression of the mycelium growth or spore germination (ED_{10}, ED_{50}, and ED_{95}, respectively) were determined by the probit analysis [53] with the involvement of a linear regression based on the data obtained in two independent experiments for each pathogen. In a regression equation ($Y = ax_n + b$) used in probit-analyses, Y is the probit value for 10, 50 or 95% growth inhibition, x values represent logarithms of these EDs, while a and b represent regression coefficients. Minimum inhibitory concentrations (MIC), i.e., the lowest 6-DMM or fungicide concentrations preventing a visible colony growth, were detected by serial two-fold dilutions of the stock solutions according to a reference CLSI method for antifungal susceptibility testing of filamentous fungi [54]. Approximation confidence values (R^2), regression coefficients, mean values, standard deviations, and standard errors as well as significant differences (at $p \leq 0.05$) between treatments and the control, determined based on the *t*-test for independent variables, were calculated using a STATISTICA 6.0 package (StatSoft Inc., Tulsa, OK, USA).

Author Contributions: Conceptualization, L.S. and V.D.; data curation, V.D.; formal analysis, L.S. and N.S.; funding acquisition, L.S.; investigation, M.K., T.P. and L.S.; methodology, L.S., M.K. and T.P.; software, N.S.; supervision, V.D.; validation, L.S., N.S. and V.D.; visualization, L.S., M.K.; writing—original draft preparation, N.S. and L.S.; writing—review and editing, N.S., V.D. and L.S. All authors have read and agreed to the published version of the manuscript.

Funding: The study was supported by the Russian Science Foundation (project no. 18-16-00084).

Acknowledgments: We acknowledge Tatyana I. Smetanina from the ARRIP Department of Potato and Vegetable Diseases for consulting on the *A. solani* spore production.

Conflicts of Interest: The authors declare no conflict of interest.

References

1. Moore, D.; Robson, G.D.; Trinci, A.P. *21st Century Guidebook to Fungi*, 2nd ed.; Cambridge University Press: Cambridge, UK, 2020; pp. 408–409.
2. Termorshuizen, A.J. Fungal and fungus-like pathogens of potato. In *Potato Biology and Biotechnology: Advances and Perspectives*; Vreugdenhil, D., Ed.; Elsevier: Oxford, UK, 2007; pp. 643–665.
3. Carson, G.R.; Edwards, N.M. Criteria of wheat and flour quality. In *Wheat Chemistry and Technology*, 4th ed.; Khan, K., Shewry, P.R., Eds.; AACC International: St. Paul, MN, USA, 2009; pp. 97–118.
4. Rosentrater, K.A.; Evers, A.D. Introduction to cereals and pseudocereals and their production. In *Kent's Technology of Cereals*, 5th ed.; Rosentrater, K.A., Evers, A.D., Eds.; Woodhead Publishing: Duxford, UK, 2018; pp. 3–75.
5. Hawkins, N.J.; Bass, C.; Dixon, A.; Neve, P. The evolutionary origins of pesticide resistance. *Biol. Rev.* **2019**, *94*, 135–155. [CrossRef] [PubMed]
6. Fisher, M.C.; Hawkins, N.J.; Sanglard, D.; Gurr, S.J. Worldwide emergence of resistance to antifungal drugs challenges human health and food security. *Science* **2018**, *360*, 739–742. [CrossRef] [PubMed]
7. Takagaki, M.; Kaku, K.; Watanabe, S.; Kawai, K.; Shimizu, T.; Sawada, H.; Kumakura, K.; Nagayama, K. Mechanism of resistance to carpropamid in *Magnaporthe grisea*. *Pest. Manag. Sci.* **2004**, *60*, 921–926. [CrossRef] [PubMed]
8. Olaya, G.; Buitrago, C.; Pearsaul, D.; Sierotzki, H.; Tally, A. Detection of resistance to QoI fungicides in *Rhizoctonia solani* isolates from rice. *Phytopathology* **2012**, *102*, 88.
9. Fairchild, K.Ł.; Miles, T.D.; Wharton, P.S. Assessing fungicide resistance in populations of *Alternaria* in Idaho potato fields. *Crop Protect.* **2013**, *49*, 31–39. [CrossRef]
10. Cheval, P.; Siah, A.; Bomble, M.; Popper, A.D.; Reignault, P.; Halama, P. Evolution of QoI resistance of the wheat pathogen *Zymoseptoria tritici* in Northern France. *Crop Protect.* **2017**, *92*, 131–133. [CrossRef]
11. Muzhinji, N.; Woodhall, J.W.; Truter, M.; van der Waals, J.E. Variation in fungicide sensitivity among *Rhizoctonia* isolates recovered from potatoes in South Africa. *Plant Dis.* **2018**, *102*, 1520–1526. [CrossRef]
12. Liu, S.; Fu, L.; Wang, S.; Chen, J.; Jiang, J.; Che, Z.; Tian, Y.; Chen, G. *Carbendazim resistance of Fusarium graminearum* from Henan wheat. *Plant Dis.* **2019**, *103*, 2536–2540. [CrossRef]
13. Fernández-Ortuño, D.; Torés, J.A.; de Vicente, A.; Pérez-García, A. Mechanisms of resistance to QoI fungicides in phytopathogenic fungi. *Int. Microbiol.* **2008**, *11*, 1–9.
14. Lucas, J.A.; Hawkins, N.J.; Fraaije, B.A. The evolution of fungicide resistance. *Adv. Appl. Microbiol.* **2015**, *90*, 29–92.
15. Chen, Z.-F.; Ying, G.-G. Occurrence, fate and ecological risk of five typical azole fungicides as therapeutic and personal care products in the environment: A review. *Environ. Int.* **2015**, *84*, 142–153. [CrossRef] [PubMed]
16. Esteve-Turrillas, F.A.; Agulló, C.; Abad-Somovilla, A.; Mercader, J.V.; Abad-Fuentes, A. Fungicide multiresidue monitoring in international wines by immunoassays. *Food Chem.* **2016**, *196*, 1279–1286. [CrossRef] [PubMed]
17. Carvalho, F.P. Pesticides, environment, and food safety. *Food Energy Secur.* **2017**, *6*, 48–60. [CrossRef]
18. Brent, K.J.; Hollomon, D.W. Fungicide Resistance in Crop Pathogens: How Can It be Managed? Fungicide Resistance Action Committee: Brussels, Belgium, 2007.

19. Álvarez-Martín, A.; Rodríguez-Cruz, M.S.; Andrades, M.S.; Sánchez-Martín, M.J. Application of a biosorbent to soil: A potential method for controlling water pollution by pesticides. *Environ. Sci. Pollut. Res. Int.* **2016**, *23*, 9192–9203. [CrossRef]
20. Campbell, B.C.; Chan, K.L.; Kim, J.H. Chemosensitization as a means to augment commercial antifungal agents. *Front. Microbiol.* **2012**, *3*, 79. [CrossRef]
21. Dzhavakhiya, V.G.; Shcherbakova, L.A.; Semina, Y.V.; Zhemchuzhina, N.S.; Campbell, B. Chemosensitization of plant pathogenic fungi to agricultural fungicides. *Front. Microbiol.* **2012**, *3*, 87. [CrossRef]
22. Kim, K.; Lee, Y.; Ha, A.; Kim, J.I.; Park, A.R.; Yu, N.H.; Son, H.; Choi, G.J.; Park, H.W.; Lee, C.W.; et al. Chemosensitization of *Fusarium graminearum* to chemical fungicides using cyclic lipopeptides produced by *Bacillus amyloliquefaciens* strain JCK-12. *Front. Plant Sci.* **2017**, *8*, 2010. [CrossRef]
23. Shcherbakova, L.A.; Syomina, Y.V.; Arslanova, L.R.; Nazarova, T.A.; Dzhavakhiya, V.G. Metabolites secreted by a nonpathogenic *Fusarium sambucinum* inhabiting wheat rhizosphere enhance fungicidal effect of some triazoles against *Parastagonospora nodorum*. *AIP Conf. Proc.* **2019**, *2063*, 030018.
24. Kim, J.H.; Chan, K.L. Augmenting the antifungal activity of an oxidizing agent with kojic acid: Control of *Penicillium* strains infecting crops. *Molecules* **2014**, *19*, 18448–18464. [CrossRef]
25. Kartashov, M.I.; Shcherbakova, L.A.; Dzhavakhiya, V.G. In vitro enhancement of the sensitivity to tebuconazole in *Bipolaris sorokiniana*, a causative agent of cereal root rots, by a microbial metabolite 6-demethylmevinolin. In *Abstract Book of 19th International Reinhardsbrunn Symposium on Modern Fungicides and Antifungal Compounds, Friedrichroda, Germany, 7–11 April 2019*; DPG Verlag: Braunschweig, Germany, 2019; p. 48.
26. Wang, F.; Wang, Z.; Zhang, B.; Zhang, Q. Degradation and adsorption of tebuconazole and tribenuron-methyl in wheat soil, alone and in combination. *Chil. J. Agric. Res.* **2017**, *77*, 281–286. [CrossRef]
27. Odintsova, T.; Shcherbakova, L.; Slezina, M.; Pasechnik, T.; Kartabaeva, B.; Istomina, E.; Dzhavakhiya, V. Hevein-like antimicrobial peptides WAMPs: Structure-function relationship in antifungal activity and sensitization of plant pathogenic fungi to tebuconazole by WAMP-2-derived peptides. *IJMS* **2020**, *21*, 7912. [CrossRef] [PubMed]
28. Smiley, R.W.; Gourlie, J.A.; Easley, S.A.; Patterson, L.M. Pathogenicity of fungi associated with the wheat crown rot complex in Oregon and Washington. *Plant Dis.* **2005**, *89*, 949–957. [CrossRef] [PubMed]
29. Gagkaeva, T.Y.; Gavrilova, O.P.; Levitin, M.M.; Novozhilov, K.V. Fusarium diseases of grain crops. *Zashchita i Karantin Rastenii* **2011**, *S5*, 69–120.
30. Matin, M.; Faruq, A. Prevalence of pathogenic fungi in soils of wheat field. *J. Exp. Biosci.* **2013**, *4*, 63–68.
31. Richer, D.L. Synergism: A patent view. *Pestic. Sci.* **1987**, *19*, 309–315. [CrossRef]
32. Paul, P.A.; Lipps, P.E.; Hershman, D.E.; McMullen, M.P.; Draper, M.A.; Madden, L.V. Efficacy of triazole-based fungicides for Fusarium head blight and deoxynivalenol control in wheat. *Phytopathology* **2008**, *98*, 999–1011. [CrossRef]
33. Klix, M.B.; Verreet, J.A.; Beyer, M. Comparison of the declining triazole sensitivity of *Gibberella zeae* and increased sensitivity achieved by advances in triazole fungicide development. *Crop Prot.* **2007**, *26*, 683–690. [CrossRef]
34. Masiello, M.; Somma, S.; Ghionna, V.; Logrieco, A.F.; Moretti, A. In vitro and in field response of different fungicides against *Aspergillus flavus* and *Fusarium* species causing ear rot disease of maize. *Toxins* **2019**, *11*, 11. [CrossRef]
35. Mullenborn, C.; Steiner, U.; Ludwig, M.; Oerke, E.C. Effect of fungicides on the complex of *Fusarium* species and saprophytic fungi colonizing wheat kernels. *Eur. J. Plant Pathol.* **2008**, *120*, 157–166. [CrossRef]
36. Ivic, D.; Sever, Z.; Kuzmanovska, B. In vitro sensitivity of *Fusarium graminearum*, *F. avenaceum* and *F. verticillioides* to carbendazim, tebuconazole, flutriafol, metconazole and prochloraz. *Pestic. Fitomedicina* **2011**, *26*, 35–42. [CrossRef]
37. Deising, H.B.; Reimann, S.; Pascholati, S.F. Mechanisms and significance of fungicide resistance. *Br. J. Microbiol.* **2008**, *39*, 286–295. [CrossRef]
38. Becher, R.; Hettwer, U.; Karlovsky, P.; Deising, H.B.; Wirsel, S.G.R. Adaptation of *Fusarium graminearum* to tebuconazole yielded descendants diverging for levels of fitness, fungicide resistance, virulence, and mycotoxin production. *Phytopathology* **2010**, *100*, 444–453. [CrossRef] [PubMed]
39. Avozani, A.; Tonin, R.B.; Reis, E.M.; Camera, J.; Ranzi, C. In vitro sensitivity of *Fusarium graminearum* isolates to fungicides. *Summa Phytopathol.* **2014**, *40*, 231–247. [CrossRef]

40. Pawar, P.; Bhosale, A.; Lolage, Y.; Arts, S.; Commerce, D. Alternaria blight of tomato (*Lycopersicon esculentum* Mill). *Int. J. Adv. Technol. Innov. Res.* **2016**, *8*, 1727–1728.
41. Arabiat, S.; Khan, M.F.R. Sensitivity of *Rhizoctonia solani* AG-2-2 from sugar beet to fungicides. *Plant Dis.* **2016**, *100*, 2427–2433. [CrossRef]
42. Castroagudin, V.L.; Fiser, S.; Cartwright, R.D.; Wamishe, Y.; Correll, J.C. Evaluation of *Rhizoctonia solani* AG 1-IA and *Rhizoctonia* species for resistance to QoI fungicides. *Phytopathology* **2013**, *103* (Suppl. 2), S2–S24.
43. Broquist, H. Slaframine and swainsonine, mycotoxins from *Rhizoctonia leguminicola*. *Toxin Rev.* **2008**, *5*, 241–252. [CrossRef]
44. Fraeyman, S.; Croubels, S.; Devreese, M.; Antonissen, G. Emerging Fusarium and Alternaria mycotoxins: Occurrence, toxicity and toxicokinetics. *Toxins* **2017**, *9*, 228. [CrossRef]
45. Dzhavakhiya, V.G.; Voinova, T.M.; Popletaeva, S.B.; Statsyuk, N.V.; Limantseva, L.A.; Shcherbakova, L.A. Effect of various compounds blocking the colony pigmentation on the aflatoxin B1 production by *Aspergillus flavus*. *Toxins* **2016**, *8*, 313. [CrossRef]
46. Fourtouni, A.; Manetas, Y.; Christias, C. Effects of UV-B radiation on growth, pigmentation, and spore production in the phytopathogenic fungus *Alternaria solani*. *Can. J. Bot.* **1998**, *76*, 2093–2099.
47. Alan, A.; Earle, E. Sensitivity of bacterial and fungal plant pathogens to the lytic peptides, MSI-99, Magainin II, and Cecropin B. *Mol. Plant Microbe Interact.* **2002**, *15*, 701–708. [CrossRef] [PubMed]
48. Ukraintseva, S.N.; Voinova, T.M.; Dzhavakhiya, V.G. Penicillium citrinum strain improvement for compactin production by induced-mutagenesis and optimization of obtained mutant cultivation conditions. In *Biotechnology and Medicine*; Zaikov, G.E., Ed.; Nova Science Publishers: New York, NY, USA, 2004; pp. 71–78.
49. Aver'yanov, A.A.; Lapikova, V.P.; Pasechnik, T.D.; Abramova, O.S.; Gaivoronskaya, L.M.; Kuznetsov, V.V.; Baker, C.J. Pre-illumination of rice blast conidia induces tolerance to subsequent oxidative stress. *Fungal Biol.* **2014**, *118*, 743–753. [CrossRef] [PubMed]
50. Canton, E.; Peman, J.; Gobernado, M.; Viudes, A.; Espinel, I.A. Synergistic activities of fluconazole and voriconazole with terbinafine against four *Candida* species determined by checkerboard, time-kill, and E-test methods. *Antimicrob. Agent Chemother.* **2005**, *49*, 1593–1596. [CrossRef] [PubMed]
51. Orhan, G.; Bayram, A.; Zer, Y.; Balci, I. Synergy tests by E test and checkerboard methods of antimicrobial combinations against *Brucella melitensis*. *J. Clin. Microbiol.* **2005**, *43*, 140–143. [CrossRef]
52. Odds, F.C. Synergy, antagonism, and what the chequerboard puts between them. *J. Antimicrob. Chemother.* **2003**, *52*, 1. [CrossRef]
53. Sparling, D.W. Modeling in ecotoxicology. In *Ecotoxicology Essentials*, 1st ed.; Donald, W., Ed.; Academic Press: London, UK, 2016; pp. 361–390.
54. Espinel-Ingroff, A.; Cantón, E.; Pemán, J. Antifungal susceptibility testing of filamentous fungi. *Curr. Fungal Infect. Rep.* **2012**, *6*, 41–50. [CrossRef]

Publisher's Note: MDPI stays neutral with regard to jurisdictional claims in published maps and institutional affiliations.

antibiotics

MDPI

Article

Antibacterial Activity of Volatile Organic Compounds Produced by the Octocoral-Associated Bacteria *Bacillus* sp. BO53 and *Pseudoalteromonas* sp. GA327

Anette Garrido [1,†], Librada A. Atencio [1,†], Rita Bethancourt [2], Ariadna Bethancourt [2], Héctor Guzmán [3], Marcelino Gutiérrez [1,*] and Armando A. Durant-Archibold [1,4,*]

1 Center for Biodiversity and Drug Discovery, Instituto de Investigaciones Científicas y Servicios de Alta Tecnología (INDICASAT AIP), Panama City 0843-01103, Panama; anecgarrido@gmail.com (A.G.); latencio@indicasat.org.pa (L.A.A.)
2 Department of Microbiology and Parasitology, College of Natural, Exact Sciences, and Technology, Universidad de Panama, Panama City 0824-03366, Panama; rita.bethancourt@up.ac.pa (R.B.); ariadna.bethancourt@up.ac.pa (A.B.)
3 Smithsonian Tropical Research Institute, Panama City 0843-03092, Panama; GuzmanH@si.edu
4 Department of Biochemistry, College of Natural, Exact Sciences, and Technology, University of Panama, Panama City 0824-03366, Panama
* Correspondence: mgutierrez@indicasat.org.pa (M.G.); adurant@indicasat.org.pa (A.A.D.-A.)
† These authors contributed equally to this work.

Received: 29 October 2020; Accepted: 16 December 2020; Published: 18 December 2020

Abstract: The present research aimed to evaluate the antibacterial activity of volatile organic compounds (VOCs) produced by octocoral-associated bacteria *Bacillus* sp. BO53 and *Pseudoalteromonas* sp. GA327. The volatilome bioactivity of both bacteria species was evaluated against human pathogenic antibiotic-resistant bacteria, methicillin-resistant *Staphylococcus aureus*, *Acinetobacter baumanni*, and *Pseudomonas aeruginosa*. In this regard, the in vitro tests showed that *Bacillus* sp. BO53 VOCs inhibited the growth of *P. aeruginosa* and reduced the growth of *S. aureus* and *A. baumanni*. Furthermore, *Pseudoalteromonas* sp. GA327 strongly inhibited the growth of *A. baumanni*, and *P. aeruginosa*. VOCs were analyzed by headspace solid-phase microextraction (HS-SPME) joined to gas chromatography-mass spectrometry (GC-MS) methodology. Nineteen VOCs were identified, where 5-acetyl-2-methylpyridine, 2-butanone, and 2-nonanone were the major compounds identified on *Bacillus* sp. BO53 VOCs; while 1-pentanol, 2-butanone, and butyl formate were the primary volatile compounds detected in *Pseudoalteromonas* sp. GA327. We proposed that the observed bioactivity is mainly due to the efficient inhibitory biochemical mechanisms of alcohols and ketones upon antibiotic-resistant bacteria. This is the first report which describes the antibacterial activity of VOCs emitted by octocoral-associated bacteria.

Keywords: octocoral-associated bacteria; antibacterial activity; volatilome; *Bacillus*; *Pseudoalteromonas*; bacteria volatile organic compounds; antibiotic-resistant bacteria

1. Introduction

For centuries, the fight against bacterial infections has been one the focus of attention of humanity. Although in the 20th century an important number of discovered antibacterial compounds have improved the quality of life of the people, it has been observed an increased antibiotic resistance prevalence among bacteria which represents the greatest challenge to human health. The main molecular mechanisms of bacteria resistance to antibiotics are due to mutations in bacterial genes; and due to the elevated number of multidrug resistance pumps (MDR pumps), which extrudes antibiotics out of the bacterial cells [1–3]. It is therefore imperative the discovery of new therapeutic compounds that overcome the bacterial resistance to antimicrobial agents. Among the major critical group of multidrug-resistant bacteria are *Staphylococcus aureus*, *Pseudomonas aeruginosa*, and *Acinetobacter baumannii* [4]. *S. aureus* is a commensal Gram-positive bacteria, which causes different types of diseases such as endocarditis, septic arthritis, necrotizing fasciitis, parotitis, pyomyositis, osteomyelitis, and skin infections [5]. *P. aeruginosa* is an opportunistic pathogen that is a leading cause of morbidity and mortality in cystic fibrosis patients and immunocompromised individuals. Moreover, it is one of the main risk factors for nosocomial infections and ventilator-associated pneumonia [6]. *A. baumannii* is a Gram-negative *Bacillus*, which is the main cause of hospital-acquired infections, leading to septicemia and pneumonia in immune-compromised hosts [7].

Natural antibacterial agents represent the main source of new drugs [8,9]. Most of the scientific investigations for the discovery of drugs with bioactivity against different illnesses have been focused on living organisms from the terrestrial ecosystem. However, in recent years an important number of researches undertaken for the discovery of new bioactive natural products have been performed on marine organisms [10,11]. In this sense, corals (Cnidaria) are aquatic invertebrates in which thousands of bacterial phylotypes coexist. Coral's microbiome association mainly depends on the corals' species on which they develop a relevant role in the biosynthesis of compounds with a high degree of bioactivities, for protection against pathogenic microorganisms. Thus, the coral microbiome is an important source of antibacterial natural products [11–14].

Many marine bacteria produce volatile organic compounds (VOCs) whose bioactivity against human pathogenic bacteria are at the early stages of investigation [15–20]. These VOCs are small molecules biosynthesized by primary and secondary metabolic pathways and include chemical classes such as alcohols, esters, aliphatic and aromatic hydrocarbons, terpenes, nitrogen, and sulfur compounds, among others. The bacteria volatile organic compounds (bVOCs), contribute to the intra- or inter-communication, and protection against other microorganisms [15,17]. Taking this into account, marine bacteria VOCs can be considered a highly potential source of drugs with antibacterial bioactivity. Despite bacterial symbionts of octocorals represent a source of potential antibiotic drugs [14], no investigation has been focused on the assessment of the antibacterial activity of bVOCs. Accordingly, the purpose of this research is to study the antibiotic activity of VOCs of bacteria isolated from two different octocoral species of the Caribbean Sea against human pathogenic bacteria *P. aeruginosa*, *A. baumannii*, and methicillin-resistant *S. aureus*.

2. Results

2.1. Identification of Marine Bacteria BO53 and GA327

The marine bacteria species BO53 and GA327, isolated in Panama from *Pseudopterogorgia acerosa* (Pallas, 1766) and *Muriceopsis sulphurea* (Donovan, 1825), respectively, were identified by 16S rRNA gene sequence analyses. In this sense, the 16S rRNA gene sequence of each species shows 99% sequence similarity with *Bacillus sp.* (BO53) and *Pseudoalteromonas* sp. (GA327) species, when compared to those in the GenBank database (https://www.ncbi.nlm.nih.gov/genbank), using the Basic Local Alignment Search Tool (BLAST). GeneBank accession numbers of the 16S rRNA sequence of BO53 and GA327 are MK291446 and KU213068, respectively.

2.2. Antibacterial Activity of VOCs Produced by Bacillus sp. BO53 and Pseudoalteromonas sp. GA327

The in vitro study to determine the antibacterial activity of volatile compounds produced by marine bacteria *Bacillus* sp. BO53 and *Pseudoalteromonas* sp. GA327 species revealed that both bacterial compounds had inhibitory activity towards *A. baumannii*, *P. aeruginosa*, and methicillin-resistant *S. aureus* growth (Table 1). The Gram-positive *Bacillus* sp. BO53 volatile organic compounds lead to significant growth inhibition of *P. aeruginosa* at 24 h, and to growth reduction of *S. aureus* and *A. baumannii* at 48 h. The volatiles released by the Gram-negative *Pseudoalteromonas* sp. GA327 lead to the growth inhibition of *A. baumannii* at 24 h, and *P. aeruginosa* at 48 h, and did not inhibit the growth of *S. aureus*. On the other hand, in the absence of bVOCs produced by marine bacteria, the growth of human pathogenic bacteria was not suppressed. The inhibition of Gram-negative pathogenic bacteria growth by marine bacteria VOCs was greater for *Pseudoalteromonas* sp. GA327, than *Bacillus* sp. BO53; while on the other hand, the bioactivity of VOCs on Gram-positive pathogenic bacteria, was higher for *Bacillus* sp. BO53 than *Pseudoalteromonas* sp. GA327.

Table 1. Antibacterial effect of *Bacillus* sp. and *Pseudoalteromonas* sp. volatile compounds against *Acinetobacter baumanni*, *Staphylococcus aureus*, and *Pseudomonas aeruginosa* pathogenic bacteria.

Species	Strain	*Bacillus* sp. BO53		*Pseudoalteromonas* sp. GA327	
		24 h	48 h	24 h	48 h
A. baumanni	ATCC 19606	-	±	+	+
S. aureus	ATCC 43300	-	±	-	-
P. aeruginosa	ATCC 10145	-	+	±	+

(+) Growth inhibition; (±) Growth reduction; (-) No inhibition.

2.3. *Identification of VOCs from Bacillus sp. BO53 and Pseudoalteromonas sp. GA327*

The analysis for identification of VOCs biosynthesized by octocoral-associated bacteria *Bacillus* sp. BO53 and *Pseudoalteromonas* sp. GA327 was performed by HS-SPME-GC-MS technique. Each marine bacteria sample was analyzed three times using the DVB/CAR/PDMS coating solid-phase microextraction fiber; 37 °C extraction temperature; and 40 min extraction time. DVB/CAR/PDMS fiber was selected due to its high capacity for the extraction of volatile and semi-volatile compounds present in samples [21–23].

In total, 19 bVOCs were identified, of which 11 were detected in *Bacillus* sp. BO53 and 13 in *Pseudoalteromonas* sp. GA327 (Table 2). Compounds that were present in the blank Luria Bertani (LB) broth, supplemented with seawater, were excluded. Ketone comprised the largest compounds detected in *Bacillus* sp. BO53, followed by alcohols, sesquiterpenes, monoterpenes, aromatics, and alkanes (Figure 1a). 5-Acetyl-2-methylpyridine (64.63%), 2-butanone (17.03%), and 2-nonanone (7.00%) were the major VOCs detected in *Bacillus* sp. BO53. In the case of *Pseudoalteromonas* sp. GA327, alcohols are the main compounds detected followed by ketones, ester, and monoterpene compounds (Figure 1b). 1-Pentanol (38.91%), 2-butanone (20.14%), and butyl formate (17.30%) were the primary VOCs detected in *Pseudoalteromonas* sp. GA327. 2,4-trimethylpentane, o-xylene, 5-acetyl-2-methylpyridine, α-cubebene, 1-undecanol, and α-longicyclene are the specific VOCs produced by *Bacillus* sp. BO53 (Figure 2b); while 1-butanol, 2-pentanone, butyl formate, 2-heptanone, 6-methyl-5-heptene-2-one, benzyl alcohol, 2-decanone, 2-undecanone were found only in *Pseudoalteromonas* sp. GA327 (Figure 2c).

Table 2. Identified mVOCs produced by the marine bacteria *Bacillus* sp. BO53 and Pseudoalteromonas sp. GA327.

			% Detected Compound on Each Species	
Compound	TRI	ERI	*Bacillus* sp. BO53	*Pseudoalteromonas* sp. GA327
2-Butanone	602	609	17.03	20.14
1-Butanol	671	676	-	4.55
2,2,4-Trimethylpentane	680	687	0.46	-
2-Pentanone	687	693	-	1.29
1-Pentanol	775	780	1.33	38.91
Butyl formate	787	793	-	17.30
o-Xylene	884	891	0.47	-

Table 2. *Cont.*

Compound	TRI	ERI	% Detected Compound on Each Species	
			Bacillus sp. BO53	*Pseudoalteromonas* sp. GA327
2-Heptanone	889	895	-	7.99
6-Methyl-5-heptene-2-one	988	995	-	0.37
p-Cymene	1021	1027	0.50	0.98
Benzyl Alcohol	1033	1040	-	3.32
2-Nonanone	1096	1102	7.00	1.38
3-Methylacetophenone	1176	1184	0.26	1.18
5-Acetyl-2-methylpyridine	1189	1193	64.63	-
2-Decanone	1190	1198	-	1.01
2-Undecanone	1291	1300	-	1.58
α-Cubebene	1354	1355	0.43	-
1-Undecanol	1370	1374	4.58	-
α-Longicyclene	1374	1380	3.30	-

TRI: Theoretical Retention Index; ERI: Experimental Retention Index.

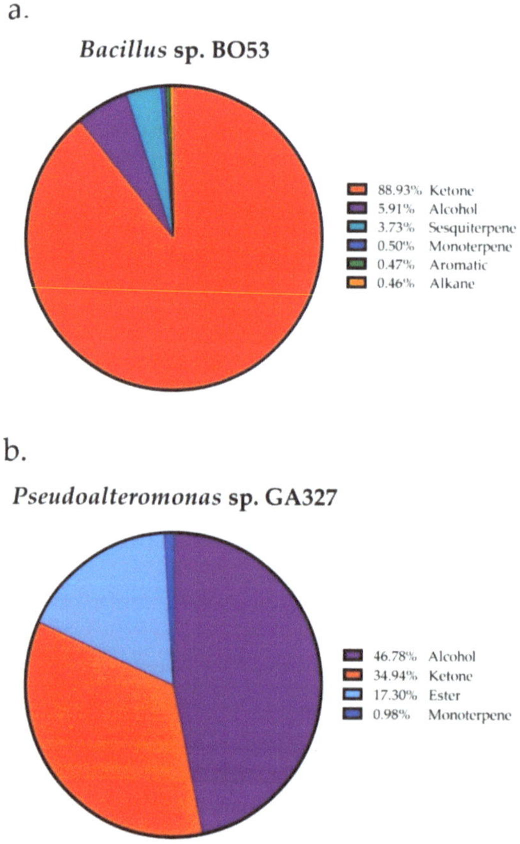

Figure 1. The proportion of the chemical families of VOCs detected in octocoral-associated bacteria *Bacillus* sp. BO53 and *Pseudoalteromonas* sp. GA327.

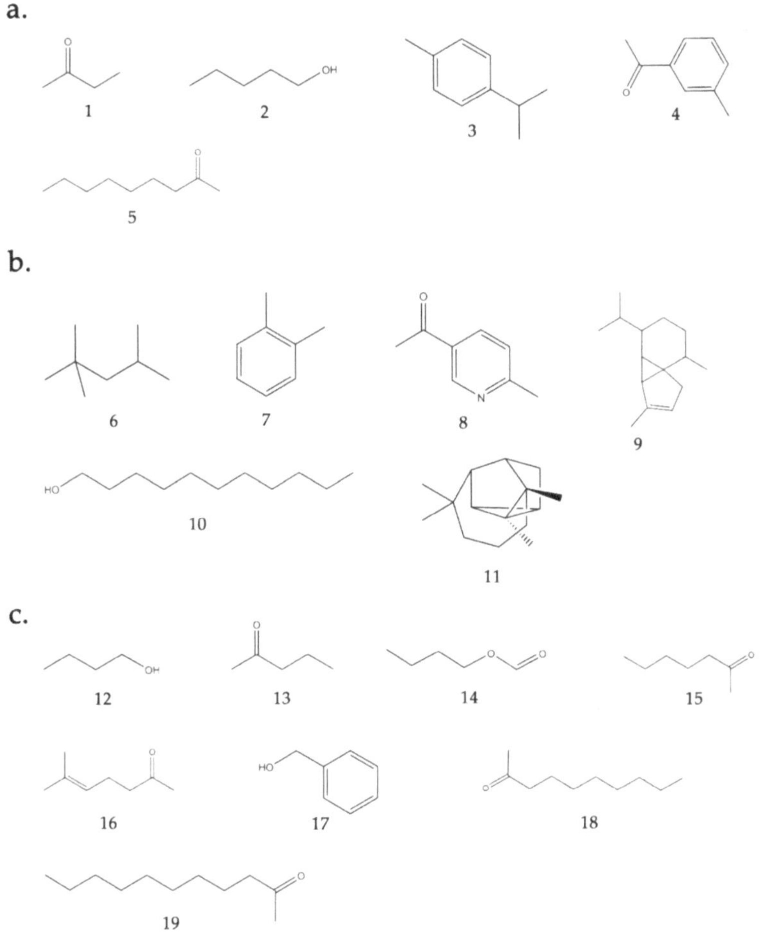

Figure 2. Molecular structure of *Bacillus* sp. BO53 and *Pseudoalteromonas* sp. GA327. detected volatile compounds. (**a**) VOCs detected on *Bacillus* sp. BO53 and *Pseudoalteromonas* sp. GA327: (1) 2-butanone, (2) 1-pentanol, (3) p-cymene, (4) 3-methylacetophenone, (5) 2-nonanone, (**b**) VOCs detected only on *Bacillus* sp. BO53: (6) 2,2,4-trimethylpentane, (7) o-xylene, (8) 5-acetyl-2-methylpyridine, (9) α-cubebene, (10) 1-decanol, (11) logicyclene, (**c**) VOCs detected only on *Pseudoalteromonas* sp. GA327: (12) 2-butanone, (13) 2-pentanone, (14) butyl formate, (15) 2-heptanone, (16) 6-methyl-5-heptene-2-one, (17) benzyl alcohol, (18) 2-decanone, (19) 2-undecanone.

3. Discussion

Human pathogenic bacteria, *A. baumannii*, methicillin-resistant *S. aureus*, and *P. aeruginosa* bacteria, are antibiotic-resistant microorganisms for which the development of research for the discovery of new antibacterial drugs have become crucial. To date, only a small number of marine bacteria have been studied for bioactive bVOCs [24].

This study aimed to determine the antibacterial activity of the volatilome produced by marine bacteria, *Bacillus* sp. BO53 and *Pseudoalteromonas* sp. GA327, isolated from octocorals. Overall, the VOCs produced by *Pseudoalteromonas* sp. GA327 lead to the inhibition of the two Gram-negative pathogenic bacteria investigated at early stages (24 h) when compared to the antibacterial activity of *Bacillus* sp. BO53. The antibacterial activity of *Bacillus* sp. BO53 VOCs against *P. aeruginosa* was higher than for *A. baumannii* and *S. aureus*, and lead to a total inhibition of *P. aeruginosa* growth within 48 h after exposure to the *Bacillus* sp. BO53 VOCs.

The HS-SPME-GC-MS analysis, lead to the identification of ketones, mainly 5-acetyl-2-methylpyridine, as the most abundant volatile compound produced by *Bacillus* sp. BO53. Among the main VOCs biosynthesized by bacteria are ketones and alcohols [25]. It is evident from this research that the volatile compound 5-acetyl-2-methylpyridine leads to the growth reduction of *A. baumannii*, *S. aureus*, and the inhibition of *P. aeruginosa*. In this sense, pyridine derivatives have shown relevant bioactivity against Gram-positive and Gram-negative bacteria [26]. The antibacterial activity of this volatile compound has to be performed to corroborate its bioactivity. On the other hand, studies have reported a relevant inhibitory activity of 2-butanone upon *S. aureus*, *P. aeruginosa*, and *E. coli* [27]. Furthermore, Arambula et al. [28] have reported an important growth inhibition of *S. aureus* and *E. coli* by 2-nonanone. The alcohol volatile compound 1-undecanol, which was determined as one of the main compounds produced by *Bacillus* sp. BO53 inhibits *S. aureus* by damaging the bacterial cell membrane [29]. This reported bioactivity also might contribute to the growth reduction of methicillin-resistant *S. aureus* observed in the current study. The lack of complete inhibition of *S. aureus* and *A. baumannii* strains can be attributed to the low amount of VOCs produced by *Bacillus* sp. BO53, which was due to its low growth rates.

The results of this investigation have revealed an effective antibacterial activity of *Pseudoalteromonas* sp. GA327 VOCs on *A. baumannii* and *P. aeruginosa* strains. These observations suggest that alcohol, ketone, and ester volatile compounds, which were the most abundant VOCs identified from *Pseudoalteromonas* sp. GA327, generate an efficient inhibitory biochemical mechanism on the Gram-negative bacteria studied. 1-Pentanol and benzyl alcohol, which were detected in high amounts in *Pseudoalteromonas* sp. GA327, affect the bacterial cell membrane, causing fluidization or interrupting the functions of the membrane proteins. The alteration of the bacterial membrane due to volatile alcohols allows other antimicrobial compounds to easily penetrate the cell membrane [30] 2-Butanone was the main antibacterial ketone identified from *Pseudoalteromonas* sp. GA327. On the other hand, it has been reported that 2-heptanone, 6-methyl-5-heptene-2-one, and 2-undecanone ketones, produced by *Pseudoalteromonas* sp. GA327, present antibacterial properties against pathogenic Gram-positive and Gram-negative bacteria [31–35]. Regarding the inhibitory activity of the volatile ester butyl formate, one of the VOCs detected in higher amounts in *Pseudoalteromonas* sp. GA327, Calvo et al. [36] reported the presence of this molecule within the antifungal VOCs produced by the bacteria *B. velezensis*. Therefore, the results of the present study suggest the potential antibacterial bioactivity of butyl formate.

4. Materials and Methods

4.1. Bacterial Isolation from Octocorals

Isolated bacteria GA327 and BO53 were obtained from two octocorals hosts located in coastal Caribbean Sea waters of Panama: GA327 from *Pseudopterogorgia acerosa* and *Muriceopsis sulphurea*. *M. sulphurea* was collected at Punta Galeta in Colon Province (9°24′16′ N 79°51′35″ W), and *P. acerosa* from San Cristobal Island in Bocas del Toro Province (9°15′31″ N 82°16′12″ W).

For isolation of the octocoral-associated bacteria, 0.5 mL of the coral mucus was inoculated on agar plates with seawater-based nutrient medium (500 mg of mannitol, 100 mg of peptone, 8 g of Noble agar, and rifampicin [5 μg/mL] in 1 L of seawater). The octocoral-associated bacteria, GA327 and BO53, were subsequently isolated from the collection plate and successively replated until the pure isolated bacteria was obtained.

4.2. Molecular Identification of Octocoral-Associated Bacteria Species

The genetic identification of the bacterial species GA237 and BO53 was performed based on the methodology described by Atencio et al. [13]. Briefly, for DNA extraction, one milliliter of the GA237 and BO53 species were cultured on Luria Bertani (LB) broth (Difco, Michigan, MI, USA), supplemented with seawater, and grown at 25 °C for 24 h. The samples were then centrifuged at 10,000 rpm for 2 min. The resulting pellet was resuspended in 500 µL of 5% Chelex-100. Each suspension was vortexed and incubated at 56 °C for 20 min, then boiled at 100 °C for 10 min, and placed on ice for 2 min. The samples were centrifuged at 13,000 rpm for 5 min. Subsequently, the supernatants containing the DNA were transferred to a new tube and stored at −20 °C.

The DNA fragment of the 16S rRNA gene was amplified by PCR using primers pairs 27F (5′-AGAGTTTGATCMTGGCTCAG-3′) and 1492R (5′-TACGGYTACCTTGTTACGACTT-3′), and sequenced using 518F (5′CCAGCAGCCGCGGTAATACG3′) and 800R (5′TACCAGGG TATCTAATCC3′) primers [37]. The obtained sequences were compared to 16S rRNA gene sequences, using the BLAST algorithm, deposited in the GenBank, keeping a maximum of 100 hits per query sequence. Moreover, 16S rRNA sequences were compared against RDP (Ribosomal Database Project) [38] and aligned against the SILVA reference database using SINA with a 98% similarity threshold [39]. The nucleotide sequence of the BO53 and GA327 species have been submitted to the GenBank database under the accession number MK291446 and KU213068, respectively.

4.3. Pathogenic Bacterial Strains

Pathogenic bacteria *A. baumannii* (ATCC 19606), *P. aeruginosa* (ATCC 10145), and *S. aureus* (ATCC 43300) were maintained on LB medium at 37 °C. Each pathogenic strain was transferred to LB broth and was grown at 37 °C overnight. These broths were used to prepare dilutions of 0.5 McFarland to use as inoculum for the antibacterial activity assays.

4.4. Antibacterial Activity of Marine bVOCs

The antibacterial activity of the VOCs of BO53 and GA327 species were determined by the double plate test method of Romoli et al. [40] with slight modifications. BO53 and GA327 species were cultured by triplicate on LB broth, supplemented with sterile seawater, and incubated at 37 °C for 24 h. A dilution of the cultured marine bacteria was made to achieve a turbidity of 0.5 McFarland, and then each dilution was inoculated on LB plates supplemented with sterile seawater (hereafter marine bacteria plate) and placed on an incubation chamber at 37 °C overnight. Afterward, the Petri dish lid was taken off and a plate with only LB medium (hereafter pathogenic bacteria plate) was placed over the marine bacteria plate. Both plates were sealed with parafilm and incubated at 37 °C for 24 h, to allow the VOCs generated by the marine bacteria to be absorbed in the pathogenic bacteria plate. Afterward, the 0.5 McFarland dilutions of each pathogenic strain were inoculated homogeneously on the pathogenic bacteria plate with a sterile cotton swab, and placed again over the marine bacteria plate, sealed with parafilm, and incubated at 37 °C for 48 h. The pathogen's growth was evaluated every 24 h. Antibacterial activities were compared to negative controls. The pathogenic bacteria growth inhibition by the marine bVOCs was judged as "+" (complete inhibition), "±" (reduced growth), and "-" (no detectable bioactivity). Each experiment was carried out in triplicate.

4.5. Marine Bacteria Volatolome Analysis

Pseudoalteromonas sp. GA327 and *Bacillus* sp. BO53 species were cultured in glass vials, by triplicate, on LB medium supplemented with sterile seawater and incubated at 37 °C for 24 h. The samples were subsequently analyzed after 48 h of incubation. Three vials containing LB medium supplemented with sterile seawater, but not inoculated, were incubated under the same conditions.

The VOCs of all samples were analyzed by headspace-solid phase microextraction-gas chromatography-mass spectrometry (HS-SPME-GC-MS) method [22,41]. A divinylbenzene-carboxen-polydimethylsiloxane (DVB/CAR/PDMS 50/30 µm) fiber (Supelco, Bellefonte, PA, USA) was exposed to the headspace of the samples for 40 min at 37 °C. The isolated VOCs were analyzed by GC-MS, on a GC 6890N coupled to a 5975C mass spectrometry detector (Agilent Technologies, Palo Alto, CA, USA). VOCs were desorbed by insertion of the SPME fiber into the GC injection port, in splitless mode, for 2 min at 250 °C. The compounds were separated on an HP-5MS capillary column (30 m length,

0.25 mm id, 0.25 μm), using He as carrier gas at 1 mL/min. The oven temperature was 50 °C for 2 min, then increased to 240 °C at 6 °C/min and held for 5 min. MS detector was operated in electron impact mode (EV = 70 eV); in scan mode from 30 to 550 *m/z*; with an ion source temperature of 250 °C.

VOCs were identified by comparing their MS spectra with Registry of Mass Spectral Data with Structures library (Wiley 7th edition, USA), and National Institute of Standards and Technology library (NIST) spectral databases, and by using authentic standards when available. Additional identification was performed by determination of the compounds Kovat's retention index (RI) by using an alkane standard solution C8-C20 (Sigma- Aldrich, Saint Louis, MO, USA). VOCs compounds identified in vials not inoculated were excluded from the data analyses. The relative quantities of the volatile compounds are expressed as percent peak areas relative to the total peak area of identified compounds from the average of the three replicates [22,42].

5. Conclusions

The antibacterial activity of octocoral-associated bacteria *Bacillus* sp. BO53 and *Pseudoalteromonas* sp. GA327 VOCs were determined for the first time. *Bacillus* sp. BO53 volatile compounds lead to complete inhibition of *P. aeruginosa* and displayed growth reduction on *A. baumannii* and methicillin-resistant *S. aureus*; while *Pseudoalteromonas* sp. GA327 VOCs exhibited a high inhibition against both Gram-negative bacteria species and were inefficient against *S. aureus* growth. HS-SPME-GC-MS methodology allowed the identification of VOCs produced by both octocoral-associated bacteria. Alcohol and ketone volatile compounds were the most abundant VOCs detected. The bacterial emission of these VOCs might explain the antibacterial activity observed. The results of this study justified future research to determine the antibacterial activity of a few of the identified VOCs to evaluate their potential bioactivity against antibiotic-resistant bacteria.

Author Contributions: M.G. and A.A.D.-A. conceived the idea; H.G. identified and collected the octocoral from the Panamanian Caribbean Sea; L.A.A. and A.G. carried out the experiments under the supervision of R.B., A.B., M.G. and A.A.D.-A. All authors approved the manuscript. All authors of the present manuscript have read and agreed to the published version of the manuscript.

Funding: This research received no external funding.

Acknowledgments: The author acknowledges K.S. Rao, Director of INDICASAT AIP, for his support. M.G. and A.A.D.-A. acknowledge the financial support given by the National Research System of Panama (SNI).

Conflicts of Interest: The authors declare that they have no conflict of interest.

References

1. Blair, J.M.A.; Webber, M.A.; Baylay, A.J.; Ogbolu, D.O.; Piddock, L.J.V. Molecular mechanisms of antibiotic resistance. *Nat. Rev. Microbiol.* **2015**, *13*, 42–51. [CrossRef]
2. Friedman, N.D.; Temkin, E.; Carmeli, Y. The negative impact of antibiotic resistance. *Clin. Microbiol. Infect.* **2016**, *22*, 416–422. [CrossRef] [PubMed]
3. Patini, R.; Mangino, G.; Martellacci, L.; Quaranta, G.; Masucci, L.; Gallenzi, P. The Effect of Different Antibiotic Regimens on Bacterial Resistance: A Systematic Review. *Antibiotics* **2020**, *9*, 22. [CrossRef] [PubMed]
4. World Health Organization. Global Priority List of Antibiotic-Resistance Bacteria to Guide Research, Discovery, and Development of New Antibiotics. 2017. Available online: https://www.who.int/medicines/publications/global-priority-list-antibiotic-resistant-bacteria/en/ (accessed on 16 May 2020).
5. Jenul, C.; Horswill, A.R. Regulation of *Staphylococcus aureus* virulence. *Microbiol. Spectr.* **2018**, *6*, 669–686. [CrossRef] [PubMed]
6. Moradali, M.F.; Ghods, S.; Rehm, B.H.A. *Pseudomonas aeruginosa* Lifestyle: A Paradigm for Adaptation, Survival, and Persistence. *Front. Cell. Infect. Microbiol.* **2017**, *7*, 39–68. [CrossRef]
7. Asif, M.; Alvi, I.A. Insight into *Acinetobacter baumannii*: Pathogenesis, global resistance, mechanisms of resistance, treatment options, and alternative modalities. *Infect. Drug Resist.* **2018**, *11*, 1249–1260. [CrossRef]
8. Singh, S.B. Confronting the challenges of discovery of novel antibacterial agents. *Bioorg. Med. Chem. Lett.* **2014**, *24*, 3683–3689. [CrossRef]

9. Katz, L.; Baltz, R.H. Natural product discovery: Past, present, and future. *J. Ind. Microbiol. Biotechnol.* **2016**, *43*, 155–176. [CrossRef]
10. Puglisi, M.P.; Sneed, J.M.; Ritson-Williams, R.; Young, R. Marine chemical ecology in benthic environments. *Nat. Prod. Rep.* **2019**, *36*, 410–429. [CrossRef]
11. Sang, V.T.; Dat, T.T.H.; Vinh, L.B.; Cuong, L.C.V.; Oanh, P.T.T.; Ha, H.; Kim, Y.H.; Anh, H.L.T.; Yang, S.Y. Coral and Coral-Associated Microorganisms: A Prolific source of Potential Bioactive Natural Products. *Mar. Drugs.* **2019**, *17*, 468. [CrossRef]
12. Hernandez-Agreda, A.; Gates, R.D.; Ainsworth, T.D. Defining the Core Microbiome in Corals' Microbial Soup. *Trends Microbiol.* **2017**, *25*, 125–140. [CrossRef] [PubMed]
13. Atencio, L.A.; Dal Grande, F.; Young, G.O.; Gavilán, R.; Guzmán, H.M.; Schmitt, I.; Mejía, L.C.; Gutiérrez, M. Antimicrobial-producing *Pseudoalteromonas* from the marine environment of Panama shows a high phylogenetic diversity and clonal structure. *J. Basic Microbiol.* **2018**, *58*, 747–769. [CrossRef] [PubMed]
14. Raimundo, I.; Silva, S.G.; Costa, R.; Keller-Costa, T. Bioactive secondary metabolites from octocoral-Associated microbes—New chances for blue growth. *Mar. Drugs* **2018**, *16*, 485. [CrossRef] [PubMed]
15. Schulz, S.; Dickschat, J.S.; Kunze, B.; Wagner-Dobler, I.; Diestel, R.; Sasse, F. Biological activity of volatiles from marine and terrestrial bacteria. *Mar. Drugs* **2010**, *8*, 2976–2987. [CrossRef] [PubMed]
16. Papaleo, M.C.; Fondi, M.; Maida, I.; Perrin, E.; Lo Giudice, A.; Michaud, L.; Mangano, S.; Bartolucci, G.; Romoli, R.; Fani, R. Sponge-associated microbial Antarctic communities exhibiting antimicrobial activity against *Burkholderia cepacia* complex bacteria. *Biotechnol. Adv.* **2012**, *30*, 272–293. [CrossRef]
17. Papaleo, M.C.; Romoli, R.; Bartolucci, G.; Maida, I.; Perrin, E.; Fondi, M.; Orlandini, V.; Mengoni, A.; Emiliani, G.; Tutino, M.L.; et al. Bioactive volatile organic compounds from Antarctic (sponges) bacteria. *New Biotechnol.* **2013**, *30*, 824–838. [CrossRef]
18. Romoli, R.; Papaleo, M.C.; De Pascale, D.; Tutino, M.L.; Michaud, L.; LoGiudice, A.; Fani, R.; Bartolucci, G. GC-MS volatolomic approach to study the antimicrobial activity of the antarctic bacterium *Pseudoalteromonas* sp. TB41. *Metabolomics* **2014**, *10*, 42–51. [CrossRef]
19. Orlandini, V.; Maida, I.; Fondi, M.; Perrin, E.; Papaleo, M.C.; Bosi, E.; de Pascale, D.; Tutino, M.L.; Michaud, L.; Lo Giudice, A.; et al. Genomic analysis of three sponge-associated *Arthrobacter* Antarctic speciess, inhibiting the growth of *Burkholderia cepacia* complex bacteria by synthesizing volatile organic compounds. *Microbiol. Res.* **2014**, *169*, 593–601. [CrossRef]
20. Sannino, F.; Parrilli, E.; Apuzzo, G.A.; de Pascale, D.; Tedesco, P.; Maida, I.; Perrin, E.; Fondi, M.; Fani, R.; Marino, G.; et al. *Pseudoalteromonas haloplanktis* produces methylamine, a volatile compound active against *Burkholderia cepacia* complex strains. *New Biotechnol.* **2017**, *35*, 13–18. [CrossRef]
21. Risticevic, S.; Lord, H.; Górecki, T.; Arthur, C.L.; Pawliszyn, J. Protocol for solid-phase microextraction method development. *Nat. Protoc.* **2010**, *5*, 122–139. [CrossRef]
22. Durant, A.A.; Rodríguez, C.; Herrera, L.; Almanza, A.; Santana, A.I.; Spadafora, C.; Gupta, M.P. Anti-malarial activity and HS-SPME-GC-MS chemical profiling of *Plinia cerrocampanensis* leaf essential oil. *Malar. J.* **2014**, *13*, 18. [CrossRef] [PubMed]
23. Laukaleja, I.; Kruma, Z. Evaluation of a headspace solid-phase microextraction with different fibres for volatile compoun d determination in specialtycoffee brews. *Res. Rural Dev.* **2019**, *1*, 215–221. [CrossRef]
24. Schmidt, R.; Cordovez, V.; De Boer, W.; Raaijmakers, J.; Garbeva, P. Volatile affairs in microbial interactions. *ISME J.* **2015**, *9*, 2329–2335. [CrossRef] [PubMed]
25. Schulz, S.; Dickschat, J.S. Bacterial volatiles: The smell of small organisms. *Nat. Prod. Rep.* **2007**, *24*, 814–842. [CrossRef] [PubMed]
26. Altaf, A.A.; Shahzad, A.; Gul, Z.; Rasool, N.; Badshah, A.; Lal, B.; Khan, E. A Review on the Medicinal Importance of Pyridine Derivatives. *J. Drug Des. Med. Chem.* **2015**, *1*, 1–11. [CrossRef]
27. Mitchell, A.M.; Strobel, G.A.; Moore, E.; Robison, R.; Sears, J. Volatile antimicrobials from Muscodor crispans, a novel endophytic fungus. *Microbiology* **2010**, *156*, 270–277. [CrossRef]
28. Arámbula, C.I.; Diaz, C.E.; Garcia, M.I. Performance, chemical composition and antibacterial activity of the essential oil of *Ruta chalepensis* and *Origanum vulgare*. *J. Phys. Conf. Ser.* **2019**, *1386*. [CrossRef]
29. Togashi, N.; Shiraishi, A.; Nishizaka, M.; Matsuoka, K.; Endo, K.; Hamashima, H.; Inoue, Y. Antibacterial activity of long-chain fatty alcohols against *Staphylococcus aureus*. *Molecules* **2007**, *12*, 139–148. [CrossRef]

30. Yano, T.; Miyahara, Y.; Morii, N.; Okano, T.; Kubota, H. Pentanol and benzyl alcohol attack bacterial surface structures differently. *Appl. Environ. Microbiol.* **2016**, *82*, 402–408. [CrossRef]
31. Hanbali, E.L.F.; Mellouki, F.; Akssira, M.; Boira, H.; Blázquez, M.A. Composition and antimicrobial activity of essential oil of anthemis tenuisecta ball. *J. Essent. Oil Bear. Plants* **2007**, *10*, 499–503. [CrossRef]
32. Arrebola, E.; Sivakumar, D.; Korsten, L. Effect of volatile compounds produced by *Bacillus* strains on postharvest decay in citrus. *Biol. Control* **2010**, *53*, 122–128. [CrossRef]
33. Popova, A.A.; Koksharova, O.A.; Lipasova, V.A.; Zaitseva, J.V.; Katkova-Zhukotskaya, O.A.; Eremina, S.I.; Mironov, A.S.; Chernin, L.S.; Khmel, I.A. Inhibitory and toxic effects of volatiles emitted by strains of *Pseudomonas* and *Serratia* on growth and survival of selected microorganisms, *Caenorhabditis elegans*, and *Drosophila melanogaster*. *Biomed. Res. Int.* **2014**, *2014*. [CrossRef] [PubMed]
34. Ould Bellahcen, T.; Cherki, M.; Sánchez, J.A.C.; Cherif, A.; EL Amrani, A. Chemical Composition and Antibacterial Activity of the Essential Oil of *Spirulina platensis* from Morocco. *J. Essent. Oil Bear. Plants* **2019**, *22*, 1265–1276. [CrossRef]
35. Benali, T.; Habbadi, K.; Khabbach, A.; Marmouzi, I. GC–MS Analysis, Antioxidant and Antimicrobial Activities of *Achillea Odorata* Subsp. *Pectinata* and *Ruta Montana* Essential Oils and Their Potential Use as Food Preservatives. *Foods* **2020**, *9*, 668. [CrossRef]
36. Calvo, H.; Mendiara, I.; Arias, E.; Gracia, A.P.; Blanco, D.; Venturini, M.E. Antifungal activity of the volatile organic compounds produced by *Bacillus velezensis* strains against postharvest fungal pathogens. *Postharvest Biol. Technol.* **2020**, *166*, 111208. [CrossRef]
37. Frank, J.A.; Reich, C.I.; Sharma, S.; Weisbaum, J.S.; Wilson, B.A.; Olsen, G.J. Critical evaluation of two primers commonly used for amplification of bacterial 16S rRNA genes. *Appl. Environ. Microbiol.* **2008**, *74*, 2461–2470. [CrossRef] [PubMed]
38. Wang, Q.; Garrity, G.M.; Tiedje, J.M.; Cole, J.R. Naïve Bayesian classifier for rapid assignment of rRNA sequences into the new bacterial taxonomy. *Appl. Environ. Microbiol.* **2007**, *73*, 5261–5267. [CrossRef]
39. Pruesse, E.; Peplies, J.; Glöckner, F.O. SINA: Accurate high-throughput multiple sequence alignment of ribosomal RNA genes. *Bioinformatics* **2012**, *28*, 1823–1829. [CrossRef]
40. Romoli, R.; Papaleo, M.C.; De Pascale, D.; Tutino, M.L.; Michaud, L.; Logiudice, A.; Fani, R.; Bartolucci, G. Characterization of the volatile profile of Antarctic bacteria by using solid-phase microextraction-gas chromatography-mass spectrometry. *J. Mass Spectrom.* **2011**, *46*, 1051–1059. [CrossRef]
41. Correa, R.; Coronado, L.M.; Garrido, A.C.; Durant-Archibold, A.A.; Spadafora, C. Volatile organic compounds associated with *Plasmodium falciparum* infection in vitro. *Parasites Vectors* **2017**, *10*, 215. [CrossRef]
42. Garrido, A.; Ledezma, J.G.; Durant-Archibold, A.A.; Allen, N.S.; Villarreal A, J.C.; Gupta, M.P. Chemical Profiling of Volatile Components of the Gametophyte and Sporophyte Stages of the Hornwort *Leiosporoceros dussii* (Leiosporocerotaceae) From Panama by HS-SPME-GC-MS. *Nat. Prod. Commun.* **2019**, *14*. [CrossRef]

antibiotics

Article

Effect of Adding the Antimicrobial L-Carnitine to Growing Rabbits' Drinking Water on Growth Efficiency, Hematological, Biochemical, and Carcass Aspects

Mohamed I. Hassan [1], Naela Abdel-Monem [2], Ayman Moawed Khalifah [1], Saber S. Hassan [3], Hossam Shahba [4], Ahmad R. Alhimaidi [5], In Ho Kim [6,7,*] and Hossam M. El-Tahan [4,6,7,*]

[1] Livestock Research Department, Arid Lands Cultivation Research Institute, City of Scientific Research and Technological Applications (SRTA-City), New Borg El-Arab 21934, Egypt
[2] Poultry Production Department, Faculty of Agriculture, Alexandria University, Alexandria 21625, Egypt
[3] Animal and Poultry Production Department, Faculty of Agriculture, Damanhour University, Damanhour 22511, Egypt
[4] Animal Production Research Institute (APRI), Agricultural Research Center (ARC), Ministry of Agriculture, Giza 12611, Egypt
[5] Department of Zoology, College of Science, King Saud University, Riyadh 11451, Saudi Arabia
[6] Department of Animal Biotechnology, Dankook University, Cheonan 31116, Republic of Korea
[7] Smart Animal Bio Institute, Dankook University, Cheonan 31116, Republic of Korea
* Correspondence: inhokim@dankook.ac.kr (I.H.K.); hossam.eltahan@dankook.ac.kr (H.M.E.-T.)

Abstract: The current study was designed to assess the impact of L-carnitine (LC) supplementation in the drinking water of growing Alexandria-line rabbits on performance and physiological parameters. Two hundred eighty-eight 35-day-old rabbits were divided into four groups of twenty-four replicates each (seventy-two rabbits/treatment). The treatment groups were a control group without LC and three groups receiving 0.5, 1, and 1.5 g/L LC in the drinking water intermittently. The results showed that the group receiving 0.5 g LC/L exhibited significant improvements in final body weight, body weight gain, feed conversion ratio, and performance index compared to the other groups. The feed intake remained unaffected except for the 1.5 g LC/L group, which had significantly decreased intake. Hematological parameters improved in all supplemented groups. Compared with those in the control group, the 0.5 g LC/L group showed significant increases in serum total protein and high-density lipoprotein, along with decreased cholesterol and low-density lipoprotein. Compared to other supplemented groups, this group also demonstrated superior carcass traits (carcass, dressing, giblets, and percentage of nonedible parts). In conclusion, intermittent supplementation of LC in the drinking water, particularly at 0.5 g/L twice a week, positively influenced the productivity, hematology, serum lipid profile, and carcass traits of Alexandria-line growing rabbits at 84 days of age.

Keywords: carnitine; Alexandria-line rabbits; productive performance; hematology; lipid profile; carcass

Citation: Hassan, M.I.; Abdel-Monem, N.; Khalifah, A.M.; Hassan, S.S.; Shahba, H.; Alhimaidi, A.R.; Kim, I.H.; El-Tahan, H.M. Effect of Adding the Antimicrobial L-Carnitine to Growing Rabbits' Drinking Water on Growth Efficiency, Hematological, Biochemical, and Carcass Aspects. *Antibiotics* **2024**, *13*, 757. https://doi.org/10.3390/antibiotics13080757

Academic Editors: Beatriz Vázquez Belda and Carlos M. Franco

Received: 11 July 2024
Revised: 5 August 2024
Accepted: 8 August 2024
Published: 11 August 2024

1. Introduction

Rabbits are an exciting field that is vital for providing humans with animal protein, especially in developing countries. Scientific institutions and the general public have recently paid close attention to the quantitative and qualitative features of rabbit meat [1,2]. The most critical period of a rabbit's life is after weaning, when rabbits turn from feeding milk to a solid diet [3]; this leads to digestive disorders, poor growth rates, poor feed efficiency, and increased mortality rates [4,5], and antibiotics are used to solve these problems. Because of the issues of human health caused by antibiotics, European Union countries have banned their use [6]. Scientists have been researching natural substitutes for antibiotics that may promote rabbit growth and ensure that rabbits are kept healthy, such as prebiotics, probiotics, phytobiotics, essential oils, and herbs [7]. Some nutritional additives, such as LC, have beneficial effects on health status and are required for growth [8–10].

LC is water soluble and plays a vital role in the metabolism of long-chain fatty acids by transferring through the inner mitochondrial membrane for β-oxidation [11,12]. It also affects carbohydrate metabolism, which regulates pyruvate dehydrogenase in the glycolysis pathway and Krebs cycle to produce energy from glucose [11,12]. Moreover, scientists widely recognize amphiphilic chemicals as effective antimicrobial substances. According to Calvani et al. [13] quaternary ammonium LC esters with long alkyl chains have a light antimicrobial effect against a wide range of Gram-positive and Gram-negative bacteria, as well as fungi. This is because LC has a polar structural motif that is common in these compounds. Antimicrobial activity was also demonstrated in synthetic acylcarnitine analogs [1,14]. Additionally, [15,16] investigated the antimicrobial effects of LC chloride using *Caenorhabditis elegans* and *Escherichia coli* strain OP50. They discovered that 3 mg/mL of LC inhibited bacterial growth. Further testing revealed that LC chloride displayed antimicrobial activity against various yeasts and bacteria, with minimal inhibitory concentrations between 2 and 8 mg/mL and minimal bactericidal concentrations ranging from 4 to over 16 mg/mL. The minimal fungicidal concentrations were between 1 and 4 mg/mL. At these concentrations, it may be feasible to use it as a secure local antimicrobial agent. The LC chloride exhibits antimicrobial activity by disrupting the cell membranes of both yeasts and bacteria, leading to increased permeability and cell lysis. It interferes with essential metabolic pathways, inhibiting enzyme production necessary for microbial growth, and reduces biofilm formation, enhancing susceptibility to other antimicrobial agents. Studies have shown its efficacy against pathogens like *E. coli*, *Staphylococcus aureus*, *Candida albicans*, and others [1]. Furthermore, LC has many biological effects, such as improving immunity and productive performance in broilers, as reported by Asadi et al. [15]; Liu et al. [16], and in fish, as shown by Wang et al. [17]. Additionally, it plays a crucial role in enhancing carcass characteristics and modulating the blood lipid profile in broiler chickens [10]. The LC plays a vital role in enhancing the balance of the microbial population in lambs [18].

After the weaning stage, rabbits do not develop LC because LC synthesis, mainly in the liver, is dependent on the amino acids (lysine and methionine), as well as cofactors of minerals and vitamins such as iron, ascorbic acid, pyridoxine, and niacin. Furthermore, four enzymes are required to complete the synthesis of LC N-tri methyl lysine dioxygenase: 3-hydroxy-N-trimethyl lysine aldolase, 4-N-trimethyl amino butyraldehyde dehydrogenase, and γ-butyrobetaine dioxygenase [19]. The vitality of LC during the weaning phase stems from the challenges associated with its formation and limited availability in grains and legume diets, which contain only 5.9 to 6.8 mg/kg dry matter of LC [19].

The current study aimed to determine the effect of adding different levels of LC twice a week (intermittently) to the drinking water of growing rabbits on productive performance, hematology, serum protein, lipid profiles, and carcass traits in response to weaning stress.

2. Results

Table 1 shows that the growth performance of growing Alexandria rabbits was influenced by water supplemented with LC. Compared with those in all the experimental groups, the final body weight (FBW) and BWG in the drinking water of the rabbits that received 0.5 g LC/L were significantly greater. However, the FBW and BWG values in the group that received 1.5 g LC/L were lower than those in the control group, although the differences were not statistically significant. The feed intake significantly decreased in rabbits that received 1.5 g LC/L in drinking water compared with those in the remaining experimental groups. Moreover, the rest of the groups exhibited no significant change in feed consumption. The FCR changed dramatically in all the experimental groups, as the group that received 0.5 g LC/L had the greatest change, followed by the group that received 1.5 g LC/L and then 1 g LC/L compared to the control group. A significant improvement in production index (PI) was found in all treated groups, especially those that received 0.5 g LC/L, followed by groups that received 1 and 1.5 g LC/L, compared to the control.

Table 1. Productive performance as influenced by L-carnitine supplementation in the drinking water of growing Alexandria-line rabbits at 84 days of age.

Item	Control	LC g/L			SEM	*p* Value
		0.5	1.0	1.5		
IBW, g	731.0	755.4	758.7	724.7	41.00	0.323
FBW, g	2095 ^{c}	2243 ^{a}	2165 ^{b}	2063 ^{c}	39.81	0.0001
BWG, g	1364 ^{c}	1488 ^{a}	1406 ^{b}	1338.3 ^{c}	30.69	0.0001
FI, g/rabbit	4065 ^{a}	4026 ^{a}	4064 ^{a}	3750 ^{b}	73.60	0.0001
FCR	2.98 ^{a}	2.71 ^{d}	2.89 ^{b}	2.80 ^{c}	0.061	0.0001
PI (%)	70.30 ^{c}	82.91 ^{a}	74.90 ^{b}	73.64 ^{b}	2.58	0.0001

Means with different superscripts in the same row are significantly different ($p < 0.05$); SEM: standard error of means; IBW: initial body weight; FBW: final body weight; BWG: body weight gain; FI: feed intake; FCR: feed conversion ratio; PI: performance index.

Table 2 shows that the hematological parameters of growing Alexandria rabbits were influenced by water supplemented with LC intermittently. The results indicated that PCV, RBC, and hemoglobin (Hb) levels were significantly improved in the groups that received varying levels of LC compared to the control group. Compared with those in the control group, MCH and MCV were significantly lower in all the LC supplementation groups.

Table 2. Hematological parameters as influenced by L-carnitine supplementation in the drinking water of growing Alexandria-line rabbits at 84 days of age.

Item	Control	LC g/L			SEM	*p* Value
		0.5	1.0	1.5		
PCV (%)	41.18 ^{b}	44.35 ^{a}	43.30 ^{a}	44.16 ^{a}	1.008	0.0001
RBCs (10^6/μL)	4.63 ^{b}	6.15 ^{a}	6.47 ^{a}	6.04 ^{a}	0.680	0.0002
Hb (g/dL)	9.03 ^{b}	9.88 ^{a}	9.94 ^{a}	9.65 ^{a}	0.554	0.0203
MCHC (g/dL)	21.93	22.26	22.94	21.82	0.831	0.079
MCH (pg)	19.52 ^{a}	16.20 ^{b}	15.65 ^{b}	16.06 ^{b}	1.411	0.0001
MCV (fL)	89.01 ^{a}	73.14 ^{b}	68.06 ^{b}	73.85 ^{b}	6.990	0.0001

Means with different superscripts in the same row are significantly different ($p < 0.05$); SEM: standard error of the mean; PCV: picked cell volume; RBCs: red blood cells; Hb: hemoglobin; MCHC: mean corpuscular Hb concentration; MCH: mean corpuscular Hb; MCV: mean corpuscular volume.

Table 3 shows the serum protein and lipid profiles of growing Alexandria rabbits treated intermittently with water supplemented with LC. The findings revealed that the group receiving 0.5 g LC/L had significantly greater total protein (TP) than the other groups. Additionally, the results revealed that TP did not significantly change in the rest of the treated groups compared with the control group. The serum ALB concentration significantly increased in the group that received 0.5 g LC/L compared to the other groups. The serum globulin and β-globulin levels were significantly greater in the group receiving 0.5 g LC/L than in the group receiving 1 g LC/L. Moreover, neither group significantly differed from the other groups. The study found no significant changes in α-globulin or γ-globulin concentrations, total lipids, triglycerides, or VLDL levels among the experimental groups. Compared with that in the control group, the serum cholesterol in the treated groups significantly decreased, especially in the group that received 1.5 g LC/L, followed by 0.5 g and 1 g LC/L. Compared with those in the control group, the LDL in all treated groups significantly decreased, especially in the groups that received 0.5 g and 1.5 g of LC, followed by the group that received 1 g of LC/L. Furthermore, HDL significantly increased only in the group that received 0.5 g LC/L compared with the other groups.

Table 3. Serum protein and lipid profile as influenced by L-carnitine supplementation in the drinking water of growing Alexandria line rabbits at 84 days of age.

Item	Control	LC g/L			SEM	*p* Value
		0.5	1.0	1.5		
			Protein profile			
Total protein (g/dL)	5.861 [b]	6.471 [a]	5.896 [b]	5.926 [b]	0.220	0.0001
Albumin (g/dL)	2.292 [b]	2.725 [a]	2.508 [ab]	2.368 [b]	0.206	0.0035
Globulin (g/dL)	3.569 [ab]	3.747 [a]	3.387 [b]	3.557 [ab]	0.258	0.106
α-Globulin (g/dL)	1.707	1.768	1.675	1.711	0.099	0.388
β-Globulin (g/dL)	1.111 [ab]	1.178 [a]	1.045 [b]	1.108 [ab]	0.078	0.0346
ɣ-Globulin (g/dL)	0.750	0.800	0.667	0.738	0.128	0.303
			Lipid profile			
Total lipid (g/dL)	360.3	357.8	361.3	361.1	5.787	0.6487
Triglycerides (mg/dL)	81.89	77.95	77.98	76.87	6.797	0.5406
Cho (mg/dL)	120.3 [a]	112.5 [b]	112.4 [b]	108.3 [c]	2.153	0.0001
LDL (mg/dL)	55.06 [a]	45.21 [c]	47.50 [b]	44.97 [c]	1.817	0.0001
HDL (mg/dL)	48.86 [bc]	51.72 [a]	49.34 [b]	47.97 [c]	0.898	0.0001
VLDL	16.38	15.59	15.59	15.38	1.360	0.5408

Means with different superscripts in the same row are significantly different ($p < 0.05$); SEM: standard error of means; Cho: total cholesterol; LDL: low-density lipoprotein; HDL: high-density lipoprotein; VLDL: very low-density lipoprotein.

Table 4 presents the effects of intermittent LC supplementation in drinking water on the carcass characteristics of growing Alexandria rabbits. The group receiving 0.5 g LC/L exhibited a highly significant increase in both carcass yield and dressing percentage compared to the control group. This group also demonstrated the highest giblets percentage and a corresponding decrease in non-edible parts. In comparison to the control, all LC-treated groups showed significantly lower percentages of non-edible parts, although this did not affect the giblets percentage. Notably, the abdominal fat percentage decreased significantly with increasing LC dosage. Furthermore, the percentage of pancreatic tissue was significantly higher in the 0.5 g LC/L group relative to other treatment groups. Similarly, the liver percentage was significantly greater in this group compared to the others. Conversely, the kidney tissue percentage was significantly lower in all treated groups compared to the control group. No significant differences were observed in the percentages of heart, lungs, spleen, or abdominal fat among the different groups.

Table 4. Carcass traits as influenced by L-carnitine supplementation in the drinking water of growing Alexandria-line rabbits at 84 days of age.

Item	Control	LC g/L			SEM	*p* Value
		0.5	1.0	1.5		
Carcass traits %						
Carcass	51.16 [c]	56.57 [a]	53.40 [b]	53.20 [b]	1.620	0.0001
Dressing	56.37 [c]	63.26 [a]	58.75 [b]	58.50 [b]	1.660	0.0001
Giblets	5.21 [b]	6.69 [a]	5.35 [b]	5.31 [b]	0.707	0.0047
Nonedible part	43.63 [a]	36.74 [c]	41.25 [b]	41.50 [b]	1.660	0.0001
pancreas	0.14 [b]	0.26 [a]	0.18 [b]	0.20 [ab]	0.067	0.032
Heart	0.34	0.31	0.38	0.37	0.042	0.074
Liver	3.39 [b]	5.09 [a]	3.77 [b]	3.77 [b]	0.610	0.0006
Kidney	0.81 [a]	0.74 [ab]	0.69 [bc]	0.64 [c]	0.075	0.008
Lungs	0.67	0.53	0.51	0.53	0.178	0.389
Spleen	0.07	0.06	0.08	0.06	0.028	0.553
Abdominal fat	0.81	0.67	0.64	0.50	0.214	0.133

Means with different superscripts in the same row are significantly different ($p < 0.05$); SEM: standard error of means.

3. Discussion

3.1. Productive Performance

The results of this study demonstrated significant improvements in final body weight (FBW) and body weight gain (BWG) in rabbits treated with LC, particularly at a concentration of 0.5 g LC/L. This aligns with previous research that has highlighted the growth-promoting effects of LC in rabbits. The primary mechanism behind these effects is LC's role in enhancing lipid metabolism, which increases energy availability by facilitating the transport of fatty acids into mitochondria for oxidation [20]. The 0.5 g LC/L dose appears to provide an optimal balance, maximizing growth benefits without causing metabolic imbalances. Studies have shown that moderate doses of LC are more effective than higher doses, which may not offer additional benefits and could potentially disrupt metabolic processes [21]. This increased energy efficiency contributes to improved growth performance and feed conversion ratios in rabbits. Similarly, Ayyat et al. [22] reported that in comparison to control rabbits, growing rabbits exposed to heat stress and receiving a 50 mg LC/kg diet exhibited significantly improved FBW and BWG. However, other authors found no effect of LC on BW or BWG [23] and reported that LC (160 mg/kg diet) did not affect FBW in broilers. Arslan et al. [24] found that when duck chicks received 200 mg of LC in drinking water, it had no significant effect on BW or BWG.

Similarly Murali et al. [25] found no significant differences in BW or BWG when LC at 900 mg/kg was incorporated into broiler diets. This finding supported our results showing a nonsignificant effect on FI and improvements in FCR and performance index in the group that received 0.5 g LC/L. Arslan et al. [26] reported that feed consumption did not significantly change in ducks that received 200 mg of LC. The FCR was significantly improved in broilers that received 200 and 300 mg LC/kg diets during the fattening period [25]. Moreover, Abdel-Fattah et al. [27] found that FCR improved in quails fed 200 and 400 mg LC/kg diets. Moreover, ref. [22] reported that heat stress in growing rabbits receiving a 50 mg LC/kg diet did not significantly influence feed consumption, while FCR was significantly enhanced in these rabbits than in control rabbits.

On the other hand, Xu et al. [28] observed that FCR was not affected by different levels of LC from 25 to 100 mg LC/kg diet in broilers. Additionally, some studies on laying hens from hatching to four weeks of age reported by Deng et al. [29] stated that different levels of LC (100 mg or 1 g/kg diet) had no significant influence on the FCR. In another study on laying hens at 22 weeks of age, Yalçin et al. [30] reported no significant variation in FCR in the group receiving 100 mg LC/kg diet. Awad et al. [31] revealed that the PI improved in groups supplemented with 300 and 400 mg LC/kg diet. According to previous results and the literature, it is clear that the improvement in PI is related to improvements in BW, BWG, and FCR without increasing feed consumption. Another explanation is that adding LC reduces the amount of methionine and lysine, which are precursors for protein production [12]. The positive effect of LC on productive performance can be explained by the ability of LC to utilize energy either through enhanced fatty acid β-oxidation or energy from glucose by activating pyruvate dehydrogenase through the glycolysis pathway and the tricarboxylic acid cycle [11,12].

3.2. Hematological Parameters

The results of this study demonstrate that intermittent supplementation of LC in drinking water positively influences the hematological parameters of growing Alexandria rabbits. Specifically, we observed significant improvements in packed cell volume (PCV), red blood cell count (RBC), and hemoglobin (Hb) levels in the LC-treated groups compared to the control. These enhancements suggest that LC supplementation may contribute to better overall blood health and increased oxygen-carrying capacity, which is crucial for optimal growth and performance in rabbits [32]. The LC plays a key role in energy metabolism by facilitating the transport of fatty acids into mitochondria for oxidation, which may enhance erythropoiesis and improve red blood cell production [33]. Additionally, the antioxidant properties of LC can protect red blood cells from oxidative damage, thus supporting improved hematological parameters [34].

However, we also observed that mean corpuscular hemoglobin (MCH) and mean corpuscular volume (MCV) were significantly lower in all LC supplementation groups compared to the control. This reduction could indicate more efficient erythrocyte function rather than a decrease in red blood cell quantity or quality. Lower MCH and MCV values may reflect a higher concentration of hemoglobin per cell and a more compact cell volume, which could be attributed to LC's effects on cellular metabolism and efficiency [34]. These findings suggest that intermittent LC supplementation can positively influence key hematological parameters, contributing to improved blood health and potentially enhancing growth performance in rabbits. The decrease in MCH and MCV warrants further investigation to understand its implications fully. Future studies should explore the long-term effects of LC supplementation on hematological profiles and overall health in rabbits.

3.3. Serum Protein and Lipid Profile

The results of the analysis of the serum TP concentration in the studied groups showed that the levels in rabbits that received 1 or 1.5 g LC/L were similar to those reported by Yalçin et al. [30], who indicated that 100 mg/kg LC in laying hens had no significant effect on the serum TP concentration. Moreover, Wang et al. [35] stated that broilers had no significant impact on serum TP, albumin (ALB), or globulin. According to Hamad et al. [36], administering 40 mg of LC orally to developing rabbits did not influence the serum TP, ALB, globulin, or the albumin-to-globulin ratio. Furthermore, Ayyat et al. [22] revealed that the serum TP concentration significantly increased while the ALB concentration did not change in New Zealand White rabbits receiving 50 mg of LC compared to those in the control group. Consistent with our findings on the TP value in the group that received 0.5 g LC/L, Rehman et al. [37] reported that broiler chicks that received a 0.5 g LC/kg diet had significantly increased serum protein levels.

The lipid profile results disagree with the results obtained by [11,12], who illustrated that LC supplementation decreased serum total lipids and triglycerides but were similar to our results in reducing cholesterol levels. In contrast, Eder [38] showed that rats fed a hyperlipidemic diet supplemented with 0.5 g LC/kg diet had considerably higher total cholesterol. Similar to our results regarding total lipids and triglycerides, as mentioned by Arslan et al.'s study [26] of ducks and Yalçin et al.'s study [30] of laying hens, they found no significant influence on serum total lipids, triglycerides, or VLDL. Moreover, they found no significant impact on serum total cholesterol. The decrease in plasma cholesterol is due to the link between acetyl coenzyme A, which is essential for cholesterol synthesis, and free carnitine to form acyl-carnitine [39]. Our results conflict with those of Parsaeimehr et al. [40] regarding serum total cholesterol, LDL, and HDL in broilers, as they concluded no significant influence on serum cholesterol, HDL, LDL, or triglycerides.

On the other hand, our study cholesterol, LDL, and HDL results agree with those of Rehman et al. [37], who reported that broiler chickens receiving a 0.5 g LC/kg diet had significantly decreased cholesterol and LDL and increased HDL. In harmony with our results regarding total cholesterol, Ayyat et al. [22] showed that growing rabbit cholesterol was significantly lower in rabbits fed a diet with high energy plus 50 mg LC/kg under heat stress than in those fed other diets. The decrease in LDL may be related to the ability of LC to reduce the activity of β-hydroxy-β-methylglutaryl-L-CoA reductase, which is involved in cholesterol generation [41].

3.4. Carcass Characteristics

The results of this study indicate that intermittent supplementation of LC in drinking water has a significant impact on the carcass characteristics of growing Alexandria rabbits. Notably, the group receiving 0.5 g LC/L exhibited the most pronounced improvements in carcass yield and dressing percentage compared to the control group. This finding aligns with previous studies suggesting that LC supplementation can enhance carcass traits by improving overall growth performance and feed efficiency [38]. The observed increase in giblet percentage and the corresponding decrease in non-edible parts in the 0.5 g LC/L group suggest that LC supplementation promotes more efficient utilization of nutrients and enhances the quality of edible parts. Similar improvements in meat quality and yield have been reported in other studies, where LC was found to enhance the overall carcass value in various animal models, including poultry and rabbits [42,43]. The significant decrease in abdominal fat percentage with increasing LC dosage is consistent with the role of LC in lipid metabolism. The LC facilitates the transport of fatty acids into mitochondria for oxidation, which can lead to reduced fat deposition [44]. This effect has been observed in rabbits, where LC supplementation resulted in decreased fat accumulation and improved body composition [45].

The increase in pancreatic and liver percentages in the 0.5 g LC/L group may reflect enhanced metabolic activity and nutrient processing, which are essential for optimal growth and health. The LC's role in improving liver function and pancreatic health has been documented in various studies, highlighting its benefits for metabolic efficiency and overall organ health [46]. Conversely, the reduction in kidney tissue percentage in the treated groups could be attributed to the overall improvement in metabolic efficiency, which may lead to a more balanced distribution of internal organ tissues. While no significant changes were observed in the percentages of heart, lungs, or spleen, the overall improvement in carcass traits suggests that LC supplementation contributes to more efficient growth and better utilization of body resources [38].

In conclusion, intermittent LC supplementation, particularly at 0.5 g/L, enhances carcass characteristics in Alexandria rabbits by improving carcass yield, reducing abdominal fat, and optimizing organ development. These findings support the potential of LC as a valuable supplement in rabbit production to improve meat quality and growth performance.

4. Materials and Methods

4.1. Ethical Statement

The study location was the El-Bostan Experimental Station, Damanhour University's Faculty of Agriculture, Al-Behera Governorate, Egypt. The authors attest to the European Parliament's and the Council's Directive 2010/63/EU of 22 September 2010, on the protection of animals and birds used for research. The ethical standards of animal research were incorporated into the present study, which was authorized by Damanhur University's Ethical Animal Care and Use Committee (Approval No: DUFA-2023-9).

4.2. Animals and Experimental Design

The Alexandria line rabbits used in this study were obtained from stocks kept at Alexandria University's Poultry Research Centre. This artificial paternal line is produced by crossing rabbits of the maternal V-line line with the paternal Black Baladi line [47,48].

A total of 288 mixed-sex Alexandria-line rabbits with an average mass of 742 ± 30.9 g at 35 days of age were randomly divided into four groups (n = 72 each). Each group was divided into 24 replicates, three rabbits each; all rabbits in all treatments provided identical standard meals prepared following NRC [49] and calculated according to AOAC [50]. The feed components and chemical proximate analysis are the same as the previous study of Hassan et al. [51]. The first group served as a control, and the subsequent groups, the second, third, and fourth groups, were supplemented with 0.5, 1.0, and 1.5 g/L LC, respectively. LC was purchased from Sigma Aldrich, Inc., St. Louis, MO, USA. LC was administered twice weekly (Monday and Thursday) via the drinking water of the growing rabbits, with the dosage adjusted to ensure proportionality to the daily water consumption specific to each treatment group. On the following day, any residual water was discarded, and the rabbits were provided with fresh water.

The rabbits were housed as three rabbits per cage in galvanized wire single cages within open-system pens ($35 \times 40 \times 50$ cm) in width, height, and length. (Italian battery). Hand feeding and an automated nipple drinker system offered constant access to fresh, clean water for every cage. The rabbits were housed under identical environmental and sanitary conditions throughout the trial.

4.3. Data Collection

4.3.1. Growth Performance

The initial and final body weights of each rabbit were recorded at 35 and 84 days of age, respectively, and were used to determine body weight gain (BWG). The BWG was calculated as the final and initial body weight difference. FI was determined by providing the rabbits with a known feed weight and subtracting the residual feed from the feed supplied after feeding. The FCR was calculated using the ratio of FI to BWG during the interval. The performance index (PI, %) was computed using the following formula: PI (%) = final live body weight (kg)/feed conversion ratio × 100.

4.3.2. Blood Hematology, Serum Protein and Lipid Profile

At 84 days of age (at the end of the experiment), twenty-four blood samples were obtained from the marginal ear veins of the rabbits of each treatment at 8 AM before the customary feeding time. Blood was drawn into clean tubes with or without heparin. Hematological parameters were measured using fresh blood samples treated with heparin. Red blood cell (RBC) and packed cell volume (PCV) counts were determined as described by Ewuola and Egbunike [52]. The Hb concentration was determined according to Drew et al. [53]. The blood constant mean corpuscular Hb concentration (MCHC), mean corpuscular Hb (MCH), and mean corpuscular volume (MCV) were calculated using the appropriate formulae [54]. Serum TP [55], serum ALB [56], and the difference between TP and total ALB were used to determine serum globulin levels [57]. By using a commercial diagnostic kit and automated electrophoresis equipment, globulins were separated on an agarose gel by zone electrophoresis (23 EL-Montazah St. Heliopolis, Cairo, Egypt, http://www.diamonddiagnostics.com accessed on 11 July 2024) according to the procedure described by the manufacturer. The total lipids were determined using the phospho-vanillin method as described by Woodman and Price [58]. The serum triglyceride concentration was measured using specific kits supplied by CAL-TECH Diagnostics, INC, Chino, California, USA, following the glycerol phosphate oxidase (GPO) method according to Frings et al. [59]. Total cholesterol levels were quantified using the Liebermann–Burchard method as described by Burstein et al. [60]. Low-density lipoprotein (LDL) and high-density lipoprotein (HDL) cholesterol concentrations were measured using the precipitation method by Wieland and Seidel [61], and the precipitation method by Bogin and Keller [62], respectively; likewise, very low-density lipoprotein (VLDL) was calculated as one-fifth of triglycerides [63].

4.3.3. Carcasses Trial

Twelve rabbits from each group were randomly selected at the end of the trial (84 days of age). They were fasted for 12 h to ensure empty gastrointestinal tracts, which is crucial for accurate carcass weight measurement. The rabbits were then individually weighed before euthanasia. Euthanasia was performed humanely using an overdose of sodium pentobarbital administered intravenously, following AVMA guidelines for the euthanasia of animals. After euthanasia, the rabbits were immediately slaughtered, and the carcasses were carefully eviscerated. This involved making an incision along the ventral midline and removing internal organs to avoid contamination and maintain sample integrity. Carcasses were weighed post-evisceration, including the head, and expressed as a percentage of the pre-slaughter body weight to determine the carcass yield.

The following tissues were dissected and weighed separately: abdominal fat, pancreas, heart, liver, kidney, lungs, and spleen. Each tissue was carefully separated using sterilized surgical instruments to prevent cross-contamination. The weights of these organs were expressed as a percentage of the live body weight. The dressing percentage was calculated by including the weight of the carcass and the giblets, which comprised the heart, liver, kidneys, and lungs. The giblets were accurately weighed and recorded, ensuring a consistent methodology across all samples. Non-edible parts were calculated as (100-dressing percentage).

4.4. Statistical Analysis

Using the general linear model (GLM) procedure of the Statistical Package for Social Sciences (SPSS®, version 20) [64], the data were statistically analyzed. The following formula was used for the one-way analysis of variance:

$$Yij = \mu + Ti + eij \quad (1)$$

where *Yij* is the observation of the statistical measurements, μ is the general overall mean, *Ti* is the treatment effect, and *eij* is the experimental random error. The significant differences among treatments were examined according to Duncan [65].

5. Conclusions

The intermittent administration of L-carnitine (LC) at a concentration of 0.5 g/L in the drinking water of growing rabbits twice weekly significantly enhanced their productivity, hematological parameters, serum lipid profiles, and carcass characteristics. This strategic supplementation method demonstrates the potential of LC to optimize growth performance and health outcomes in rabbit production systems.

Author Contributions: M.I.H., N.A.-M., A.M.K., S.S.H. and H.S.: conceptualization, methodology, software, formal analysis, investigation, resources, and writing—original draft; A.R.A., I.H.K. and H.M.E.-T.: conceptualization, validation, data curation, writing—review and editing, visualization, supervision, and project administration. All authors have read and agreed to the published version of the manuscript.

Funding: This research received no external funding.

Institutional Review Board Statement: This study followed the instructions of Damanhur University's Ethical Animal Care and Use Committee (Approval No: DUFA-2023-9).

Informed Consent Statement: Not applicable.

Data Availability Statement: All data generated or analyzed during this study are included in this published article.

Acknowledgments: The authors sincerely acknowledge the Researcher Support Project (RSP-2024/R232) at King Saud University, Riyadh, Saudi Arabia, for funding this work.

Conflicts of Interest: The authors declare no conflicts of interest related to this manuscript.

References

1. Olgun, A.; Kisa, O.; Yildiran, S.T.; Tezcan, S.; Akman, S.; Erbil, M.K. Antimicrobial efficacy of L-carnitine. *Ann. Microbiol.* **2004**, *54*, 95–101.
2. El-Hanoun, A.; El-Komy, A.; El-Sabrout, K.; Abdella, M. Effect of bee venom on reproductive performance and immune response of male rabbits. *Physiol. Behav.* **2020**, *223*, 112987. [CrossRef] [PubMed]
3. Krieg, R.; Vahjen, W.; Awad, W.; Sysel, M.; Kroeger, S.; Zocher, E.; Hulan, H.W.; Arndt, G.; Zent, J. Performance, digestive disorders and the intestinal microbiota in weaning rabbits are affected by a herbal feed additive. *World Rabbit Sci.* **2010**, *17*, 87–95. [CrossRef]
4. Gidenne, T. Dietary fibres in the nutrition of the growing rabbit and recommendations to preserve digestive health: A review. *Animal* **2015**, *9*, 227–242. [CrossRef] [PubMed]
5. Attia, Y.A.; Hamed, R.S.; Abd El-Hamid, A.E.; Al-Harthi, M.A.; Shahba, H.A.; Bovera, F. Performance, blood profile, carcass and meat traits and tissue morphology in growing rabbits fed mannanoligosaccharides and zinc-bacitracin continuously or intermittently. *Anim. Sci. Pap. Rep.* **2015**, *33*, 85–101.
6. Pradella, G.; Anadon, A.; Klose, V.; Plail, R.; Mohni, M.; Schatzmayr, G.; Spring, P.; Montesissa, C.; Calini, F. Workshop III: 2006 EU Ban on Antibiotics as Feed Additives—Consequences and Perspectives. *J. Vet. Pharmacol. Ther.* **2006**, *29*, 41–46. [CrossRef]
7. El-Sabrout, K.; Khalifah, A.; Mishra, B. Application of botanical products as nutraceutical feed additives for improving poultry health and production. *Vet. World* **2023**, *16*, 369–379. [CrossRef] [PubMed]

8. Attia, Y.A.; El-Hanoun, A.M.; Bovera, F.; Monastra, G.; El-Tahawy, W.S.; Habiba, H.I. Growth performance, carcass quality, biochemical and haematological traits and immune response of growing rabbits as affected by different growth promoters. *J. Anim. Physiol. Anim. Nutr.* **2014**, *98*, 128–139. [CrossRef] [PubMed]
9. Ismail, E.; Ouda, M. Impact of Dietary L-Carnitine as a Feed Additive on Performance, Carcass Characteristics and Blood Biochemical Measurements of Broiler. *J. Anim. Poult. Prod.* **2020**, *11*, 21–25. [CrossRef]
10. Hassan, S.; Abou-Shehema, B.; Shahba, H.; Hassan, M.; Boriy, E.; Rozan, M. Impact of dietary vitamin (E) and Eruca sativa seeds powder on broiler productivity, health, carcass characteristics, and meat quality. *Anim. Biotechnol.* **2023**, *34*, 5037–5054. [CrossRef] [PubMed]
11. El-Azeem, N.A.A.; Abdo, M.S.; Madkour, M.; El-Wardany, I. Physiological and histological responses of broiler chicks to in ovo injection with folic acid or L-carnitine during embryogenesis. *Glob. Vet.* **2014**, *13*, 544–551.
12. Ringseis, R.; Keller, J.; Eder, K. Basic mechanisms of the regulation of L-carnitine status in monogastrics and efficacy of L-carnitine as a feed additive in pigs and poultry. *J. Anim. Physiol. Anim. Nutr.* **2018**, *102*, 1686–1719. [CrossRef] [PubMed]
13. Calvani, M.; Critelli, L.; Gallo, G.; Giorgi, F.; Gramiccioli, G.; Santaniello, M.; Scafetta, N.; Tinti, M.O.; De Angelis, F. L-Carnitine esters as "soft", broad-spectrum antimicrobial amphiphiles. *J. Med. Chem.* **1998**, *41*, 2227–2233. [CrossRef] [PubMed]
14. Gülçin, İ. Antioxidant and antiradical activities of l-carnitine. *Life Sci.* **2006**, *78*, 803–811. [CrossRef] [PubMed]
15. Asadi, H.; Sadeghi, A.A.; Eila, N.; Aminafshar, M. Carcass traits and immune response of broiler chickens fed dietary L-carnitine, coenzyme Q10 and ractopamine. *Rev. Bras. Cienc. Avic.* **2016**, *18*, 677–682. [CrossRef]
16. Liu, Y.; Yang, Y.; Yao, R.; Hu, Y.; Liu, P.; Lian, S.; Lv, H.; Xu, B.; Li, S. Dietary supplementary glutamine and L-carnitine enhanced the anti-cold stress of Arbor Acres broilers. *Arch. Anim. Breed.* **2021**, *64*, 231–243. [CrossRef] [PubMed]
17. Wang, S.; Yao, Q.; Wang, N.; Zhang, D.; Wang, X.; Zheng, N.; Zhang, B.; Liu, H.; Wan, J.; Chen, Y.; et al. Effects of L-carnitine supplementation on growth performance, histomorphology, antioxidant and immune function of Rhynchocypris lagowskii fed dietary fish oil replaced with corn oil. *Aquac. Res.* **2022**, *53*, 1981–1994. [CrossRef]
18. Martín, A.; Giráldez, F.J.; Montero, O.; Andrés, S. Dietary administration of L-carnitine during the fattening period of early feed restricted lambs modifies liver transcriptomic and plasma metabolomic profiles. *Anim. Feed Sci. Technol.* **2022**, *292*, 1115426. [CrossRef]
19. Vaz, F.M.; Wanders, R.J.A. Carnitine biosynthesis in mammals. *Biochem. J.* **2002**, *361*, 417–429. [CrossRef] [PubMed]
20. Abu-Alya, I.S.; Alharbi, Y.M.; Abdel-Rahman, H.A.; Zahran, I.S. Effect of l-carnitine and/or calf thymus gland extract supplementation on immunity, antioxidant, duodenal histomorphometric, growth, and economic performance of japanese quail (*Coturnix coturnix japonica*). *Vet. Sci.* **2021**, *8*, 251. [CrossRef] [PubMed]
21. Wu, C.; Zhu, M.; Lu, Z.; Zhang, Y.; Li, L.; Li, N.; Yin, L.; Wang, H.; Song, W.; Xu, H. L-carnitine ameliorates the muscle wasting of cancer cachexia through the AKT/FOXO3a/MaFbx axis. *Nutr. Metab.* **2021**, *18*, 98. [CrossRef] [PubMed]
22. Ayyat, M.S.; Abd El-Latif, K.M.; Helal, A.A.; Al-Sagheer, A.A. Interaction of supplementary L-carnitine and dietary energy levels on feed utilization and blood constituents in New Zealand White rabbits reared under summer conditions. *Trop. Anim. Health Prod.* **2021**, *53*, 279. [CrossRef] [PubMed]
23. Sarica, S.; Corduk, M.; Kilinc, K. The effect of dietary L-carnitine supplementation on growth performance, carcass traits, and composition of edible meat in japanese quail (Coturnix coturnix japonica). *J. Appl. Poult. Res.* **2005**, *14*, 709–715. [CrossRef]
24. Arslan, C.; Tufan, T. Effects of chitosan oligosaccharides and L-carnitine individually or concurrent supplementation for diets on growth performance, carcass traits and serum composition of broiler chickens. *Rev. Med. Vet.* **2018**, *169*, 130–137.
25. Murali, P.; George, S.K.; Ally, K.; Dipu, M.T. Effect of L-carnitine supplementation on growth performance, nutrient utilization, and nitrogen balance of broilers fed with animal fat. *Vet. World* **2015**, *8*, 482–486. [CrossRef] [PubMed]
26. Arslan, C.; Çitil, M.; Saatci, M. Effects of L-carnitine administration on growth performance, carcass traits, blood serum parameters and abdominal fatty acid composition of ducks. *Arch. Anim. Nutr.* **2003**, *57*, 381–388. [CrossRef] [PubMed]
27. Abdel-Fattah, S.A.; El-Daly, E.F.; Ali, N.G.M. Growth performance, immune response, serum metabolites and digestive enzyme activities of Japanese quail fed supplemental L-carnitine. *Glob. Vet.* **2014**, *12*, 277–286.
28. Xu, Z.R.; Wang, M.Q.; Mao, H.X.; Zhan, X.A.; Hu, C.H. Effects of L-carnitine on growth performance, carcass composition, and metabolism of lipids in male broilers. *Poult. Sci.* **2003**, *82*, 408–413. [CrossRef] [PubMed]
29. Deng, K.; Wong, C.W.; Nolan, J.V. Long-term effects of early-life dietary L-carnitine on lymphoid organs and immune responses in Leghorn-type chickens. *J. Anim. Physiol. Anim. Nutr.* **2006**, *90*, 81–86. [CrossRef] [PubMed]
30. Yalçin, S.; Ergün, A.; Özsoy, B.; Yalçin, S.; Erol, H.; Onbaşilar, I. The effects of dietary supplementation of L-carnitine and humic substances on performance, egg traits and blood parameters in laying hens. *Asian-Australas. J. Anim. Sci.* **2006**, *19*, 1478–1483. [CrossRef]
31. Awad, A.L.; Fahim, H.N.; Beshara, M.M. Effect of Dietary L-Carnitine Supplementation on Productive Performance and Carcass Quality of Local Duck Breeds in Summer Season. *Egypt. Poult. Sci. J.* **2016**, *36*, 11–27. [CrossRef]
32. Rebouche, C.J. Kinetics, pharmacokinetics, and regulation of L-Carnitine and acetyl-L-carnitine metabolism. *Ann. N. Y. Acad. Sci.* **2004**, *1033*, 30–41. [CrossRef] [PubMed]
33. Al-Dhuayan, I.S. Biomedical role of L-carnitine in several organ systems, cellular tissues, and COVID-19. *Braz. J. Biol.* **2022**, *82*, e267633. [CrossRef] [PubMed]

34. Lee, B.J.; Lin, J.S.; Lin, Y.; Lin, P.-T. Effects of L-carnitine supplementation on oxidative stress and antioxidant enzymes activities in patients with coronary artery disease: A randomized, placebo-controlled trial. *Nutr. J.* **2014**, *13*, 79. Available online: http://www.nutritionj.com/content/pdf/1475-2891-13-79.pdf (accessed on 4 August 2014). [PubMed]
35. Wang, Y.W.; Ning, D.; Peng, Y.Z.; Guo, Y.M. Effects of dietary L-carnitine supplementation on growth performance, organ weight, biochemical parameters and ascites susceptibility in broilers reared under low-temperature environment. *Asian-Australas. J. Anim. Sci.* **2013**, *26*, 233. [CrossRef] [PubMed]
36. Hamad, M.; Ashour, G.; Gabr, S.; Younan, G.; Kamel, D. Growth Performance, Liver and Kidney Function, Lipid Metabolism and Thyroid Hormones of Growing Rabbits Treated with Different Types of Metabolic Agents. *J. Anim. Poult. Prod.* **2016**, *7*, 447–456. [CrossRef]
37. Rehman, Z.; Chand, N.; Khan, R.U.; Naz, S.; Alhidary, I.A. Serum biochemical profile of two broiler strains supplemented with vitamin E, raw ginger (*Zingiber officinale*) and L-carnitine under high ambient temperatures. *S. Afr. J. Anim. Sci.* **2019**, *48*, 935. [CrossRef]
38. Eder, K. L-carnitine supplementation and the lipid metabolism of rats fed a hyperlipidaemic diet. *J. Anim. Physiol. Anim. Nutr.* **2000**, *83*, 132–140. [CrossRef]
39. Bell, F.P.; Raymond, T.L.; Patnode, C.L. The influence of diet and carnitine supplementation on plasma carnitine, cholesterol and triglyceride in WHHL (Watanabe-heritable hyperlipidemic), Netherland dwarf and New Zealand rabbits (oryctolagus cuniculus). *Comp. Biochem. Physiol. Part B Comp. Biochem.* **1987**, *87*, 587–591. [CrossRef] [PubMed]
40. Parsaeimehr, K.; Afrouziyeh, M.; Hoseinzadeh, S. The Effects of L—Carnitine and Different Levels of Animal Fat on Performance, Carcass Characteristics, some Blood Parameters and Immune Response in Broiler Chicks. *Iran. J. Appl. Anim. Sci.* **2014**, *4*, 561–566.
41. Mondola, P.; Santillo, M.; De Mercato, R.; Santangelo, F. The effect of l-Carnitine on cholesterol metabolism in rat (*Rattus bubalus*) hepatocyte cells. *Int. J. Biochem.* **1992**, *24*, 1047–1050. [CrossRef] [PubMed]
42. Stephens, F.B.; Constantin-teodosiu, D.; Greenhaff, P.L. New insights concerning the role of carnitine in the regulation of fuel metabolism in skeletal muscle. *J. Physiol.* **2007**, *581*, 431–444. [CrossRef] [PubMed]
43. Sarica, S.; Corduk, M.; Ensoy, U.; Basmacioglu, H.; Karatas, U. Effects of dietary supplementation of L-carnitine on performance, carcass and meat characteristics of quails. *S. Afr. J. Anim. Sci.* **2007**, *37*, 189–201. [CrossRef]
44. Khan, B.A.; Anwar, S.; Maqbool, R.M.; Amin, M.M.; Javaid, M.M.; Amin, M.M.; Iqbal, M.K.; Aziz, A. Assessing allelopathic potential of *Sonchus oleraceus* L. (milk thistle) on germination and seedling growth of Oryza puncta. *J. Weed Sci. Res.* **2021**, *27*, 1–12. [CrossRef]
45. Kwiecień, M.; Jachimowicz-Rogowska, K.; Krupa, W.; Winiarska-Mieczan, A.; Krauze, M. Effects of Dietary Supplementation of L-Carnitine and Mannan-Oligosaccharides on Growth Performance, Selected Carcass Traits, Content of Basic and Mineral Components in Liver and Muscle Tissues, and Bone Quality in Turkeys. *Animals* **2023**, *13*, 770. [CrossRef] [PubMed]
46. Ribas, G.S.; Vargas, C.R.; Wajner, M. L-carnitine supplementation as a potential antioxidant therapy for inherited neurometabolic disorders. *Gene* **2014**, *533*, 469–476. [CrossRef] [PubMed]
47. El-Raffa, A.M. Formation of a rabbit synthetic line (Alexandria line) and primary analysis of its productive and reproductive performance. *Egypt. Poult. Sci.* **2007**, *27*, 321–334.
48. El-Sabrout, K.; Aggag, S.; El-Raffa, A. Comparison of milk production and milk composition for an exotic and a local synthetic rabbit lines. *Vet. World* **2017**, *10*, 526–529. [CrossRef] [PubMed]
49. National Research Council. *Nutrients Requirements of Poultry*, 7th ed.; National Academy Press: Washington, DC, USA, 1977; Volume 42.
50. AOAC. *Official Methods of Analysis of AOAC International*; AOAC: Rockville, MD, USA, 1995.
51. Hassan, S.S.; Shahba, H.A.; Mansour, M.M. Influence of Using Date Palm Pollen or Bee Pollen on Some Blood Biochemical-metabolites, Semen Characteristics Andsubsequentreproductive Performance of V-Line Male Rabbits. *J. Rabbit Sci.* **2022**, *32*, 19–39.
52. Ewuola, E.O.; Egbunike, G.N. Haematological and serum biochemical response of growing rabbit bucks fed dietary fumonisin B1. *Afr. J. Biotechnol.* **2008**, *7*, 4304–4309.
53. Drew, P.; Charles, R.; Trevor, B.; Johnn, L. *Oxford Handbook of Clinical Haematology*, 2nd ed.; Oxford University Press: New York, NY, USA, 2004.
54. Jain, N.C. Scanning electron micrograph of blood cells. In *Schalm's Veterinary Haematology*; Wiley: Hoboken, NJ, USA, 1983; pp. 63–70.
55. Armstrong, W.D.; Carr, C.W. *Physiological Chemistry Laboratory Directions*, 3rd ed.; Burges Publishing Co.: Clayton, NC, USA, 1964.
56. Doumas, B.T.; Watson, W.A.; Biggs, H.G. Albumin standards and the measurement of serum albumin with bromocresol green. *Clin. Clin. Chem. Acta* **1977**, *31*, 87–96. [CrossRef] [PubMed]
57. Sturkie, P. Body fluids: Blood. In *Avian Physiology*; Springer: New York, NY, USA, 1986; pp. 103–121.
58. Woodman, D.D.; Price, C.P. Estimation of serum total lipids. *Clin. Chim. Acta* **1972**, *38*, 39–43. [CrossRef] [PubMed]
59. Frings, C.S.; Fendley, T.W.; Dunn, R.T.; Queen, C.A. Improved Determination of Total Serum Lipids by the Sulfo-Phospho-Vanillin Reaction. *Clin. Chem.* **1972**, *18*, 673–674. [CrossRef] [PubMed]
60. Burstein, M.; Scholnick, H.R.; Morfin, R. Rapid method for the isolation of lipoproteins from human serum by precipitation with polyanions. *J. Lipid Res.* **1970**, *11*, 583–595. [CrossRef] [PubMed]

61. Wieland, H.; Seidel, D. Improved assessment of plasma lipoprotein patterns: IV. Simple preparation of a lyophilized control serum containing intact human plasma lipoproteins. *Clin. Chem.* **1982**, *28*, 1335–1337. [CrossRef] [PubMed]
62. Bogin, E.; Keller, P. Application of clinical biochemistry to medically relevant animal models and standardization and quality control in animal biochemistry. *J. Clin. Chem. Clin. Biochem* **1987**, 873–878.
63. Friedewald, W.T.; Levy, R.I.; Fredrickson, D.S. Estimation of the Concentration of Low-Density Lipoprotein Cholesterol in Plasma, Without Use of the Preparative Ultracentrifuge. *Clin. Chem.* **1972**, *18*, 499–502. [CrossRef] [PubMed]
64. *IBM SPSS Statistics for Windows, Version 20*; IBM Corp.: New York, NY, USA, 2011.
65. Duncan, D.B. Multiple range and Multiple F tests. *Biometrics* **1955**, *11*, 1–42. [CrossRef]

antibiotics

MDPI

Article

Polygonum hydropiper Compound Extract Inhibits *Clostridium perfringens*-Induced Intestinal Inflammatory Response and Injury in Broiler Chickens by Modulating NLRP3 Inflammasome Signaling

Jinwu Zhang [1], Chunzi Peng [1], Maojie Lv [1], Shisen Yang [1], Liji Xie [2], Jiaxun Feng [3], Yingyi Wei [1], Tingjun Hu [1], Jiakang He [1], Zhixun Xie [2,*] and Meiling Yu [1,2,*]

1 Guangxi Key Laboratory of Animal Breeding, Disease Control and Prevention, College of Animal Science and Technology, Guangxi University, Nanning 530004, China; 2018302041@st.gxu.edu.cn (J.Z.); 2218393073@st.gxu.edu.cn (S.Y.)
2 Guangxi Key Laboratory of Veterinary Biotechnology, Guangxi Veterinary Research Institute, Nanning 530001, China
3 College of Life Science and Technology, Guangxi University, Nanning 530004, China
* Correspondence: xiezhixun@126.com (Z.X.); yumeiling@gxu.edu.cn (M.Y.); Tel.: +86-0771-3235635 (M.Y.)

Abstract: *Necrotic enteritis* (*NE*) is a critical disease affecting broiler health, with *Clostridium perfringens* as its primary pathogen. *Polygonum hydropiper* compound extract (PHCE), formulated based on traditional Chinese veterinary principles, contains primarily flavonoids with antibacterial, anti-inflammatory, and antioxidant properties. However, PHCE's efficacy against *Clostridium perfringens*-induced *NE* and its underlying mechanism remain unclear. This study employed network pharmacology and molecular docking to predict PHCE's potential mechanisms in treating NE, followed by determining its minimum inhibitory concentration (MIC) and minimum bactericidal concentration (MBC) against *Clostridium perfringens* (*C. perf*). Subsequently, the effects of various PHCE doses on intestinal damage, antioxidant capacity, and inflammatory factors in *C. perf*-infected broilers were assessed. Network pharmacology and molecular docking suggested that PHCE's therapeutic mechanism for *NE* involves the NOD-like receptor thermal protein domain associated protein 3 (NLRP3) inflammasome signaling pathway, with flavonoids such as quercetin, kaempferol, and isorhamnetin as key active components. PHCE exhibited an MIC of 3.13 mg/mL and an MBC of 12.5 mg/mL against *C. perf.* High PHCE doses effectively reduced intestinal damage scores in both the jejunum and ileum, accompanied by attenuated intestinal pathological changes. Additionally, the high dose significantly increased superoxide dismutase (SOD) levels while decreasing malondialdehyde (MDA), hydrogen peroxide (H_2O_2), tumor necrosis factor-alpha (TNF-α), interleukin-1 beta (IL-1β), and interleukin-6 (IL-6) in the jejunum and ileum ($p < 0.01$ or $p < 0.05$). PHCE also modulated the expression of caspase-1, IL-1β, gasdermin D (GSDMD), and NLRP3 mRNA, key components of the NLRP3 inflammasome signaling pathway, in both intestinal segments. These findings collectively indicate that PHCE protects against *C. perf*-induced oxidative stress and inflammatory damage in *NE*. By enhancing antioxidant capacity, PHCE likely reduces oxidative stress and inflammatory responses, subsequently modulating NLRP3 inflammasome signaling pathway key factor expression. Overall, this research provides valuable insights into the protective mechanism of the herbal compound PHCE and its potential benefits for avian health.

Keywords: *Polygonum hydropiper* compound extract; *Necrotic enteritis*; network pharmacology; oxidative stress; NLRP3 inflammasome

Citation: Zhang, J.; Peng, C.; Lv, M.; Yang, S.; Xie, L.; Feng, J.; Wei, Y.; Hu, T.; He, J.; Xie, Z.; et al. *Polygonum hydropiper* Compound Extract Inhibits *Clostridium perfringens*-Induced Intestinal Inflammatory Response and Injury in Broiler Chickens by Modulating NLRP3 Inflammasome Signaling. *Antibiotics* **2024**, *13*, 793. https://doi.org/10.3390/antibiotics13090793

Academic Editors: Carlos M. Franco and Beatriz Vázquez Belda

Received: 15 July 2024
Revised: 19 August 2024
Accepted: 20 August 2024
Published: 23 August 2024

1. Introduction

Necrotic enteritis (*NE*) is a significant intestinal disease in poultry, manifesting in either acute clinical or chronic subclinical forms. The latter is prevalent in the global poultry industry, representing a substantial threat to broiler production and resulting in approximately $6 billion in economic losses annually [1]. *NE* is caused by pathogenic *Clostridium perfringens* (*C. perf*), a spore-forming, anaerobic, Gram-positive bacterium, which can also be a component of the normal microbiota in humans and animals [2,3]. The pathogenesis of *NE* is primarily attributed to toxins produced by *C. perf*, including α-toxin (CPA), β-toxin (CPB), ε-toxin (ETX), ι-toxin (ITX), enterotoxin (CPE), and NetB toxin (NetB) [4]. The conserved *CPA* gene encodes phospholipase C, sphingomyelinase, and a zinc-dependent metalloenzyme with hemolytic activity. This metalloenzyme is capable of hydrolyzing phospholipids within cell membranes, thereby inducing membrane dysfunction and subsequent cell death [2]. Known for its hemolytic, skin necrotic, and lethal properties, CPA plays a pivotal role in gangrene [5]. While PFO is leukotoxic at high doses, it stimulates the production of intracellular adhesion molecule 1 (ICAM-1) and adhesion glycoprotein CD11b/CD18 in endothelial cells at sublethal concentrations, resulting in leukocyte stasis in vessels adjacent to gangrene [6,7]. *C. perf* types A and C are the primary causative agents of *NE* in poultry, with type A strains producing toxins such as CPA and PFO [8]. Although *C. perf* exhibits low pathogenicity in the intestinal tract of healthy poultry, environmental changes or feed-induced stress can trigger its overgrowth, leading to food poisoning, *NE*, and other diseases [9]. Antibiotics were once the mainstay of *NE* treatment, but concerns over residues, contamination, and resistance have prompted their ban in many countries. Consequently, *NE* cases surged, necessitating urgent development of antibiotic alternatives [10]. Traditional Chinese medicine has emerged as a promising replacement due to its low toxicity, minimal residue, and reduced likelihood of inducing drug resistance [11].

In recent years, clinical applications of *Polygonum hydropiper* L. and its compounds have steadily increased, with studies demonstrating their efficacy against various bacterial and fungal diseases. Extracts of *Polygonum hydropiper* L. exhibit protective effects against *Vibrio parahaemolyticus*-infected *Litopenaeus vannamei*, and the minimum inhibitory concentration (MIC) against *Aeromonas hydrophila*, a pathogen affecting weather fish, is less than 10 μL/mL, indicating potent bacteriostatic activity [12,13]. *Polygonum hydropiper* compound extract (PHCE) is formulated based on traditional Chinese veterinary medicine principles, incorporating six herbs, including *Polygonum hydropiper* L. and patchouli. As the core component, *Polygonum hydropiper* L., with its heat-clearing and detoxifying properties, constitutes the largest proportion of PHCE and plays a primary role. The remaining herbs—*Patchouli, Pulsatilla, Astragalus, Ligustrum lucidum*, and *licorice*—enhance the therapeutic efficacy of *Polygonum hydropiper* L. while mitigating potential adverse effects [14]. PHCE, a commonly used herb in Guangxi, contains abundant flavonoids and is employed in treating dysentery, gastroenteritis, diarrhea, and other gastrointestinal disorders. Plant-derived flavonoids possess diverse biological activities, including anti-inflammatory, antiviral, antioxidant, and antimicrobial properties [15–17]. Liu et al. demonstrated that licorice chalcone A, extracted from licorice, reduced pro-inflammatory TNF-α levels in broiler chicken small intestinal tissues and effectively treated *C. perf*-induced *NE* with associated intestinal pathology [18]. In addition, the traditional Chinese medicine compound composed of *Polygonum hydropiper* L. also has great efficacy in treating animal diseases. Shen reported the efficacy of a compound *Polygonum hydropiper* powder containing *Polygonum hydropiper* L., cinnabar, and tea in preventing and treating contagious porcine gastroenteritis [19].

Network pharmacology, a novel discipline rooted in systems biology, analyzes biological system networks to identify specific signaling nodes for targeted drug molecule design [20]. Its aim is to systematically elucidate complex scientific challenges at multiple levels [21]. A key characteristic of network pharmacology is its predictive nature, emphasizing multi-channel signaling pathway regulation to enhance therapeutic efficacy, reduce adverse effects, and improve clinical trial success rates, thereby lowering drug research and development costs [22]. Molecular docking, a drug design method based on receptor characteristics and drug–receptor interactions, is a theoretical simulation technique for predicting binding patterns and affinities. It has become a crucial technology in computer-aided drug research [23]. To optimize the development and utilization of PHCE, this study integrated network pharmacology and molecular docking to explore potential network targets and mechanisms of action in treating chicken *NE*. PHCE and its constituents were extracted using decoction and alcohol extraction methods and subsequently applied to *C. perf* type A. The in vitro bacteriostatic effects of PHCE on *C. perf* type A were evaluated by determining the MIC, minimum bactericidal concentration (MBC), and bacteriostatic curves. To further elucidate PHCE's protective effects and mechanisms against *NE* in vivo, different PHCE doses were incorporated into broiler chicken diets. This experimental approach investigated PHCE's protective effects and mechanisms in the jejunum and ileum, providing a theoretical foundation for PHCE application in preventing and controlling *C. perf*-induced *NE* in broilers.

2. Results

2.1. Compound-Target Network

As illustrated in Figure 1, a Traditional Chinese Medicine Systems Pharmacology (TCMSP) search identified 140 active ingredients within PHCE: 13 from *Polygonum hydropiper* L., 7 from patchouli, 8 from pulsatilla, 16 from astragalus, 8 from *Ligustrum lucidum*, and 88 from licorice, along with 216 potential targets. The PHCE compound–target network, visualized in Cytoscape, comprised 339 nodes and 2291 edges. The top 10 drug components were selected based on degree, betweenness centrality, and closeness centrality, with higher degree values indicating greater network significance (Table 1). Within this network, active ingredients such as quercetin, kaempferol, isorhamnetin, and lignocerol exhibited high degree values, suggesting their potential as key components of PHCE for *NE* treatment.

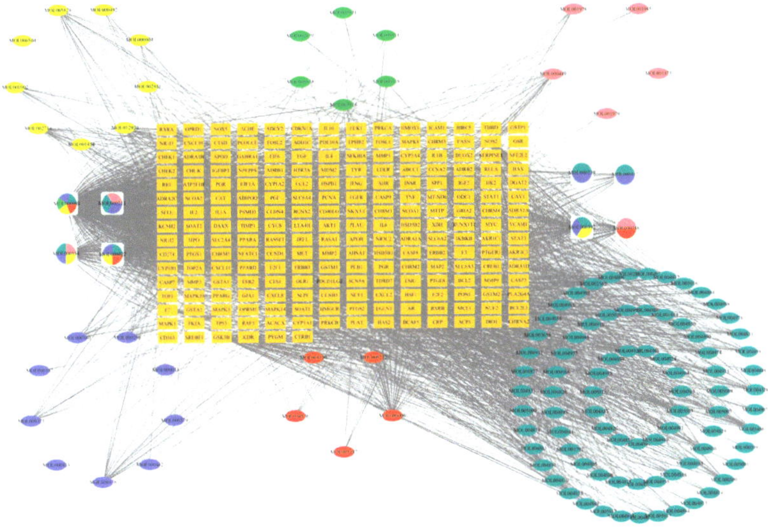

Figure 1. Herb–compound–target network of PHCE. Nine yellow ellipses and four yellow sectors represent *Polygonum hydropiper* L. components, six green ellipses and one green sector represent patchouli components, five pink ellipses and three pink sectors represent pulsatilla components, nine blue ellipses and seven blue sectors represent astragalus components, five red ellipses and thee red sectors represent *Ligustrum lucidum* components, and eighty-one cyan ellipses and seven cyan sectors represent licorice components. Orange rectangles represent the 216 potential targets of PHCE. Circles represent compounds shared among different herbs.

Table 1. Top ten compounds information of PHCE network.

Mol ID	Compound	Degree	Closeness Centrality	Betweenness Centrality	OB	DL
MOL000098	quercetin	605	0.51057	0.39401	46.43	0.28
MOL000422	kaempferol	188	0.41936	0.07805	41.88	0.24
MOL000354	isorhamnetin	84	0.39302	0.02112	49.6	0.31
MOL000392	formononetin	71	0.39211	0.03092	69.67	0.21
MOL000358	β-sitosterol	50	0.39579	0.02882	36.91	0.75
MOL000006	luteolin	47	0.41523	0.08765	36.16	0.25
MOL000378	7-*O*-methylisomucronulatol	30	0.40143	0.01798	74.69	0.3
MOL003896	7-Methoxy-2-methyl isoflavone	28	0.39953	0.01247	42.56	0.2
MOL004328	Calycosin	27	0.39302	0.08517	47.75	0.24
MOL000417	naringenin	26	0.38585	0.00097	59.29	0.21

Note: OB (Oral Bioavailability) refers to the rate and extent to which a drug is absorbed into the bloodstream from a dosage form. DL (drug-likeness) indicates the similarity of a compound to known drugs.

2.2. Protein–Protein Interaction (PPI) Network Construction

As shown in Figure 2A, PHCE exhibited 216 targets, while NE displayed 2774 targets, with 141 shared targets between the two. Figure 2B illustrates the PHCE and NE common target PPI network. In this network, node size and darkness correlated with degree value, indicating the likelihood of a target being a core target, such as NLRP3 and CASP1.

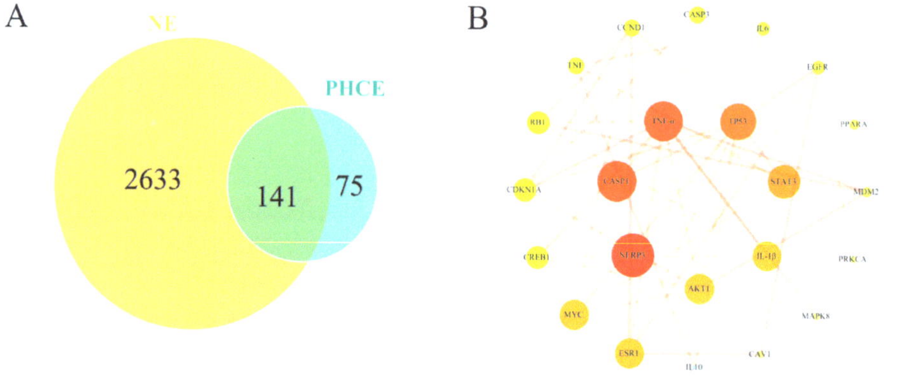

Figure 2. Network pharmacological analysis of PHCE and NE. (**A**) Venn diagram illustrating the overlap between PHCE targets and NE targets. (**B**) Protein–protein interaction (PPI) network of shared PHCE and NE targets. Nodes represent proteins, with a color gradient from yellow to red, indicating increasing protein interaction strength. Edges represent protein–protein association.

2.3. Results of Functional Enrichment Analysis

As illustrated in Figure 3A, GO enrichment analysis identified 414 GO entries ($p < 0.01$), from which the top ten terms were selected for biological processes, cellular components, and molecular functions, respectively. These terms included positive regulation of transcription from the RNA polymerase II promoter, response to hypoxia, and inflammatory response. Figure 3B presents the results of the KEGG pathway analysis, revealing 174 pathways associated with common targets. The top 30 pathways were ranked based on significant enrichment and gene count, with the NLRP3 inflammasome signaling pathway, TNF signaling pathway, and pathways in cancer identified as the most prominent. Notably, the NLRP3 inflammasome signaling pathway exhibited lower p-values and higher gene counts compared to other pathways, suggesting its potential as a primary mechanism of action for PHCE in treating *NE*.

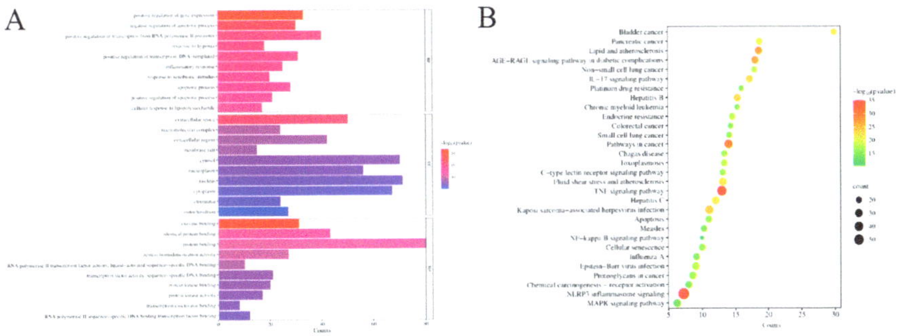

Figure 3. Network pharmacology prediction for PHCE treatment of *NE*. Top 10 GO terms of hub genes (**A**) and top 30 KEGG pathway of hub genes (**B**). The cut-off values of count and −log10(pvalue) for (**A**) are 5.3 and 8, respectively, while the cut-off values of count and −log10(pvalue) for (**B**) are 10.3 and 16, respectively.

2.4. Molecular Docking

Two primary targets, NLRP3 and CASP1, exhibiting the highest degree values within the network diagram, were selected, along with the top components, for molecular docking analysis. The absolute values of docking scores reflect the affinity and conformational stability of component–target interactions. A score exceeding 4.25 indicates a certain binding activity, while values greater than 5.0 and 7.0 suggest good and strong binding activities, respectively [24]. As shown in Table 2, all docking results were negative, and the absolute binding energies of NLRP3 and CASP1 to the ligand surpassed 5.0, indicating favorable ligand–receptor binding activity. Figure 4 illustrates the specific binding site of each protein and the calculated specific binding energy for each target, with those exhibiting superior binding energies highlighted.

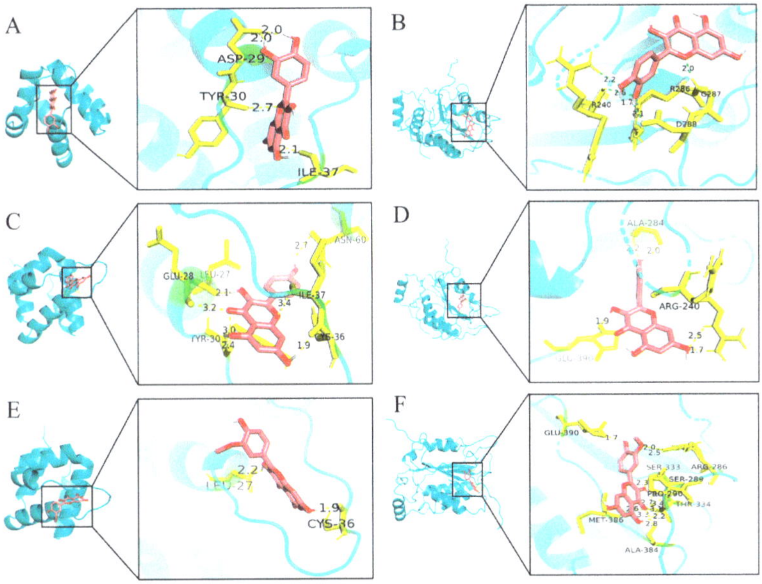

Figure 4. Molecular docking results of main chemical components of PHCE. Quercetin-NLRP3 (**A**), Quercetin-CASP1 (**B**), Kaempferol-NLRP3 (**C**), Kaempferol-CASP1 (**D**), Isorhamnetin-NLRP3 (**E**) and Isorhamnetin-CASP1 (**F**).

Table 2. Docking results of core target proteins and core active components.

Active Components	NLRP3	CASP1
quercetin	–5.71	–5.72
kaempferol	–6.28	–5.31
isorhamnetin	–5.79	–6.15

2.5. MIC and MBC of PHCE

As presented in Table 3, the MIC of PHCE was 3.13 mg/mL. For individual components, patchouli exhibited an MIC of 12.5 mg/mL, *Polygonum hydropiper* L. 6.25 mg/mL, astragalus and licorice 50 mg/mL each, and pulsatilla and Ligustrum lucidum 6.25 mg/mL each. Following PHCE treatment, the bacterial solution was uniformly spread on a sterile TSC solid medium. A colony count below five indicated the MBC, which was found to be 12.5 mg/mL for PHCE, as shown in Table 4. Comparative analysis of PHCE and its constituent fractions revealed that PHCE demonstrated superior bacteriostatic efficacy against *C. Perf.*

Table 3. MIC of Chinese herbs against *C. perf.*

Drug	Drug Concentration (mg/mL)										Negative	Positive
	200	100	50	25	12.5	6.25	3.13	1.56	0.78	0.39		
PHCE	–	–	–	–	–	–	–	+	+	+	–	+
Patchouli	–	–	–	–	–	+	+	+	+	+	–	+
Polygonum hydropiper L.	–	–	–	–	–	–	+	+	+	+	–	+
Astragalus	–	–	–	+	+	+	+	+	+	+	–	+
Pulsatilla	–	–	–	–	–	–	+	+	+	+	–	+
Ligustrum lucidum	–	–	–	–	–	–	+	+	+	+	–	+
Licorice	–	–	–	+	+	+	+	+	+	+	–	+

Note: "+" indicates bacterial growth; "–" indicates no bacterial growth.

Table 4. Determination of *C. perf* MBC by PHCE.

	0.78 mg/mL	1.56 mg/mL	3.13 mg/mL	6.25 mg/mL	12.5 mg/mL
C. perf	+	+	+	+	–

Note: "+" indicates more colonies; "–" indicates that the number of colonies is less than five.

2.6. Antibacterial Activity Curve of PHCE against C. perf

As shown in Figure 5, different PHCE concentrations induced varying growth curve alterations in *C. perf.* At 1/4× MIC (0.78 mg/mL) and 1/2× MIC (1.56 mg/mL) PHCE concentrations, *C. perf* bacterial concentration declined. While growth rates slowed at 1 and 4 h, respectively, bacteria eventually entered a stable phase following rapid logarithmic growth, with final growth approximating that of the 0 mg/mL PHCE control group. Conversely, at MIC and 2× MIC PHCE concentrations, *C. perf* exhibited no logarithmic growth phase, and colony numbers decreased over time, falling below the initial colony count.

2.7. Levels of SOD, H_2O_2, and MDA in Jejunal and Ileal Tissues

As shown in Figure 6, compared to the control group, SOD enzyme activity in jejunal tissues of model group broilers was significantly lower ($p < 0.05$), while MDA and H_2O_2 levels were significantly higher ($p < 0.05$). H_2O_2 levels were significantly elevated in jejunal tissues of both the low-dose and medium-dose group broilers ($p < 0.05$), with no significant changes in SOD enzyme activity or MDA levels ($p > 0.05$). The levels of H_2O_2 in the jejunal tissues of broilers in both the high-dose group and ampicillin group were significantly lower than those in the control group ($p < 0.01$). There were no significant differences in SOD enzyme activity or MDA levels in the high-dose, high-dose control, or ampicillin groups ($p > 0.05$). Compared to the model group, SOD enzyme activity was significantly or highly significantly increased ($p < 0.01$ or $p < 0.05$) in jejunal tissues of high-dose, high-dose control, and positive control group broilers, accompanied by significantly or highly significantly decreased MDA and H_2O_2 levels ($p < 0.01$ or $p < 0.05$). There was no significant change in SOD enzyme activity or MDA level in the low-dose, medium-dose, or medium-dose prophylaxis groups ($p > 0.05$).

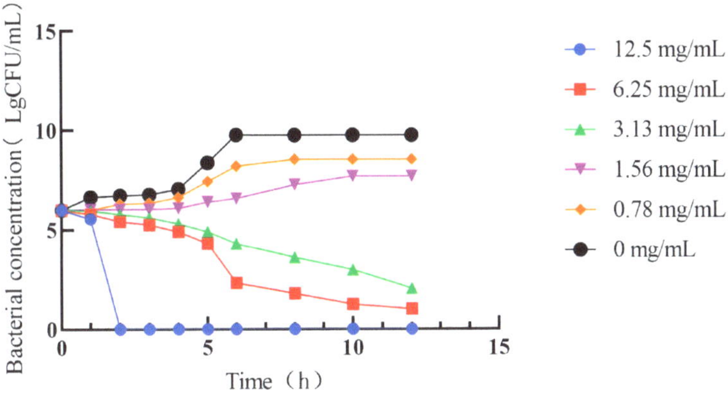

Figure 5. Antibacterial activity curve of PHCE against *C. perf*.

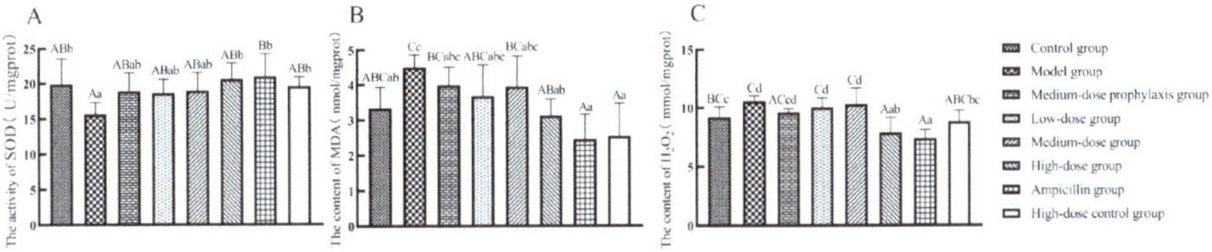

Figure 6. Levels of oxidative stress-related factors in the jejunum: SOD (**A**), MDA (**B**), and H_2O_2 (**C**). Data represent mean ± standard error of the mean ($n = 5$). A total of 200 one-day-old healthy broilers were randomly assigned to eight groups, with five replicates per treatment group and five broilers per replicate. The experiment lasted 17 days. Bars labeled with different uppercase letters indicate highly significant differences ($p < 0.01$), while bars labeled with different lowercase letters indicate significant differences ($p < 0.05$). Bars labeled with the same letter indicate no significant difference ($p > 0.05$).

Compared to the control group, the model group exhibited significantly decreased SOD activity ($p < 0.05$) and significantly or highly significantly increased MDA and H_2O_2 levels ($p < 0.01$ or $p < 0.05$). No significant changes in SOD activity were observed in the ileal tissues of low-dose and medium-dose groups, while MDA and H_2O_2 levels were significantly or highly significantly elevated in the medium-dose group ($p < 0.01$ or $p < 0.05$). Relative to the model group, the high-dose, high-dose control, and ampicillin groups displayed significantly or highly significantly increased SOD activity and decreased MDA and H_2O_2 levels ($p < 0.01$ or $p < 0.05$). No significant alterations in SOD activity, MDA, or H_2O_2 levels were observed in the ileal tissues of low-dose, medium-dose, or medium-dose prevention groups ($p > 0.05$) (Figure 7).

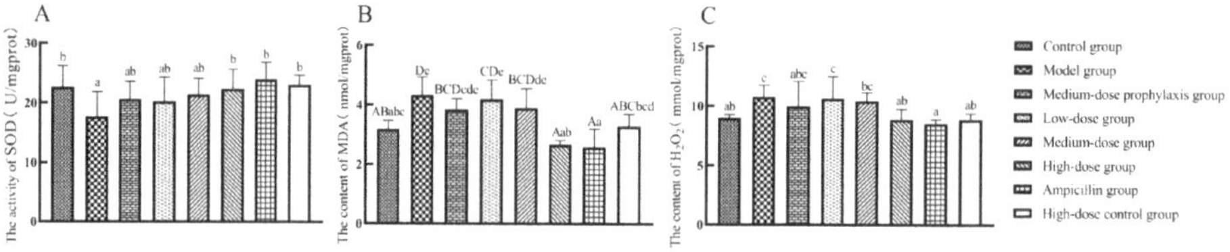

Figure 7. Levels of oxidative stress-related factors in the ileum: SOD (**A**), MDA (**B**), and H_2O_2 (**C**). Data represent mean ± standard error of the mean (n = 5). A total of 200 one-day-old healthy broilers were randomly assigned to eight groups, with five replicates per treatment group and five broilers per replicate. The experiment lasted 17 days. Bars labeled with different uppercase letters indicate highly significant differences ($p < 0.01$), while bars labeled with different lowercase letters indicate significant differences ($p < 0.05$). Bars labeled with the same letter indicate no significant difference ($p > 0.05$).

2.8. Levels of TNF-α, IL-1β, and IL-6 in Jejunal and Ileal Tissues

As shown in Figure 8, compared to the control group, TNF-α secretion levels in jejunal tissues were significantly elevated in model and low-dose group broilers ($p < 0.05$), while IL-1β and IL-6 secretion levels were significantly or highly significantly increased in model, low-dose, and medium-dose group broilers ($p < 0.01$ or $p < 0.05$). IL-1β secretion levels were significantly increased in jejunal tissues of medium-dose prevention group broilers ($p < 0.05$). No significant changes in TNF-α, IL-1β, or IL-6 secretion levels were observed in jejunal tissues of high-dose, high-dose control, or ampicillin groups ($p > 0.05$). Compared to the model group, TNF-α, IL-1β, and IL-6 secretion levels were significantly or extremely significantly reduced in jejunal tissues of high-dose, high-dose control, and ampicillin group broilers ($p < 0.01$ or $p < 0.05$). No significant changes in TNF-α, IL-1β, or IL-6 secretion levels were observed in jejunal tissues of low-dose, medium-dose, or medium-dose prevention groups ($p > 0.05$).

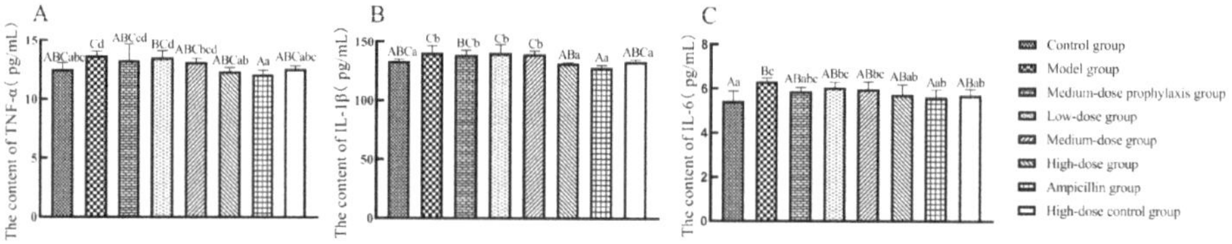

Figure 8. Secretion level of inflammatory factors in jejunum. TNF-α (**A**), IL-1β (**B**), and IL-6 (**C**). Data represent mean ± standard error of the mean (n = 5). A total of 200 one-day-old healthy broilers were randomly assigned to eight groups, with five replicates per treatment group and five broilers per replicate. The experiment lasted 17 days. Bars labeled with different uppercase letters indicate highly significant differences ($p < 0.01$), while bars labeled with different lowercase letters indicate significant differences ($p < 0.05$). Bars labeled with the same letter indicate no significant difference ($p > 0.05$).

As shown in Figure 9, compared to the control group, ileal tissue TNF-α, IL-1β, and IL-6 secretion levels were significantly or highly significantly elevated ($p < 0.01$ or $p < 0.05$) in model group broilers. Low-dose and medium-dose group broilers exhibited significantly increased IL-6 levels ($p < 0.05$) in ileal tissues. The high-dose and ampicillin groups displayed significantly or highly significantly reduced IL-1β secretion ($p < 0.01$ or $p < 0.05$) in ileal tissues, with no significant changes in TNF-α or IL-6 levels ($p > 0.05$). Compared to the model group, ileal tissue TNF-α, IL-1β, and IL-6 secretion levels were significantly or highly significantly decreased ($p < 0.01$ or $p < 0.05$) in high-dose, high-dose control, and ampicillin group broilers. No significant changes in TNF-α, IL-1β, or IL-6 secretion levels were observed in ileal tissues of low-dose, medium-dose, or medium-dose prevention group broilers ($p > 0.05$).

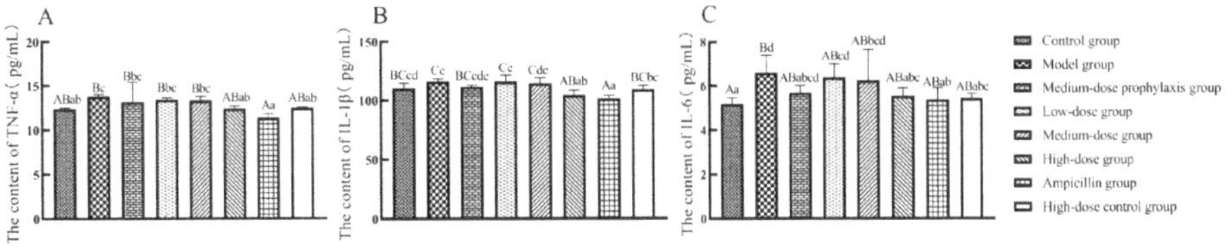

Figure 9. Secretion level of inflammatory factors in ileum. TNF-α (**A**), IL-1β (**B**), and IL-6 (**C**). Data represent mean ± standard error of the mean ($n = 5$). A total of 200 one-day-old healthy broilers were randomly assigned to eight groups, with five replicates per treatment group and five broilers per replicate. The experiment lasted 17 days. Bars labeled with different uppercase letters indicate highly significant differences ($p < 0.01$), while bars labeled with different lowercase letters indicate significant differences ($p < 0.05$). Bars labeled with the same letter indicate no significant difference ($p > 0.05$).

2.9. *Injury Scoring of Jejunum and Ileum*

Compared to the control group, jejunal and ileal tissue damage scores were significantly higher ($p < 0.01$) in the model and low-dose groups, while no significant differences were observed in the medium-dose prophylaxis, medium-dose, high-dose, high-dose control, and ampicillin groups ($p > 0.05$). Relative to the model group, no significant changes were detected in the medium-dose prophylaxis, low-dose group, and medium-dose groups ($p > 0.05$), whereas jejunal and ileal tissue damage scores were significantly or highly significantly lower ($p < 0.01$ or $p < 0.05$) in the high-dose, high-dose control, and ampicillin groups (Table 5).

Table 5. Effect of PHCE on jejunal and ileal lesion scores.

Treatments	Rank Score Means	
	Jejunum	**Ileum**
Control group	5.5 ^{Aa}	6.0 ^{Aa}
Model group	36.2 ^{Bb}	34.8 ^{Bb}
Medium dose Prophylaxis group	27.0 ^{ABab}	26.2 ^{ABab}
Low-dose group	32.1 ^{Bb}	32.7 ^{Bb}
Medium-dose group	27.4 ^{ABab}	27.7 ^{ABab}
High-dose group	11.3 ^{ABa}	11.8 ^{ABa}
Ampicillin group	10.6 ^{ABa}	9.4 ^{ABa}
High-dose control group	8.9 ^{Aa}	9.4 ^{ABa}

Note: Data represent rank score means ($n = 5$). Rank score means and difference in rank score means were calculated by the Kruskal–Wallis test. Data with different uppercase letters superscripts differ highly significantly ($p < 0.01$), data with varying letters of lowercase superscripts differ significantly ($p < 0.05$), and data with common superscripts denote no significant difference ($p > 0.05$).

2.10. *Histopathologic Lesions of the Jejunum and Ileum*

As shown in Figure 10, enterochromaffin structures in the jejunum and ileum of broilers from the model, low-dose, medium-dose, and medium-dose prophylactic groups exhibited breakage with necrotic shedding or disintegration. In contrast, enterochromaffin cells in the jejunum and ileum of the high-dose and ampicillin groups displayed relatively intact structures with only minor detachment at the cellular ends. The jejunal and ileal intestinal villi of broilers in the high-dose control and control groups maintained structural integrity without apparent pathological damage.

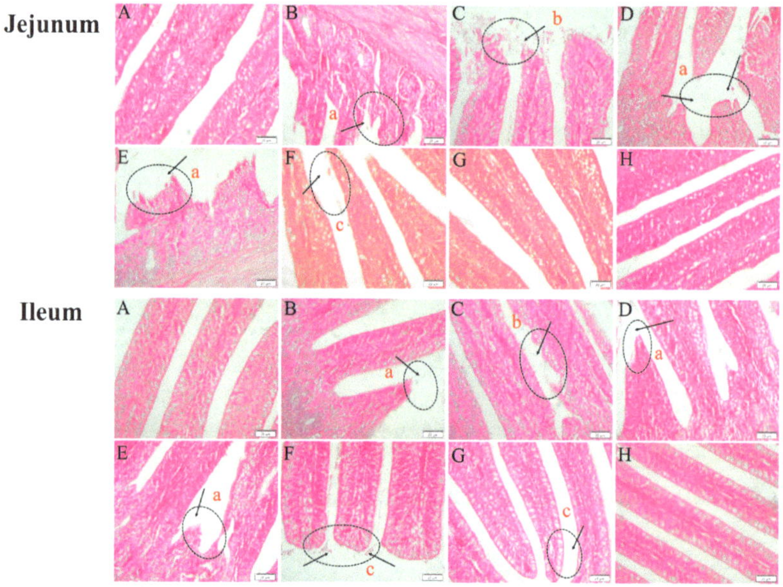

Figure 10. Histopathological changes in the jejunum and ileum of broilers (400× magnification). Control group (**A**), model group (**B**), medium-dose prophylactic group (**C**), low-dose group (**D**), medium-dose group (**E**), high-dose group (**F**), ampicillin group (**G**), and high-dose control group (**H**). Histopathological abnormalities include intestinal villi rupture (a), extensive necrotic shedding of intestinal villi (b), and minor shedding at the tips of intestinal villi (c).

2.11. The mRNA Levels of NLRP3 Inflammasome Signaling Pathway-Related Factors in Jejunal and Ileal Tissues

As shown in Figure 11, mRNA expression levels of TNF-α, IL-1β, IL-6, caspase-1, NLRP3, and GSDMD were significantly elevated ($p < 0.01$) in jejunal tissues of broilers from the model, low-dose, medium-dose, and medium-dose prophylactic groups compared to the control group. In contrast, mRNA expression levels of these same cytokines remained unchanged in jejunal tissues of broilers from the high-dose, high-dose control, and ampicillin groups ($p > 0.05$). Compared to the model group, mRNA expression levels of TNF-α, IL-1β, IL-6, caspase-1, NLRP3, and GSDMD were significantly reduced ($p < 0.01$) in jejunal tissues of broilers from the high-dose, high-dose control, and ampicillin groups. No significant changes in mRNA expression levels of these cytokines were observed in jejunal tissues of broilers from the low-dose or medium-dose prevention groups ($p > 0.05$).

As shown in Figure 12, ileal tissue mRNA expression levels of TNF-α, IL-1β, IL-6, caspase-1, NLRP3, and GSDMD were significantly elevated in broilers from the model, low-dose, medium-dose, and medium-dose prophylaxis groups compared to the control group ($p < 0.01$). While IL-6 mRNA levels were significantly increased in the high-dose group ($p < 0.05$), TNF-α, IL-1β, caspase-1, NLRP3, and GSDMD mRNA levels remained unchanged in the high-dose, high-dose drug control, and ampicillin groups ($p > 0.05$). Compared to the model group, ileal tissue mRNA expression levels of TNF-α, IL-1β, IL-6, caspase-1, NLRP3, and GSDMD were significantly reduced in the high-dose, high-dose control, and ampicillin groups ($p < 0.01$). No significant changes in mRNA expression levels of these cytokines were observed in the low-dose, medium-dose, or medium-dose prophylaxis groups ($p > 0.05$).

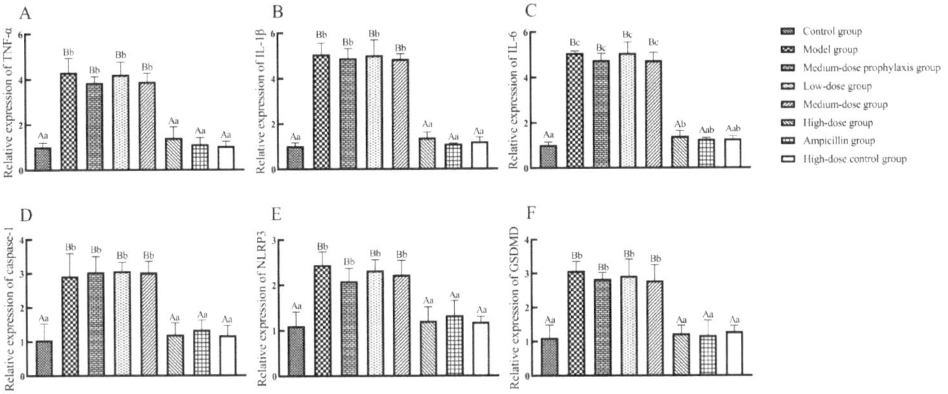

Figure 11. The expression of inflammasome signal pathway-related factors in the jejunum. TNF-α (**A**), IL-1β (**B**), IL-6 (**C**), caspase-1 (**D**), NLRP3 (**E**), and GSDMD (**F**). Data represent mean ± standard error of the mean (n = 5). A total of 200 one-day-old healthy broilers were randomly assigned to eight groups, with five replicates per treatment group and five broilers per replicate. The experiment lasted 17 days. Bars labeled with different uppercase letters indicate highly significant differences ($p < 0.01$), while bars labeled with different lowercase letters indicate significant differences ($p < 0.05$). Bars labeled with the same letter indicate no significant difference ($p > 0.05$).

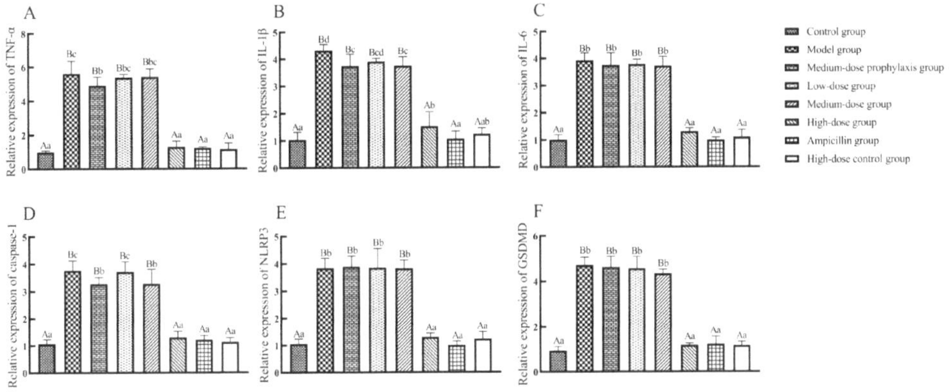

Figure 12. The expression of inflammasome signal pathway-related factors in ileum. TNF-α (**A**), IL-1β (**B**), IL-6 (**C**), caspase-1 (**D**), NLRP3 (**E**), and GSDMD (**F**). Data represent mean ± standard error of the mean (n = 5). A total of 200 one-day-old healthy broilers were randomly assigned to eight groups, with five replicates per treatment group and five broilers per replicate. The experiment lasted 17 days. Bars labeled with different uppercase letters indicate highly significant differences ($p < 0.01$), while bars labeled with different lowercase letters indicate significant differences ($p < 0.05$). Bars labeled with the same letter indicate no significant difference ($p > 0.05$).

3. Discussion

PHCE, formulated based on traditional Chinese veterinary principles, comprises six Chinese herbs including *Polygonum hydropiper* L., *Patchouli, Pulsatilla, Astragalus, Ligustrum lucidum*, and *Licorice*, with *Polygonum hydropiper L.* serving as the primary therapeutic agent [14]. Primarily affecting broiler chicks aged 2–6 weeks, *NE* is induced by *C. perf* and represents a significant bacterial intestinal disease in poultry [25]. Analysis of the PHCE herb–compound–target network identified flavonoids, such as quercetin, kaempferol, and isorhamnetin, as primary active components. Activation of the NLRP3 inflammasome stimulates caspase-1, leading to the release of inflammatory cytokines IL-1β and IL-18 and subsequent cellular pyroptosis. However, NLRP3 inflammasome hyperactivation can induce pathological inflammation [26]. Studies have demonstrated that flavonoids, including quercetin, kaempferol, and isorhamnetin, effectively inhibit NLRP3 inflammasome activation, ameliorating various inflammatory conditions by targeting the NLRP3 inflammasome and suppressing its mediated inflammatory responses [27–29]. Interestingly, PPI network analysis, GO analysis, and KEGG pathway enrichment analysis identified NLRP3 and CASP1 as key PHCE targets against *NE*, suggesting the NLRP3 inflammasome pathway as a potential mechanism of action. Molecular docking analysis revealed strong binding affinities between PHCE's primary active components (quercetin, kaempferol, and isorhamnetin) and the target proteins NLRP3 and CASP1, indicating that PHCE may counteract *NE* by inhibiting NLRP3 inflammasome activation.

Herbal medicines, with their antimicrobial, antioxidant, and anti-inflammatory properties, and their reduced propensity for bacterial resistance, are promising alternatives to antibiotics in preventing and treating bacterial diseases [30]. PHCE, composed of *Polygonum hydropiper* L., *Patchouli*, *Pulsatilla*, *Astragalus*, *Ligustrum lucidum*, and *Licorice*, aligns with traditional Chinese veterinary medicine, with *Polygonum hydropiper* L. serving as the primary therapeutic agent. Zhou et al. [31] reported the potent bacteriostatic effects of *hydropiper* L. decoction against *Salmonella dysentery*, *Escherichia coli*, and *Staphylococcus* aureus (MICs of 15.62 g/L), as well as *Candida albicans* and *Pseudomonas aeruginosa* (MICs of 31.25 g/L). Zhu et al. [32] employed a compound preparation containing *Polygonum hydropiper* L., *Sapium sebiferum* leaves, and neem to treat red crucian carp fingerling disease, achieving a 100% worm kill rate within 24 h at a dose of 8 g/L. These studies collectively suggest that PHCE or compound formulations centered around *Polygonum hydropiper* L. possess significant antibacterial potential and hold promise for preventing and treating animal diseases. The current study determined MIC and MBC values of 3.13 mg/mL and 12.5 mg/mL, respectively, against *C. perf* type A, confirming PHCE's potent antibacterial activity. However, given the potential discrepancies between in vitro and in vivo antimicrobial activities of traditional Chinese medicines or their compounds, further in vivo investigations are necessary to evaluate PHCE's efficacy against *C. perf*-induced *NE*.

Network pharmacology analysis revealed that most PHCE active ingredients are flavonoids, possessing antioxidant properties through antioxidant enzyme activation and reduced oxygen-containing free radical levels. SOD, MDA, and H_2O_2 play crucial roles in maintaining redox balance in animals. SOD, an antioxidant enzyme, catalyzes the disproportionation of superoxide anions into H_2O_2 and molecular oxygen. H_2O_2, in the presence of iron chelates, reacts with oxygen to generate harmful hydroxyl radicals. MDA exhibits cytotoxicity, while the antioxidant enzyme CAT scavenges H_2O_2 [33]. Flavonoids exert antioxidant effects by increasing antioxidant enzyme activity and chelating metal ions [34]. Wang et al. [35] demonstrated that *Angelica dahurica* (polymethoxylated flavonoids) attenuated oxidative stress damage in astrocytes by reducing ROS and MDA production in a hypoxia-induced oxidative stress model. Tao et al. [36] reported that a compound containing *Polygonum hydropiper* L. and astragalus enhanced broiler chicken antioxidant capacity by increasing SOD enzyme activity when added to the basal diet. In this in vivo experiment, high-dose and ampicillin groups exhibited significantly or highly significantly increased SOD enzyme activity and decreased MDA and H_2O_2 levels in jejunal and ileal tissues compared to the model group, aligning with previous findings. While the ampicillin group demonstrated a stronger effect on increasing SOD enzyme activity and reducing MDA and H_2O_2 levels in broiler and ileal tissues compared to the medium- and low-dose groups, the high-dose group exhibited comparable efficacy. These results suggest that high-dose PHCE can alleviate oxidative stress damage induced by *C. perf NE* by enhancing antioxidant capacity.

Oxidative stress and inflammation can contribute to intestinal injury in poultry, with a reciprocal relationship between the two processes. Inflammatory cells release substantial quantities of reactive substances at inflammation sites, exacerbating oxidative stress. Conversely, reactive oxygen and nitrogen species (ROS/RNS) can initiate intracellular signaling cascades and upregulate pro-inflammatory gene expression [37–40]. Flavonoids possess diverse biological activities, including antioxidant and anti-inflammatory properties. *Angelica dahurica* attenuated hepatic injury in a mouse model of hepatic ischemia by enhancing antioxidant capacity and reducing TNF-α, IL-1β, and IL-6 mRNA expression [41]. Similarly, compound formulations containing *Polygonum hydropiper* L. have demonstrated comparable effects. Yang et al. [42] reported that *Polygonum hydropiper* L.-based Changyanning compound ameliorated ulcerative colitis in rats by increasing antioxidant enzyme activity and reducing MDA and TNF-α levels. In the current in vivo experiment, the high-dose and ampicillin groups exhibited the most pronounced effects, significantly or markedly reducing TNF-α, IL-1β, and IL-6 secretion and mRNA expression in the jejunum and ileum of broilers compared to the model group.

Inflammatory cytokines contribute to host defense against bacterial and other microbial invasions, but their overexpression disrupts biological homeostasis, triggering an inflammatory cascade that compromises intestinal integrity and barrier function [43,44]. Conversely, these cytokines protect the intestinal tract from pathogen-induced damage. Liu et al. [18] demonstrated that licorice chalcone A reduced intestinal damage scores and preserved intestinal villus structure with minimal pathological changes in *C. perf*-induced *NE* in broiler chickens. In the present study, all treatment groups (low-dose, medium-dose, high-dose, medium-dose prophylaxis, high-dose control, and ampicillin control) reduced intestinal damage scores. However, only the high-dose, high-dose control, and ampicillin groups exhibited relatively intact intestinal villi structures in jejunal and ileal histological sections, with minimal necrotic detachment at intestinal villus apices in the high-dose group. Pathological damage in the jejunum and ileum of ampicillin-treated broilers was less severe than in the low-dose and medium-dose groups, resembling that of the control group. The high-dose and high-dose control groups displayed similar levels of jejunal and ileal pathological damage as the ampicillin group. The high-dose control group, lacking antibacterial treatment and receiving only high-dose PHCE, exhibited effects on antioxidant capacity, inflammatory cytokine secretion and expression, and intestinal damage similar to the control group, indicating the safety and non-toxicity of high-dose PHCE in broilers. Moreover, these findings suggest that high-dose PHCE effectively treats *C. perf*-induced *NE*.

Inflammation is a biological process safeguarding the body from pathogen invasion and cellular stress signals. The inflammatory response is triggered by inflammasome activation (intracellular protein complexes). Stress-induced danger signals encompass various molecules, including cellular debris (e.g., DNA fragments, ATP), pathogen-associated molecular patterns (e.g., bacterial LPS, viral dsRNA), and cytokines/chemokines released by damaged or stressed cells [45]. These molecules are recognized by the immune system and cellular receptors, activating inflammasomes and initiating the inflammatory response. PAMPs and DAMPs stimulate NLRP3 inflammasome activation, which subsequently activates caspase-1. Caspase-1 cleaves pro-IL-1β and pro-IL-18 into active IL-1β and IL-18, releasing them extracellularly. Additionally, caspase-1 cleaves GSDMD, a member of the gasdermin family, resulting in cell perforation and pyroptosis. However, excessive inflammasome activation can induce pathological inflammation [26]. Studies have demonstrated the anti-inflammatory effects of flavonoids through inflammasome inhibition. For instance, Tsai et al. [46] observed that the flavonoid gallocatechin gallate (EGCG) attenuated lupus nephritis in mice by inhibiting caspase-1 activation through decreased NLRP3 mRNA expression, reducing IL-1β and IL-18 production. Jiang et al. [47] investigated the effects of quercetin on mice with sodium urate crystal-induced gouty arthritis, finding that quercetin ameliorated symptoms by inhibiting NLRP3, caspase-1, and IL-1β mRNA expression in knee joints. ROS, as upstream regulators of NLRP3 inflammasome activation, are inhibited by flavonoids, which also reduce MDA production [48]. For example, lidocaine suppresses ROS production, inhibiting NLRP3 inflammasome activation and preventing THP-1 cell pyroptosis [49]. Apigenin and fusaricin decrease MDA levels in Adriamycin-induced renal injury and glyoxylate-induced renal tissue injury, respectively [50,51]. H_2O_2 and ·OH, both

ROS, contribute to MDA formation through lipid oxidation [52]. Compared to the model group, the high-dose PHCE group exhibited significantly or highly significantly lower H_2O_2 and MDA levels in jejunal and ileal tissues, suggesting reduced ROS production. Concurrently, NLRP3, caspase-1, IL-1β, and GSDMD mRNA expression was significantly or highly significantly elevated in jejunal and ileal tissues of high-dose, high-dose control, and ampicillin groups, but not in low-dose, medium-dose, or medium-dose prophylaxis groups. The high-dose control group (no antibacterial treatment) and ampicillin group (antibiotic effective against *C. perf*) served as controls. These findings suggest that high-dose PHCE protects against *C. perf*-induced NE by inhibiting NLRP3 inflammasome activation through reduced ROS production, subsequently attenuating intestinal damage.

4. Materials and Methods

4.1. Bacterial Strains and Drugs

C. perf type A (ATCC13124) was purchased from the American type culture collection (ATCC, Rockville, MD, USA), and anaerobically cultured in Brain heart infusion (BHI) broth (Qingdao Hopebio Co., Qingdao, China) or on Tryptose sulfite–cyloserine (TSC) agar base (Qingdao Hopebio Co. Qingdao, China) at 37 °C. The raw materials for the test drug, such as *Polygonum hydropiper L., Patchouli, Pulsatilla, Astragalus, Ligustrum lucidum,* and *Licorice,* were purchased from Gaoxiong Chinese Medicine (Kaohsiung, China).

4.2. Screening the Active Ingredients and Targets of PHCE

The compounds and the corresponding targets of *Polygonum hydropiper* L., *Patchouli, Pulsatilla, Astragalus, Ligustrum lucidum,* and *Licorice* in PHCE were obtained from the databases TCMSP (https://old.tcmsp-e.com/tcmsp.php, accessed on 15 June 2024) and CNN (https://www.cnki.net/, accessed on 15 June 2024). We retrieved all components of the PHCE compounds from TCMSP and CNN, which had oral bioavailability (OB) $\geq$ 30% and drug-likeness (DL) $\geq$ 0.18 as screening conditions to retrieve the active components and targets of action of PHCE [53,54].

4.3. NE Target Acquisition

To identify potential therapeutic targets for *NE*, the GeneCards database (https://www.genecards.org/, accessed on 16 June 2024) and PharmGKB database (https://www.pharmgkb.org/, accessed on 16 June 2024) were queried using the keyword "Necrotic Enteritis". Extracted disease targets from both databases were then merged and subjected to de-weighting to account for potential redundancies. Subsequently, the de-weighted list of potential targets was imported into the UniPort database (https://www.uniprot.org/, accessed on 16 June 2024) for name standardization to ensure a consistent nomenclature for NE targets.

4.4. Construction of Active Compound–Target Networks

Cytoscape is a network biology visualization and analysis software that enables the visualization of molecular interactions and biological processes [55]. The PHCE chemical composition and target files were imported into Cytoscape 3.9.1 to construct a compound–target network. In this network, each compound or target is represented by a node, and the relationships between them are depicted as connecting lines.

4.5. Construction of PPI Network

To identify potential PHCE targets for treating *NE*, the active ingredients of PHCE and known *NE* targets were imported into a bioinformatics platform (https://www.bioinformatics.com.cn/, accessed on 19 June 2024) to generate a Venn diagram. Overlapping genes within the diagram were considered potential targets for PHCE action against *NE*. These intersecting genes were subsequently uploaded to the STRING database (https://cn.string-db.org/, accessed on 20 June 2024) with a high confidence interaction score threshold of 0.9 [56]. Irrelevant nodes were hidden, and default settings were maintained for other parameters. The resulting interaction network data were exported in ".tsv" format and imported into Cytoscape 3.9.1 software. Network node properties, including Degree, Betweenness Centrality, and Closeness Centrality, were calculated within Cytoscape [57]. Nodes exceeding the median values for these properties were identified as key targets and used to construct the final PPI network.

4.6. Functional Enrichment Analysis

Genes identified at the intersection of PHCE's mechanism of action and *NE* were imported into the DAVID database (https://david.ncifcrf.gov/summary.jsp, accessed on 21 June 2024) for GO analysis with background enrichment against the KEGG pathway database [58]. The top 10 significantly enriched (p-value < 0.01) GO terms in each category (biological process, cellular component, and molecular function) were selected, along with the top 30 pathways ranked by gene enrichment [59]. Subsequently, GO and KEGG data were uploaded to the microbiometrics platform for visual analysis.

4.7. Molecular Docking Studies

Leveraging network pharmacology, the top three key components of PHCE effective against *NE* were identified and subjected to molecular docking simulations with the two most promising target proteins from the PPI network. Here, the key components act as ligands, and the target proteins function as receptors. The 3D structures of the target proteins were downloaded in PDB format from the RCSB PDB database (http://www.rcsb.org/, accessed on 22 June 2024) while ligand structures were retrieved in SDF format from the TCMSP and PubChem databases (https://pubchem.ncbi.nlm.nih.gov/, accessed on 23 June 2024). Open Babel 2.4.1 was used to convert ligand SDF files into MOL2 format, which is suitable for docking simulations. Finally, PyMoL 5.2.7 and AutoDock Vina [60] were employed to perform the docking simulations between the key PHCE ingredients and core target proteins. The minimum binding efficiency served as the metric for assessing ligand–receptor binding activity.

4.8. Preparation of PHCE

PHCE was prepared using a hydrodecoction–alcohol method. A 20 g mixture of six herbs (*Polygonum hydropiper* L., *Patchouli*, *Pulsatilla*, *Astragalus*, *Ligustrum lucidum*, and *Licorice*), sieved through a 100-mesh sieve, was decocted three times in ten volumes of distilled water at 100 °C for 1.5 h each. The combined filtrates (passed through 8-layer gauze) were centrifuged at 3000 rpm for 30 min, and the supernatant was concentrated using a rotary evaporator at 60 °C. Subsequently, 95% ethanol was added to achieve an 80% ethanol concentration, and the solution was left to stand for 24 h. After filtration through eight layers of gauze and centrifuging at 3000 rpm for 30 min, the ethanol was recovered using a rotary evaporator, and the solid obtained (20 g) was suspended in the corresponding medium to have 20 mL of solution (conc. 1 g/mL). The prepared PHCE was stored at 4 °C for subsequent use.

4.9. Determination of MIC and MBC

PHCE, *Polygonum hydropiper* L., patchouli, pulsatilla, astragalus, *Ligustrum lucidum*, and licorice medicinal liquids were autoclaved at 100 °C for 30 min for sterilization. Subsequently, 4 mL of each drug solution was pipetted into 1 mL of BHI medium, yielding an 800 mg/mL concentration. Serial two-fold dilutions were performed using BHI liquid medium to prepare 10 drug concentrations ranging from 0.78 mg/mL to 400 mg/mL. The *C. perf* bacterial solution in the logarithmic growth phase was diluted to about 2×10^6 CFU/mL in BHI medium. A sterile 96-well plate was then used to introduce 100 μL of the bacterial solution to each well, followed by the addition of 100 μL of the prepared drug solutions at varying concentrations, resulting in final drug concentrations ranging from 0.39 mg/mL to 200 mg/mL and a final bacterial concentration of 1×10^6 CFU/mL. Positive (100 μL BHI medium + 100 μL bacterial solution) and negative (200 μL BHI medium) controls were established in triplicate. In 96-well plates, each experimental group was repeated four times, and three 96-well plates were used for 12 replicates. All 96-well plates were incubated for 12–18 h at 37 °C in a 2.5 L round-bottomed vertical anaerobic culture bag (Qingdao Hopebio Co.) containing a 2.5 L anaerobic gas-generating bag and oxygen indicator. Following incubation, the cultures were mixed, and the OD600 nm value was measured using an enzyme-labeling instrument. An absorbance change of less than 0.05 indicated effective bacterial growth inhibition, defining the MIC. An absorbance change of less than 0.05 at OD600 nm was considered the absence of bacterial growth, allowing for the determination of the MIC of PHCE, *Polygonum hydropiper* L., patchouli, pulsatilla, astragalus, *Ligustrum lucidum*, and licorice medicinal liquids against *C. perf* based on the measured absorbance values.

To determine the MBC of PHCE against *C. perf*, five PHCE drug concentrations (0.78 mg/mL, 1.56 mg/mL, 3.13 mg/mL, 6.25 mg/mL, and 12.5 mg/mL) were prepared. One hundred microliters of bacterial suspension were evenly inoculated onto a sterile TSC solid medium and placed within a 2.5 L round-bottom vertical anaerobic culture bag. A 2.5 L anaerobic gas production package and oxygen indicator were added before incubation at 37 °C overnight. The PHCE concentration resulting in fewer than five colonies on the medium was defined as the MBC against *C. perf*.

4.10. Determination of Antimicrobial Activity

C. perf liquid was diluted to 1×10^6 CFU/mL in BHI medium and distributed into tubes. PHCE was introduced at final concentrations of 0 mg/mL, 0.78 mg/mL, 1.56 mg/mL, 3.13 mg/mL, 6.25 mg/mL, and 12.5 mg/mL, respectively. Tubes were incubated statically at 37 °C, and 100 μL bacterial aliquots were collected at 0, 1, 2, 3, 4, 6, 8, 10, and 12 h for bacterial count determination via plaque assay, with triplicate samples at each time point. Time–kill curves were generated, plotting time (hours) on the horizontal axis and colony number (log) on the vertical axis.

4.11. Animals and Treatment

Two hundred one-day-old healthy yellow-feathered broilers were purchased from Guangxi Shengde Poultry; the cages and equipment used in the test were thoroughly disinfected, and the coops were fumigated. Broilers were routinely vaccinated, fed, and watered ad libitum, and the coops were cleaned and disinfected regularly. Environmental temperature control was as follows: 32–34 °C in the first week, the temperature was reduced by 2–3 °C every week, and the humidity of the chicken house was controlled at 50–60%. The basal feed (BF) was the same corn–soybean meal type, formulated with reference to the Chicken Feeding Standard (2004), and its composition and nutrient content are shown in Table 6 [61]. The present experiment was carried out at the hatchery facility of the Laboratory of Histoembryology, College of Animal Science and Technology of Guangxi University, and the broiler rearing facility of Guangxi Key Laboratory of Animal Breeding, Disease Control and Prevention, Nan-Ning City, China. All animal procedures were approved by the Animal Care and Welfare Committee of Guangxi University, China (Protocol GXU-2020-008).

Two hundred one-day-old healthy broilers were randomly assigned to eight groups: control, model, medium-dose prophylaxis, low-, medium-, and high-dose ampicillin, and high-dose drug control. Each treatment group contained five replicates of five broilers. Broilers received a basal diet throughout the experiment (Table 7). On day 17, chicks were slaughtered via CO_2 asphyxiation. Under aseptic conditions, the abdominal cavity was opened to observe and score jejunal and ileal pathological changes. One-centimeter segments from the mid-sections of the jejunum and ileum were collected, fixed in precooled 4% paraformaldehyde, washed with sterilized phosphate-buffered saline (PBS) to remove intestinal contents, and stored at −80 °C in liquid nitrogen for rapid freezing.

Table 6. Basal feed composition and nutritional level (air-dried basis).

Ingredients	Contents (%)	Nutrient Level	Contents (%)
Grain	51.73	Crude protein	21.50
Soybean meal	40.73	Lys	1.17
Soybean oil	3.36	Met	0.59
Calcium phosphate	1.92	Calcium	1.00
Limestone	1.16	Available phosphorus	0.45
Salt	0.35	Met + Cystine	0.90
DL-Met	0.26		
50% Choline chloride	0.25		
Mineral premix	0.20		
Vitamin premix	0.24		
Total	100.00		

Note: (1) Mineral premix provided the following per kg of diet: Mn, 100 mg; Zn, 75 mg; Fe, 80 mg; Cu, 8 mg; I, 0.35 mg; Se, 0.15 mg. (2) Vitamin premix provided the following per kg of diet: VA 12 500 IU, VD_3 2500 IU, VE 30 IU, VK 2.65 mg, VB_1 2 mg, VB_2 6 mg, VB_{12} 0.025 mg, biotin 0.0325 mg, folic acid 1.25 mg, niacin 12 mg, and niacin 50 mg. (3) Metabolism, 3.35 Mcal/kg.

Table 7. Experimental group classification by dosage and stimulation protocols.

Group	1–6 d	7–13 d	14–16 d
Control group	BF	BF	PBS
Model group	BF	BF	*C. perf*
Medium-dose Prophylaxis group	BF	BF + 1 g/kg PHCE	*C. perf* + 1 g/kg PHCE
Low-dose group	BF	BF	*C. perf* + 0.5 g/kg PHCE
Medium-dose group	BF	BF	*C. perf* + 1 g/kg PHCE
High-dose group	BF	BF	*C. perf* + 2 g/kg PHCE
Ampicillin group	BF	BF	*C. perf* + 1 g/kg ampicillin
High-dose control group	BF	BF	2 g/kg PHCE

Note: BF: basal feed. Except for the medium-dose prophylaxis group, the other seven groups were fed basal feed from 1 to 13 days. The control group and model group were fed 1 mL sterile PBS and 1×10^9 CFU/mL *C. perf* bacterial solution once a day from 14 to 16 d, respectively, with the same basal feed. The medium-dose prophylaxis group was fed a basal feed from 1 to 6 d, 1 g/kg PHCE was added to the basal feed from 7 to 16 d, and 1 mL 1×10^9 CFU/mL *C. perf* bacterial solution was fed once a day from 14 to 16 d. The high-dose control group was fed a basal feed from 1 to 13 days, and 2 g/kg PHCE was added to the basal feed from 14 to 16 days. The low-dose group, medium-dose group, high-dose group, and ampicillin group were fed basal feed from 1 to 13 days, 1 mL of 1×10^9 CFU/mL *C. perf* bacterial solution from 14 to 16 d, and 0.5 g/kg, 1 g/kg, 2 g/kg PHCE, and 1 g/kg ampicillin were added to the basal diet, respectively, once a day. All groups were sampled at 17 d.

4.12. Determination of Oxidation and Antioxidant Indexes

Jejunal and ileal tissues were retrieved from −80 °C storage and minced into small pieces using autoclaved scissors. Approximately 0.1 g of tissue from each group was weighed and homogenized in 0.9 mL PBS containing sterile grinding beads using a high-speed tissue grinder. The homogenate was centrifuged at 5000× *g* for 10 min, and the supernatant was carefully transferred to a new centrifuge tube for subsequent analysis. SOD activity and H_2O_2 and MDA levels were quantified according to kit instructions. The H_2O_2 (20220509), MDA (20220428), and SOD (20220507) kits were procured from Nanjing Jiancheng Bioengineering Institute (Nanjing, China).

4.13. Determination of TNF-α, IL-1β and IL-6

TNF-α, IL-1β, and IL-6 levels were quantified in jejunal and ileal tissue homogenate supernatants using ELISA kits (TNF-α [MM-093801], IL-1β [MM-3691001], and IL-6 [MM-052101]) procured from Jiangsu Enzyme Immunity Industry Co., Ltd. (Taizhou, China), following the manufacturer's protocol.

4.14. Scoring of Jejunal and Ileal Lesions

Ocular pathologic changes were observed within a 5 cm segment of the jejunum and the anterior ileum. Intestinal injury was scored using a 0–6 point scale adapted from Shojadoost et al. [62]: 0 (no apparent injury), 1 (intestinal wall thinning and brittleness),

2 (1–5 necrotic foci), 3 (6–15 necrotic foci), 4 (16 or more necrotic foci), 5 (a 2–3 cm necrotic sheet), and 6 (a large, diffuse necrotic area).

4.15. Observation of Histopathologic Changes in the Jejunum and Ileum

Jejunal and ileal tissues were fixed in 4% paraformaldehyde for 24 h, followed by dehydration, clearing, wax embedding, sectioning, and hematoxylin–eosin (HE) staining to produce histopathological sections. Microscopic examination under low magnification assessed the morphology and structure of jejunum and ileum tissues, while high-magnification examination focused on intestinal villi morphology and structure.

4.16. RNA Extraction and Gene Expression Analysis

Approximately 50–100 mg of jejunum and ileum tissues, stored at −80 °C, were transferred to sterile, enzyme-free microcentrifuge tubes. Sterile, enzyme-free grinding beads were added, followed by 1 mL of RNA isolator reagent. The samples were homogenized using a high-speed tissue grinder, subsequently centrifuged at 12,000× g for 5 min at 4 °C, and the supernatant was transferred to a sterile, enzyme-free microcentrifuge tube. Total RNA from intestinal tissues was extracted using the RNA isolator Total RNA Extraction Reagent kit following sample processing.

RNA purity was assessed by spectrophotometric determination of OD260/280 values, while integrity was verified using 1% agarose gel electrophoresis. Reverse transcription employed the All-In-One 5× RT MasterMix kit (Nanjing Ai Bi MnegBiological Material Co., Ltd., Nanjing, China), generating cDNA from 1 μL RNA template, 4 μL MasterMix, and 15 μL ddH_2O through incubation at 37 °C for 15 min, 60 °C for 10 min, and 95 °C for 3 min. cDNA was stored at −20 °C. RT-qPCR primers targeting GADPH, caspase-1, IL-1β, NLRP3, IL-6, GSDMD, and TNF-α genes were designed using Primer Express 6.0 software based on GenBank sequences and synthesized by Shanghai Sangon Biotech Co., Shanghai, China (Table 8). Real-time PCR reactions were conducted using BlasTaq™ 2× qPCR MasterMix (Nanjing Ai Bi MnegBiological Material Co., Ltd.), comprising 10 μL MasterMix, 0.5 μL of each primer (10 μM), 1 μL cDNA, and 8 μL ddH_2O. A Light Cycler 96 real-time fluorescence quantitative PCR instrument (Roche, Basel, Switzerland) was used with the following cycling conditions: initial denaturation at 95 °C for 3 min, followed by 40 cycles of denaturation at 95 °C for 15 s and annealing/extension at 60 °C for 60 s.

Table 8. The sequence of primers.

Gene	Primer Name	Primer Sequence (5′→3′)
GADPH	GADPH-F	5′-CTGGCAAAGTCCAAGTGGTG-3′
	GADPH-R	5′-AGCACCACCCTTCAGATGAG-3′
caspase-1	caspase-1-F	5′-TTCCTTCAACACCATCTACG-3′
	caspase-1-R	5′-GGTGAGCTTCTCTGGTTTTA-3′
IL-1β	IL-1β-F	5′-AGCAGCCTCAGCGAAGAGACC-3′
	IL-1β-R	5′-GTCCACTGTGGTGTGCTCAGAATC-3′
NLRP3	NLRP3-F	5′-AGCTACCACACATCTAGGAT-3′
	NLRP3-R	5′-GGTGTCCAAATCCTCAATCT-3′
IL-6	IL-6-F	5′-AAGTTCACCGTGTGCGAGAA-3′
	IL-6-R	5′-TCAGGCATTTCTCCTCGTCG-3′
GSDMD	GSDMD-F	5′-ACTGAGGTCCACAGCCAAGAGG-3′
	GSDMD-R	5′-GCCACTCGGAATGCCAGGATG-3′
TNF-α	TNF-α-F	5′-CTTCCTGCTGGGGTGCATAG-3′
	TNF-α-R	5′-AAGAACCAACGTGGGCATTG-3′

4.17. Statistics Analysis

SPSS 23 software was used for statistical analysis. One-way analysis of variance (ANOVA) was used to test the main effect. When the differences were significant ($p < 0.05$), the group means were further compared using Tukey's test. Data were presented as mean ± standard deviation, with significance levels set at $p < 0.05$ and high significance at $p < 0.01$. The analysis for lesion score was performed using the nonparametric Kruskal–Wallis test. Graphical representations were generated using GraphPad Prism 9.

5. Conclusions

In summary, network pharmacology and molecular docking analyses indicated that PHCE's primary active components, including flavonoids such as quercetin, kaempferol, and isorhamnetin, potentially counteract *NE* by modulating the NLRP3 inflammasome signaling pathway. In vitro experiments demonstrated that both PHCE and its constituent herbs exhibit antibacterial properties against *Clostridium perfringens*, with PHCE demonstrating superior efficacy. In vivo studies align with network pharmacology findings, revealing that PHCE (2 g/kg) is safe and effectively mitigates oxidative stress in NE-infected broilers by enhancing intestinal antioxidant capacity. Moreover, PHCE inhibits NLRP3 inflammasome activation, potentially through reduced ROS production, leading to decreased expression of NLRP3 inflammasome signaling pathway-related factors. Consequently, PHCE attenuates jejunal and ileal inflammation, preserving intestinal integrity. The findings of this study provide a theoretical basis for the clinical application of PHCE. The active ingredients and content in PHCE still need further study, and the mechanism of PHCE against *NE* should be verified by knockout and over-expression of genes.

Author Contributions: Conceptualization, M.Y., T.H. and Z.X.; methodology, J.Z., C.P., S.Y. and J.H.; software, J.Z., M.L. and L.X.; validation, J.Z., C.P., M.L. and Y.W.; formal analysis, L.X. and J.H.; investigation, J.Z., C.P., S.Y. and J.F.; resources, M.Y., T.H. and Y.W.; data curation, J.Z., S.Y. and L.X.; writing—original draft preparation, J.Z., C.P., S.Y. and M.Y.; writing—review and editing, J.Z., C.P., S.Y., L.X., J.F., Y.W., T.H., J.H., Z.X. and M.Y.; visualization, J.Z. and Z.X.; supervision, J.H.; project administration, M.Y.; funding acquisition, M.Y. All authors have read and agreed to the published version of the manuscript.

Funding: This research was funded by the Guangxi BaGui Scholars Program Foundation [Grant number: 2019A50], the National Natural Science Foundation of China [Grant number: 32202850], Guangxi Natural Science Foundation [Grant number: 2021GXNSFAA196004], and Guangxi University Natural Science and Technology Innovation Development Multiplier Program Project [Grant number: 2024BZRC013].

Institutional Review Board Statement: The animal study protocol was approved by the Ethics Committee of Guangxi University (protocol code GXU-2020-008 and 15 March 2020).

Informed Consent Statement: Not applicable.

Data Availability Statement: All data are available from the corresponding author by request.

Conflicts of Interest: The authors declare no conflicts of interest.

References

1. Mehdizadeh Gohari, I.; Navarro, M.A.; Li, J.; Shrestha, A.; Uzal, F.; McClane, B.A. Pathogenicity and virulence of *Clostridium perfringens*. *Virulence* **2021**, *12*, 723–753. [CrossRef] [PubMed]
2. Brynestad, S.; Granum, P.E. *Clostridium perfringens* and foodborne infections. *Int. J. Food Microbiol.* **2002**, *74*, 195–202. [CrossRef] [PubMed]
3. Buiatte, V.; Dominguez, D.; Lesko, T.; Jenkins, M.; Chopra, S.; Lorenzoni, A.G. Inclusion of high-flavonoid corn in the diet of broiler chickens as a potential approach for the control of necrotic enteritis. *Poult. Sci.* **2022**, *101*, 102–104. [CrossRef] [PubMed]
4. Rood, J.I.; Adams, V.; Lacey, J.; Lyras, D.; Mcclane, B.A.; Melville, S.B.; Moore, R.J.; Popoff, M.R.; Sarker, M.R.; Songer, J.G. Expansion of the *Clostridium perfringens* toxin-based typing scheme. *Anaerobe* **2018**, *107*, 599–601. [CrossRef]
5. Jun, S.; Masahiro, N.; Masataka, O. *Clostridium perfringens* Alpha-Toxin: Characterization and Mode of Action. *J. Biochem.* **2004**, *136*, 569–574.
6. Bryant, A.E.; Stevens, D.L. Phospholipase C and perfringolysin O from *Clostridium perfringens* upregulate endothelial cell-leukocyte adherence molecule 1 and intercellular leukocyte adherence molecule 1 expression and induce interleukin-8 synthesis in cultured human umbilical vein endothelial cells. *Infect. Immun.* **1996**, *64*, 358–362.

7. Bryant, A.E.; Bergstrom, R.; Zimmerman, G.A.; Salyer, J.L.; Hill, H.R.; Tweten, R.K.; Sato, H.; Stevens, D.L. *Clostridium perfringens* invasiveness is enhanced by effects of theta toxin upon PMNL structure and function: The roles of leukocytotoxicity and expression of CD11/CD18 adherence glycoprotein. *FEMS Immunol. Med. Microbiol.* **1993**, *7*, 321–336. [CrossRef]
8. Uzal, F.A.; Vidal, J.E.; Mcclane, B.A.; Gurjar, A.A. Clostridium Perfringens Toxins Involved in Mammalian Veterinary Diseases. *Open Toxinology J.* **2010**, *2*, 24. [CrossRef]
9. Branton, S.L.; Reece, F.N.; Hagler, W.M. Influence of a wheat diet on mortality of broiler chickens associated with necrotic enteritis. *Poult. Sci.* **1987**, *66*, 1326–1330. [CrossRef]
10. Silva, P.; Rohloff, N.; Catoia, M.R.R.; Kaufmann, C.; Tesser, G.L.S.; Weber, S.H.; Campos, F.P.; Silva, L.; Ferreira, A.; Nunes, R.V.; et al. Alternative to antimicrobial growth promoters in the diets of broilers challenged with subclinical necrotic enteritis. *Poult. Sci.* **2024**, *103*, 103986. [CrossRef]
11. Li, S.; Zhang, K.; Bai, S.; Wang, J.; Zeng, Q.; Peng, H.; Lv, H.; Mu, Y.; Xuan, Y.; Li, S.; et al. Extract of Scutellaria baicalensis and Lonicerae flos improves growth performance, antioxidant capacity, and intestinal barrier of yellow-feather broiler chickens against *Clostridium perfringens*. *Poult. Sci.* **2024**, *103*, 103718. [CrossRef]
12. Gao, H.J.; Chen, C.F. In vitro bacterial inhibition of Aeromonas hydrophila by 21 Chinese herbs. *Reserv. Fish.* **1996**, *4*, 16–17.
13. Wei, Z.; Chen, X.Y.; Wang, S.; Ye, X.L.; Huang, L.P.; Wei, Y.Y.; Liu, M.; Hu, T.; Yu, M. Protective effect of Chinese herbal compound preparation on hepatopancreas necrosis of Penaeus vannamei induced by Vibrio parahaemolyticus. *Heilongjiang Anim. Sci. Vet. Med.* **2022**, *16*, 108–115.
14. Hu, T.J. Preparation Process of Compound Polygonum hydropiper L. Chinese Herbal Liquid Preparation. Guangxi University, as-signee. Chinese Patent No. CN105055541A, 11 January 2019.
15. Ahn-Jarvis, J.H.; Parihar, A.; Doseff, A.I. Dietary Flavonoids for Immunoregulation and Cancer: Food Design for Targeting Disease. *Antioxidants* **2019**, *8*, 202. [CrossRef]
16. Dai, H.; Lv, Z.; Huang, Z.; Ye, N.; Li, S.; Jiang, J.; Cheng, Y.; Shi, F. Dietary hawthorn-leaves flavonoids improves ovarian function and liver lipid metabolism in aged breeder hens. *Poult. Sci.* **2021**, *100*, 101499. [CrossRef]
17. Zhou, Y.; Mao, S.; Zhou, M. Effect of the flavonoid baicalein as a feed additive on the growth performance, immunity, and antioxidant capacity of broiler chickens. *Poult. Sci.* **2019**, *98*, 2790–2799. [CrossRef]
18. Liu, Z.Z. Therapeutic Effect and Application of Licochalcone A against *Clostridium perfringens* Infection. Ph.D. Thesis, Jilin University, Changchun, China, 2019.
19. Shen, X.Y. Treatment of infectious gastroenteritis in pigs with self-proposed *Polygonum hydropiper* L. compound formula. *J. Tradit. Chin. Vet. Med.* **1993**, *4*, 42–43.
20. Hopkins, A.L. Network pharmacology. *Nat. Biotechnol.* **2007**, *25*, 1110–1111. [CrossRef]
21. Hopkins, A.L. Network pharmacology: The next paradigm in drug discovery. *Nat. Chem. Biol.* **2008**, *4*, 682–690. [CrossRef]
22. Li, S.; Zhang, B. Traditional Chinese medicine network pharmacology: Theory, methodology and application. *Chin. J. Nat. Med.* **2013**, *11*, 110–120. [CrossRef]
23. Pinzi, L.; Rastelli, G. Molecular docking: Shifting paradigms in drug discovery. *Int. J. Mol. Sci.* **2019**, *20*, 4331. [CrossRef]
24. Hsin, K.Y.; Ghosh, S.; Kitano, H. Combining machine learning systems and multiple docking simulation packages to improve docking prediction reliability for network pharmacology. *PLoS ONE* **2013**, *8*, e83922. [CrossRef]
25. Long, J.R.; Pettit, J.R.; Barnum, D.A. Necrotic enteritis in broiler chickens. II. Pathology and proposed pathogenesis. *Can. J. Comp. Med. Rev. Can. Med. Comp.* **1974**, *38*, 467–474.
26. Jo, E.K.; Kim, J.K.; Shin, D.M.; Sasakawa, C. Molecular mechanisms regulating NLRP3 inflammasome activation. *Cell. Mol. Immunol.* **2016**, *13*, 148–159. [CrossRef]
27. Ahn, H.; Lee, G.S. Isorhamnetin and hyperoside derived from water dropwort inhibits inflammasome activation. *Phytomedicine Int. J. Phytother. Phytopharm.* **2017**, *24*, 77–86. [CrossRef]
28. Ruiz-Miyazawa, K.W.; Staurengo-Ferrari, L.; Mizokami, S.S.; Domiciano, T.P.; Vicentini, F.; Camilios-Neto, D.; Pavanelli, W.R.; Pinge-Filho, P.; Amaral, F.A.; Teixeira, M.M.; et al. Quercetin inhibits gout arthritis in mice: Induction of an opioid-dependent regulation of inflammasome. *Inflammopharmacology* **2017**, *25*, 555–570. [CrossRef] [PubMed]
29. Tian, H.; Lin, S.; Wu, J.; Ma, M.; Yu, J.; Zeng, Y.; Liu, Q.; Chen, L.; Xu, J. Kaempferol alleviates corneal transplantation rejection by inhibiting NLRP3 inflammasome activation and macrophage M1 polarization via promoting autophagy. *Exp. Eye Res.* **2021**, *208*, 108627. [CrossRef] [PubMed]
30. Abou Baker, D.H. An ethnopharmacological review on the therapeutical properties of flavonoids and their mechanisms of actions: A comprehensive review based on up to date knowledge. *Toxicol. Rep.* **2022**, *9*, 445–469. [CrossRef]
31. Zhou, L.Y.; Tao, J.Y.; Yu, M.L.; Hu, T.J. Study on antibacterial effect of the extracts from different polar of Laliao decotion in vitro. *Mod. J. Anim. Husb. Vet. Med.* **2020**, *9*, 1–4.
32. Zhu, D.G. Study on control effect of Compound Chinese Herbal medicine on Dactylogyriasis in Ornamental Fish. *J. Anhui Agric. Sci.* **2010**, *38*, 2977–2978.
33. Lucy, J.A. Free radicals in biology and medicine. *FEBS Lett.* **1985**, *188*, 331–332. [CrossRef]
34. Shen, N.; Wang, T.; Gan, Q.; Liu, S.; Wang, L.; Jin, B. Plant flavonoids: Classification, distribution, biosynthesis, and antioxidant activity. *Food Chem.* **2022**, *383*, 132531. [CrossRef]
35. Wang, D.; Gao, F.; Hu, F.; Wu, J. Nobiletin Alleviates Astrocyte Activation and Oxidative Stress Induced by Hypoxia In Vitro. *Molecules* **2022**, *27*, 14–16. [CrossRef] [PubMed]

36. Tao, J.Y.; Li, X.F.; Hu, T.J. Effect of compound Polygonum hydropiper L. on growth performance and antioxidant capacity of broiler chickens. *China Anim. Husb. Vet. Med.* **2013**, *5*, 490–493.
37. Collins, T. Acute and chronic inflammation. *Robbins Pathol. Basis Dis.* **1999**, *36*, 56–57.
38. Anderson, M.T.; Staal, F.J.; Gitler, C.; Herzenberg, L.A.; Herzenberg, L.A. Separation of oxidant-initiated and redox-regulated steps in the NF-kappa B signal transduction pathway. *Proc. Natl. Acad. Sci. USA* **1994**, *91*, 11527–11531. [CrossRef]
39. Flohé, L.; Brigelius-Flohé, R.; Saliou, C.; Traber, M.G.; Packer, L. Redox regulation of NF-kappa B activation. *Free Radic. Biol. Med.* **1997**, *22*, 1115–1126. [CrossRef] [PubMed]
40. Wan, S.S.; Li, X.Y.; Liu, S.R.; Tang, S. The function of carnosic acid in lipopolysaccharides-induced hepatic and intestinal inflammation in poultry. *Poult. Sci.* **2023**, *103*, 103415. [CrossRef]
41. Dusabimana, T.; Kim, S.R.; Kim, H.J.; Park, S.W.; Kim, H. Nobiletin ameliorates hepatic ischemia and reperfusion injury through the activation of SIRT-1/FOXO3a-mediated autophagy and mitochondrial biogenesis. *Exp. Mol. Med.* **2019**, *51*, 1–16. [CrossRef]
42. Yang, X. Efficacy Evaluation and Quality Control on the Compound Chinese Medicine Wumei Changyanning for the Treatment of Ulcerative Colitis. Ph.D. Thesis, Chongqing University, Chongqing, China, 2012.
43. Liu, C.; Chu, D.; Kalantar-Zadeh, K.; George, J.; Young, H.A.; Liu, G. Cytokines: From Clinical Significance to Quantification. *Adv. Sci.* **2021**, *8*, e2004433. [CrossRef]
44. McKay, D.M.; Baird, A.W. Cytokine regulation of epithelial permeability and ion transport. *Gut* **1999**, *44*, 283–289. [CrossRef] [PubMed]
45. Schaefer, L. Complexity of danger: The diverse nature of damage-associated molecular patterns. *J. Biol. Chem.* **2014**, *289*, 35237–35245. [CrossRef] [PubMed]
46. Tsai, P.Y.; Ka, S.M.; Chang, J.M.; Chen, H.C.; Shui, H.A.; Li, C.Y.; Hua, K.F.; Chang, W.L.; Huang, J.J.; Yang, S.S.; et al. Epigallocatechin-3-gallate prevents lupus nephritis development in mice via enhancing the Nrf2 antioxidant pathway and inhibiting NLRP3 inflammasome activation. *Free Radic. Biol. Med.* **2011**, *51*, 744–754. [CrossRef] [PubMed]
47. Jiang, W.; Huang, Y.; Han, N.; He, F.; Li, M.; Bian, Z.; Liu, J.; Sun, T.; Zhu, L. Quercetin suppresses NLRP3 inflammasome activation and attenuates histopathology in a rat model of spinal cord injury. *Spinal Cord* **2016**, *54*, 592–596. [CrossRef] [PubMed]
48. Man, S.M.; Kanneganti, T.D. Regulation of inflammasome activation. *Immunol. Rev.* **2015**, *265*, 6–21. [CrossRef]
49. Zou, Y.; Luo, X.; Feng, Y.; Fang, S.; Tian, J.; Yu, B.; Li, J. Luteolin prevents THP-1 macrophage pyroptosis by suppressing ROS production via Nrf2 activation. *Chem.-Biol. Interact.* **2021**, *345*, 109573. [CrossRef]
50. Wu, Q.; Li, W.; Zhao, J.; Sun, W.; Yang, Q.; Chen, C.; Xia, P.; Zhu, J.; Zhou, Y.; Huang, G.; et al. Apigenin ameliorates doxorubicin-induced renal injury via inhibition of oxidative stress and inflammation. *Biomed. Pharmacother. Biomed. Pharmacother.* **2021**, *137*, 111308. [CrossRef]
51. Ding, T.; Zhao, T.; Li, Y.; Liu, Z.; Ding, J.; Ji, B.; Wang, Y.; Guo, Z. Vitexin exerts protective effects against calcium oxalate crystal-induced kidney pyroptosis in vivo and in vitro. *Phytomedicine Int. J. Phytother. Phytopharm.* **2021**, *86*, 153562. [CrossRef]
52. Gutteridge, J.M. Biological origin of free radicals, and mechanisms of antioxidant protection. *Chem.-Biol. Interact.* **1994**, *91*, 133–140. [CrossRef]
53. Ru, J.; Li, P.; Wang, J.; Zhou, W.; Li, B.; Huang, C.; Li, P.; Guo, Z.; Tao, W.; Yang, Y.; et al. TCMSP: A database of systems pharmacology for drug discovery from herbal medicines. *J. Cheminformatics* **2014**, *6*, 13. [CrossRef]
54. Zhu, F.F.; Zhao, Y.M.; Chen, B.P.; Niu, H.Y.; Ren, S.Z. Mechanism of Polygonum Hydropiper L. in the Treatment of Gastric Ulcer Based on Network Pharmacology. *J. Emerg. Tradit. Chin. Med.* **2023**, *32*, 11–14.
55. Otasek, D.; Morris, J.H.; Bouças, J.; Pico, A.R.; Demchak, B. Cytoscape Automation: Empowering workflow-based network analysis. *Genome Biol.* **2019**, *20*, 185. [CrossRef] [PubMed]
56. Szklarczyk, D.; Morris, J.H.; Cook, H.; Kuhn, M.; Wyder, S.; Simonovic, M.; Santos, A.; Doncheva, N.T.; Roth, A.; Bork, P.; et al. The STRING database in 2017: Quality-controlled protein-protein association networks, made broadly accessible. *Nucleic Acids Res.* **2017**, *45*, D362–D368. [CrossRef] [PubMed]
57. Gan, X.X.; Zhong, L.K.; Shen, F.; Feng, J.H.; Li, Y.Y.; Li, S.J.; Cai, W.S.; Xu, B. Network Pharmacology to Explore the Molecular Mechanisms of Prunella vulgaris for Treating Hashimoto's Thyroiditis. *Front. Pharmacol.* **2021**, *12*, 700896. [CrossRef]
58. Huang, D.W.; Sherman, B.T.; Lempicki, R.A. Systematic and integrative analysis of large gene lists using DAVID bioinformatics resources. *Nat. Protoc.* **2009**, *4*, 44–57. [CrossRef]
59. Consortium, T.G.O. Expansion of the Gene Ontology knowledgebase and resources. *Nucleic Acids Res.* **2017**, *45*, D331–D338.
60. Trott, O.; Olson, A.J. AutoDock Vina: Improving the speed and accuracy of docking with a new scoring function, efficient optimization, and multithreading. *J. Comput. Chem.* **2010**, *31*, 455–461. [CrossRef]
61. *NY/T 33-2004*; Feeding Standard of Chicken. China Agriculture Press: Beijing, China, 2004.
62. Shojadoost, B.; Vince, A.R.; Prescott, J.F. The successful experimental induction of necrotic enteritis in chickens by *Clostridium perfringens*: A critical review. *Vet. Res.* **2012**, *43*, 74. [CrossRef]

antibiotics

Article

Antimicrobial Properties of Fennel By-Product Extracts and Their Potential Applications in Meat Products

Marica Egidio [1], Loriana Casalino [2], Filomena De Biasio [3], Marika Di Paolo [1,*], Ricardo Gómez-García [4], Manuela Pintado [4], Alma Sardo [1] and Raffaele Marrone [1]

1 Department of Veterinary Medicine and Animal Production, University of Naples Federico II, 80138 Naples, Italy; marica.egidio@libero.it (M.E.); alma.sardo@unina.it (A.S.); raffaele.marrone@unina.it (R.M.)
2 Department of Economic and Legal Sciences, Universitas Mercatorum, 00186 Rome, Italy; loriana.casalino@libero.it
3 EVRA S.r.l. Società Benefit, 85044 Lauria, Italy; filomena.debiasio@evra-ingredients.com
4 CBQF—Centro de Biotecnologia e Química Fina—Laboratório Associado, Escola Superior de Biotecnologia, Universidade Católica Portuguesa, Rua Diogo Botelho 1327, 4169-005 Porto, Portugal; rgarcia@ucp.pt (R.G.-G.); mpintado@ucp.pt (M.P.)
* Correspondence: marika.dipaolo@unina.it

Abstract: Background: Beef burgers are perishable meat products, and to extend their shelf life, EU Regulation 1129/11 permits the use of certain additives. **Objectives:** However, given the concerns of health-conscious consumers and the potential toxicity of synthetic substances, this study aimed to explore the use of fennel waste extracts as natural preservatives. **Methods:** This study characterized the bioactive compounds (phenolic content), the antioxidant activity ($ABTS^+$ and DPPH assay), and the antimicrobial properties (against *Salmonella enterica* serotype Enteritidis, *Escherichia coli*, *Staphylococcus aureus*, *Bacillus cereusi*, and *Pseudomonas aeruginosa*) of different fennel waste extracts (LF, liquid fraction; SF, solid fraction and PF, pellet fraction). Additionally, the potential use of the best fennel extract was evaluated for its impact on beef burger shelf life (up to 18 days at 4 ± 1 °C) in terms of microbiological profile, pH, and activity water (a_w). **Results:** The PF extract, which was rich in flavones, hydroxybenzoic, and hydroxycinnamic acids, demonstrated the highest antioxidant and antimicrobial activities. Microbiological analyses on beef burgers with PF identified this extract as a potential antimicrobial substance. The a_w and pH values did not appear to be affected. **Conclusions:** In conclusion, fennel extracts could be proposed as natural compounds exploitable in beef burgers to preserve their quality and extend their shelf-life.

Keywords: minced meat products; additives; natural compounds; fennel extracts

Citation: Egidio, M.; Casalino, L.; De Biasio, F.; Di Paolo, M.; Gómez-García, R.; Pintado, M.; Sardo, A.; Marrone, R. Antimicrobial Properties of Fennel By-Product Extracts and Their Potential Applications in Meat Products. *Antibiotics* **2024**, *13*, 932. https://doi.org/10.3390/antibiotics13100932

Academic Editor: William N. Setzer

Received: 3 August 2024
Revised: 26 September 2024
Accepted: 28 September 2024
Published: 1 October 2024

1. Introduction

Meat and meat products represent a significant source of energy and various nutrients in the human diet, such as high-value proteins containing essential amino acids, essential fatty acids, minerals (iron or zinc), and vitamins (B1, B2, B6, and B12) [1]. Their consumption is widespread globally, and they are commonly found in restaurants, fast food chains, and stores [2]. However, due to some physicochemical characteristics such as pH, water content, oxygen sensibility [3], and the production process, they are highly perishable products with a shorter shelf life compared to whole-muscle meat [4]. Microorganisms responsible for rapid microbiological spoilage lead to changes in flavor, odor, texture, color, and a reduction in shelf life [5]. In addition, microorganisms responsible for microbiological deterioration not only cause economic losses to the food industry, but some also pose health risks to consumers (*Salmonella* spp., *Staphylococcus aureus*, *Escherichia coli*, *Campylobacter jejuni*, *Listeria monocytogenes*, *Clostridium perfringes*, *Yersinia enterocolitica*, and *Aeromonas hydrophila*) [6]. Moreover, during storage, burgers also undergo natural chemical spoilage, characterized by lipid oxidation and protein degradation, resulting in nutrient loss and the development of unpleasant odors and flavors, further reducing their shelf life [7]. Therefore, given the economic importance of these meat products and high consumer demand, the primary concern of the food industry is to reduce and control microbial proliferation and chemical changes, thereby enhancing shelf life and overall safety [8]. Traditional meat preservation techniques include the use of synthetic chemical antioxidant and/or antimicrobial additives and preservatives, such as sulfur dioxide–sulphates, acetic acid, sorbates, and benzoates. These substances "prolong the shelf-life of foods by protecting them against deterioration caused by microorganisms and/or which protect against growth of pathogenic microorganisms." [9]. In this regard, the addition of nitrates and nitrites to meat products offers several benefits, including enhanced quality characteristics and improved microbiological safety. These compounds, in fact, play a crucial role in developing the distinctive flavor, maintaining the red color stability, and protecting against lipid oxidation in cured meats [10]. Nitrites, in particular, exhibit significant bacteriostatic and bactericidal properties against various spoilage bacteria and foodborne pathogens commonly found in meat products and prevent the growth and toxin production of *Clostridium botulinum*. However, the International Agency for Research on Cancer (IARC) has recently determined that ingested nitrates or nitrites can be probable human carcinogens under conditions that promote endogenous nitrosation. Although the use of synthetic preservations offers numerous technological benefits and are thus of great importance to the meat industry, the safety of these chemical agents is questionable [11]. Concerns persist about the potential risks associated with their consumption, such as allergic reactions, gastrointestinal issues, carcinogenicity, asthma, and behavioral disorders like hyperactivity [12–14]. For this reason, the development and application of novel technologies involving natural compounds with antimicrobial and antioxidant properties have become a new strategy adopted by the meat industry to produce healthier products. These products are free from synthetic chemical additives or preservatives yet still offer appealing colors and flavors [15–18]. Numerous natural antimicrobial substances are already in use today, including bacteriophages and their lysins [19,20], bacteriocins [21], and plant extracts (including essential oils) [22,23]. In this regard, extracts from medicinal and aromatic plants, rich in bioactive compounds (e.g., phenolics, terpenoids, carotenoids) with potent activity against microbial growth and oxidative reactions, are increasingly being used as natural additives and quality enhancers in minced meat processing [24–26].

Fennel (*Foeniculum vulgare* Mill.) is one of these medicinal and aromatic plants, renowned for its anticancer, antidementia, antiplatelet, antihirsutism, hepatoprotective, and anti-hyperlipidemic properties due to its chemical composition [27,28]. Fennel extracts, particularly from fennel leaves, are important sources of vitamins (A, C, thiamine, riboflavin, and niacin), minerals (potassium, sodium, calcium, and manganese) [29], and n-3 fatty acids (primarily found in the leaves) [30]. Rich in bioactive compounds (phenolic acids, tocopherols, flavonoids, terpenoids, carotenoids) and essential oils [31–33], these extracts have significant potential as preservatives (antioxidative and antimicrobial) in the food processing sector [34–36]. Notably, Yanan Sun et al. [36], in a study focused on the meat industry, demonstrated the antimicrobial effectiveness of fennel extracts against aerobic mesophilic bacteria in ground pork. The study showed that fennel extracts could be used to improve the quality of ground pork and extend its shelf life.

Recently, new attention has shifted towards the search for sustainable solutions for recycling and valorizing food by-products for their reintegration into industrial supply chains [37]. However, despite the potential of by-products of fruits and vegetables, little attention has been given to studying and utilizing this waste and its potential. Therefore, this study aimed to explore the use of fennel waste extracts as natural preservatives. This study characterized the bioactive compounds, the antioxidant activity, and the antimicrobial properties (against *Salmonella enterica* serotype Enteritidis, *Escherichia coli*, *Staphylococcus aureus*, *Bacillus cereus*, and *Pseudomonas aeruginosa*) of different fennel waste extracts. Additionally, the potential use of the best fennel extract was evaluated for its impact on beef burger shelf life (up to 18 days at 4 ± 1 °C) in terms of microbiological profile, pH, and activity water (a_w).

2. Results and Discussion

2.1. Total Phenolic and HPLC Analyses of the Phenolic Compounds in the Fennel Extracts

The results for the total phenolic content (TPC) and HPLC-DAD analysis are presented in Table 1. The data show that the fennel extract PF had the highest TPC, with 949 mg GAE/100 g DE, followed by the LF and SF fennel extracts, with 369.49 and 346.72 mg GAE/100 g DE, respectively. The total phenolic compounds measured based on the HPLC-DAD analysis were different and lower than those measured using the spectrophotometer based on the Folin–Ciocalteu assay (Table 1). These discrepancies can be attributed to the detection limits for certain phenolic compounds and to the fact that the Folin–Ciocalteu reagent can react with other molecules (such as sugars and proteins) present in the extracts, leading to measurement interference and an overestimation of total phenolic content. According to the HPLC-DAD analysis, four different chemical classes of polyphenols, including flavonols, flavones, and hydroxybenzoic and hydroxycinnamic acids, and seventeen individual phenolic compounds were identified in the three different fennel extracts (SF, LF, and PF), showing a high complexity and richness of such well-known bioactive compounds. Specifically, for each fennel fraction, at least nine of them (gallic acid, protocatechuic acid, catechin, dihydroxycaffeic acid, 4-hydroxybenzoic acid, caffeic acid, p-coumaric, ferulic acid, and rutin) were quantified using HPLC (Table 1). In this study, the most relevant phenolic compound was the protocatechuic acid hydroxybenzoic acid, whose highest concentration was found in the PF (126.89 mg/100 g DE). Moreover, 4-hydroxybenzoic acid and catechin (87.17 and 79.54 mg/100 g DE, respectively), which are bioactive polyphenols with antimicrobial, anti-inflammatory, and antioxidant properties attributed to their high free radical-scavenging capacity, were the most prominent compounds in the PF fennel extract. These higher PF values could be attributed to the presence of more lipophilic compounds, such as flavonoids (naringinin-7-glucoside, rutin, quercetin, and luteolin), which are associated with insoluble fiber and were separated during the milling and centrifugation process. Our data are consistent with previous research, such as the study by Roby et al. [38], which identified fifteen individual compounds, including quercetin, chlorogenic acid, and ferulic acid, and reported a lower total polyphenol content (TPC) in methanolic fennel extracts—340 mg GAE/100 g DE—compared to what was observed in this study. Similarly, Salami et al. [39] reported a high concentration of phenolic compounds (including chlorogenic acid, caffeic acid, p-coumaric acid, rutin, ferulic acid, 1,5-dicaffeoylquinic acid, quercetin, and apigenin) in fennel leaves and seeds, with TPC values of 200 mg GAE/100 g DE and 262 mg GAE/100 g DE, respectively.

2.2. Antioxidant Activity of the Fennel Extracts

The antioxidant activity of the free radical-scavenging fennel extracts is illustrated in Table 1. The results show that SF and PF had the highest antioxidant capacities compared to the LF extract. Specifically, SF exhibited strong scavenging effects on DPPH radicals, with values of 653.22 ± 14.66 µM TE/100 g DE, and on ABTS radicals, with values of 323.10 ± 25.11 mg AAE/100 g DE (dry extract). The PF extract, on the other hand, demonstrated the highest antioxidant activity on ABTS radicals, with values of 383.00 ± 18.07 mg AAE/100 g DE (dry extract), and on DPPH radicals, with values of 638.28 ± 46.70 µM TE/100 g DE. In contrast, the LF extract exhibited lower antioxidant activity, with scavenging effects on ABTS and DPPH radicals showing values of 328.85 ± 14.92 mg AAE/100 g DE (dry extract) and 143.64 ± 35.17 µM TE/100 g DE, respectively. These antioxidant activity results can be partly attributed to the high polyphenol content in the fennel extracts. Polyphenols are well-known for their potent ability to reduce oxidative compounds and scavenge free radicals, often surpassing the effectiveness of vitamins and carotenoids found in plants. Various studies have reported differing outcomes on this matter; while some authors have identified a correlation between phenolic content and antioxidant activity [40,41], others have found no such relationship, suggesting that other compounds may be responsible for the antioxidant effects [42]. Overall, the high antioxidant capacity observed in our study's extracts highlights the importance of studying fennel by-products as a cost-effective and rich source of natural antioxidants.

Table 1. HPLC-DAD analysis of individual phenolic compounds and antioxidant activity of the hydroalcoholic extracts from the fennel by-product fractions.

	Compound	Chemical Class	RT	Concentration (mg/100 g DE)		
				SF	LF	PF
1	Gallic acid	Hydroxybenzoic acid	14.07	22.45 ± 3.66	30.63 ± 1.58	30.05 ± 1.14
2	3,4 dihydroxybenzoic acid	Hydroxybenzoic acid	17.49	42.88 ± 10.90	24.17 ± 1.24	nd
3	Protocatechuic acid	Hydroxybenzoic acid	17.92	83.86 ± 36.54	32.11 ± 1.92	126.89 ± 7.56
4	Catechin	Flavonol	23.53	10.38 ± 3.62	27.34 ± 1.57	79.54 ± 4.58
5	Neochlorogenic acid	Hydroxycinnamic acids	25.57	ud	ud	ud
6	Chlorogenic acid	Hydroxycinnamic acids	28.00	ud	ud	ud
7	Dihydroxycaffeic acid	Hydroxycinnamic acids	29.29	45.98 ± 6.76	27.43 ± 1.39	79.80 ± 4.05
8	4-Hydroxybenzoic acid	Hydroxybenzoic acid	34.03	nd	90.90 ± 5.15	87.17 ± 4.58
9	Caffeic acid	Hydroxycinnamic acids	37.19	40.16 ± 0.30	9.53 ± 0.60	33.18 ± 2.05
10	*p*-Coumaric	Hydroxybenzoic acid	43.80	38.2 ± 6.41	36.35 ± 2.27	56.71 ± 3.37
11	Ferulic acid	Hydroxybenzoic acid	48.63	ud	22.81 ± 1.12	26.13 ± 0.90
12	Myricetin	Flavonoid	50.53	ud	nd	ud
13	Naringinin-7-glucoside	Flavones	52.45	ud	ud	ud
14	Naringinin	Flavones	53.45	ud	ud	ud
15	Rutin	Flavonoid	54.66	15.28 ± 12.51	2.09 ± 0.15	6.08 ± 0.45
16	Quercetin	Flavonol	55.13	ud	ud	ud
17	Luteolin	Flovones	56.46	ud	ud	ud
Total by HPLC				298.69 ± 24.16	303.37 ± 17.00	525.25 ± 29.05
TPC (mg GAE/100 g DE)				346.72 ± 17.67	369.49 ± 12.76	949.29 ± 114.37
Antioxidant activity						
ABTS (mg AAE/100 g DE)				323.10 ± 25.11	328.85 ± 14.92	383.00 ± 18.07
DPPH (µM TE/100 g DE)				653.22 ± 14.66	143.64 ± 35.17	638.28 ± 46.70

RT: retention time; LF, liquid fraction used directly; SF, solid fraction extract; PF, pellet fraction extract; TPC, total phenolic content; GAE, gallic acid equivalents; DE, dry extract; AAE, ascorbic acid equivalents; TE, trolox equivalents; nd: not detected; ud; under detection limit. All determinations were carried out in triplicate, and results are shown as mean value ± standard deviation.

2.3. Antimicrobial Screening of Fennel Fraction Extracts

The results illustrated in Figure 1 and Table 2 show the effects of the four fennel extracts (liquid fraction used directly, LF; liquid fraction at 40 °C, LF*; solid fraction, SF; and pellet fraction, PF) on the growth of *Staphylococcus aureus* and *Escherichia coli* in the first preliminary screening. Neither the directly used LF nor the concentrated LF* exhibited any inhibitory activity against the tested microorganisms. In fact, no inhibition zones were observed on the agar plates (Figure 1). Although a different concentration of the extracts was used in the screening test, making it difficult to compare antimicrobial activity, an inhibitory effect was observed for the SF and for PF extracts at concentrations of 110 mg DE/mL and 20 mg DE/mL, respectively, but only against *Staphylococcus aureus* (Figure 1a), with inhibition zones of 10 mm and 10.5 mm (Table 2). These data revealed less antimicrobial activity against *Staphylococcus aureus* compared to other studies. For example, Ghasemian et al., [43] reported that fennel essential oils exhibited an inhibition zone of 19 mm, while Barrahi et al., [44] found inhibition zones ranging from 10 to 20 mm. However, Lemiasheuski et al., [45], who also worked with essential oils extracted from various parts of the fennel plant (such as seeds, leaves, or aerial parts), reported inhibition zones of 5–6 mm, which were even smaller than those obtained in our study. It is important to note that the cited studies were focused on essential oils extracted directly from fennel, unlike our research, which examined whole extracts derived from fennel waste already used for other purposes. This difference could potentially explain the slight variations in our results compared to other studies in the field. In fact, it is well known that the antibacterial activity of a fennel extract is influenced by several different factors, including the fennel variety [33], the extraction method, and the part of the plant used for the extraction. For example, Ghafarizadeh et al., [46] observed that in the case of extracts obtained from fennel leaves, the ethanolic extract showed the highest inhibitory effect. Additionally, Rafieian et al., [47] demonstrated that extracts obtained from fennel flowers showed better antibacterial properties compared to the extracts obtained from other parts of the plant.

Moreover, in this first preliminary screening, none of the four extracts (LF, LF*, SF, PF) showed inhibitory effects against *Escherichia coli* (Figure 1b). Thus, the best result was obtained for the PF extract with the lowest concentration (20 mg DE/mL) and the higher inhibition (10.5 mm) only against *Staphylococcus aureus*, a Gram-positive bacterium. This result agrees with several previous studies [43,48] and has shown that fennel scrap extracts like fennel essential oils exhibit stronger antimicrobial activity against Gram-positive bacteria than Gram-negative bacteria. This difference in efficacy could be attributed to the distinct structural characteristics of their cell membranes. Gram-negative bacteria have a relatively thin peptidoglycan cell wall, which is further shielded by an outer membrane containing lipopolysaccharides. In contrast, Gram-positive bacteria lack this outer membrane but possess a much thicker peptidoglycan layer than that of Gram-negative bacteria [49,50].

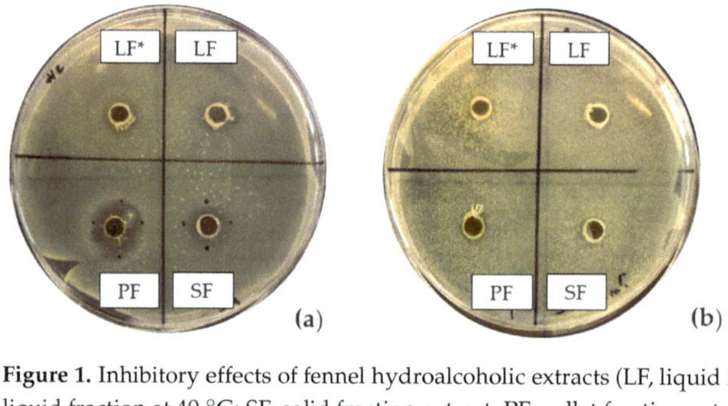

Figure 1. Inhibitory effects of fennel hydroalcoholic extracts (LF, liquid fraction used directly; LF*, liquid fraction at 40 °C; SF, solid fraction extract; PF, pellet fraction extract) against *Staphylococcus aureus* growth (**a**) and *Escherichia coli* (**b**).

Table 2. Inhibitory potentials of fennel extracts on *Staphylococcus aureus* and *Escherichia coli*.

Fennel Extract	Positive Effect (×)		Concentration (mg DE/mL)	Inhibition Zone Diameter (mm)
	S. aureus	*E. coli*		
LF	-	-	38.7	-
LF*	-	-	160	-
SF	×	-	110	10.00 ± 0.21
PF	×	-	20	10.50 ± 0.52

LF, liquid fraction used directly; LF*, liquid fraction at 40 °C; SF, solid fraction extract; PF, pellet fraction extract. (-) no inhibition zone was observed; (×), positive effect was observed.

After identifying the fennel extracts with antimicrobial properties (SF and PF) and their respective concentrations (50, 100, and 150 mg/mL for the SF; 10 and 20 mg DE/mL for PF), a second screening using a microplate-based assay was conducted. The data revealed that the fennel extracts (SF and PF) had inhibitory effects on the growth of all tested indicator strains (*Salmonella enterica* serotype Enteritidis, *Escherichia coli*, *Pseudomonas aeruginosa*, *Staphylococcus aureus*, and *Bacillus cereus*). Microbial growth, derived from absorbance data, decreased as the fennel extract concentration increased. As shown in Table 3, higher concentrations of the extracts resulted in greater growth inhibition, suggesting a proportional relationship between the amount of fennel extract used and its antimicrobial activity, as noted by Di Napoli et al., [51]. In summary, PF exhibited the best antimicrobial properties against all the tested microorganisms due to its effectiveness at lower concentrations (MIC = 2%) compared to the SF extract (MIC = 10%). In fact, an almost total inhibition (72–81%) of microbial growth was observed, particularly for the Gram-negative bacteria *Salmonella enterica* serovar Enteritidis, *Escherichia coli*, and *Pseudomonas aeruginosa*. Our data are consistent with other studies, which have confirmed the bactericidal activity of the fennel extract against these pathogens [51–53]. Moreover, in agreement with Ozcan et al., [54], the PF has shown its biocidal power against the Gram-positive *Bacillus cereus*, with an inhibition percentage (76%) similar to that observed for Gram-negative bacteria. Finally, an action, albeit minimal, was also observed on Gram-positive *Staphylococcus aureus*, with an inhibition of 28%. As the extracts were obtained from fresh fennel waste, this antimicrobial effectiveness could be attributed to the bioactive compounds present in the vegetable waste, such as hydroxybenzoic acids (gallic acid; 3,4 dihydroxybenzoic acid; protocatechuic acid; 4-hydroxybenzoic acid; *p*-coumaric and ferulic acid), hydroxycinnamic acids (dihydroxycaffeic acid and caffeic acid), flavonols (catechin), and flavonoids (rutin). Specifically, flavonoids are often reported to possess membrane-disrupting and inhibition activities of cell envelope synthesis, nucleic acid synthesis, electron transport chain, ATP synthesis, and biofilm formation [55], while the antimicrobial mechanisms of phenolic acids (hydroxybenzoic and hydroxycinnamic acids) are not yet fully understood. However, the leading theory suggests that these compounds destabilize microbial cell surfaces and cytoplasmic membranes, causing irreversible damage to the cell wall and various intracellular organelles. This process may lead to the coagulation of cellular components and the inhibition of intracellular enzymes. Additionally, once the cell wall is compromised, phenolic compounds may interact with intracellular components and DNA. The disruption of internal membranes can also release free radicals, which can further damage DNA and induce lipid oxidation [56].

Table 3. Antimicrobial activity of fennel solid fraction extract, SF (concentrations: 50 mg DE/mL = 5%, 100 mg DE/mL = 10%, and 150 mg DE/mL = 15%) and pellet fraction extract, PF (concentrations: 10 mg DE/mL = 1% and 20 mg DE/mL = 2%) on *Salmonella enterica* serotype Enteritidis, *Escherichia coli*, *Pseudomonas aeruginosa*, *Staphylococcus aureus*, and *Bacillus cereus* growth and comparison with the positive control (PC).

Indicator Strains	SF	PF
Salmonella enterica serotype Enteritidis	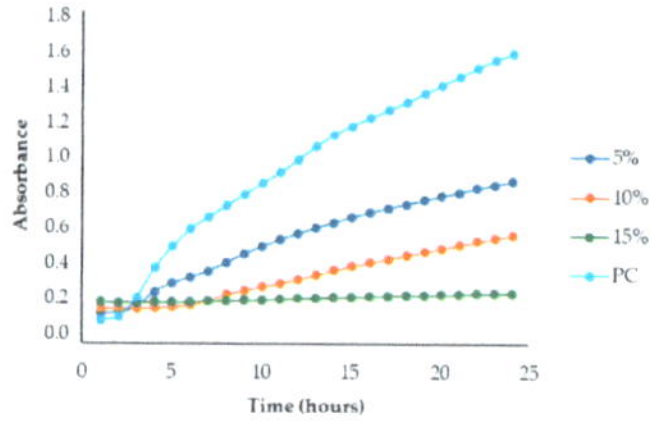	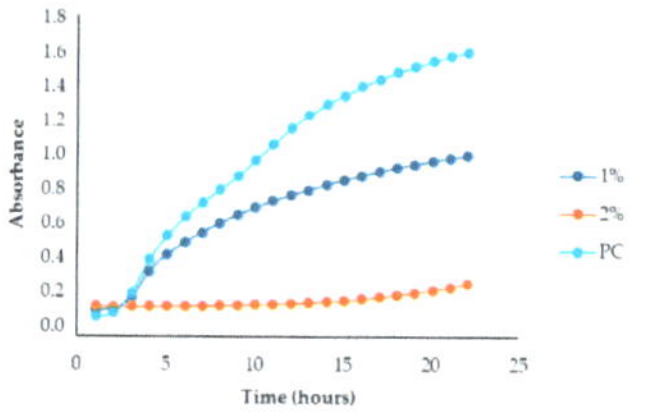
Escherichia coli	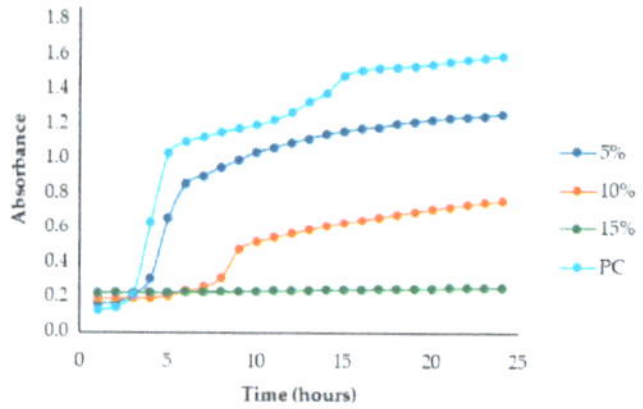	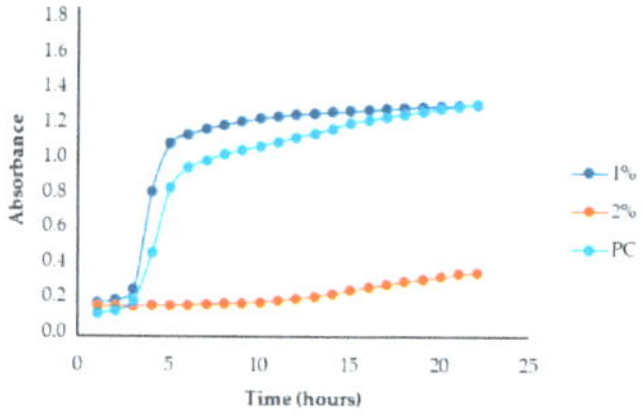
Pseudomonas aeruginosa	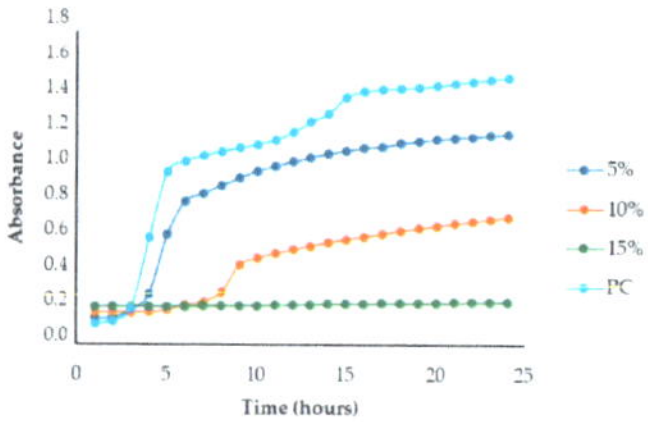	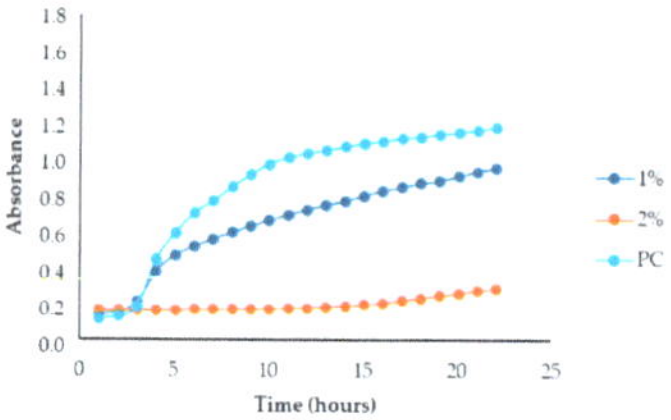
Staphylococcus aureus	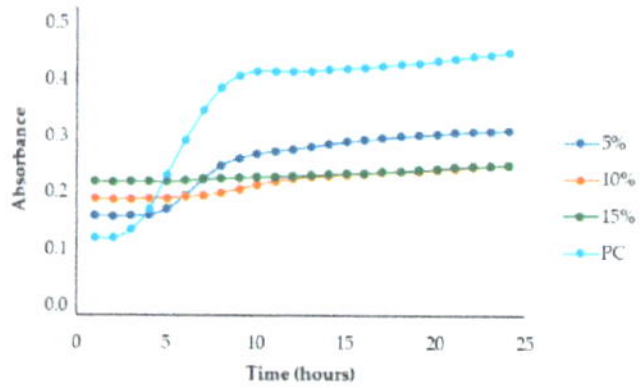	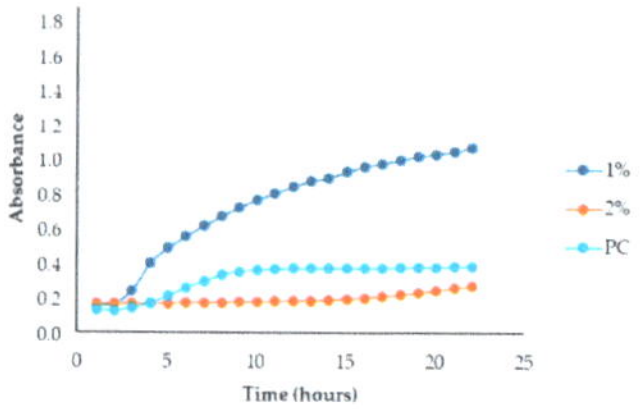
Bacillus cereus	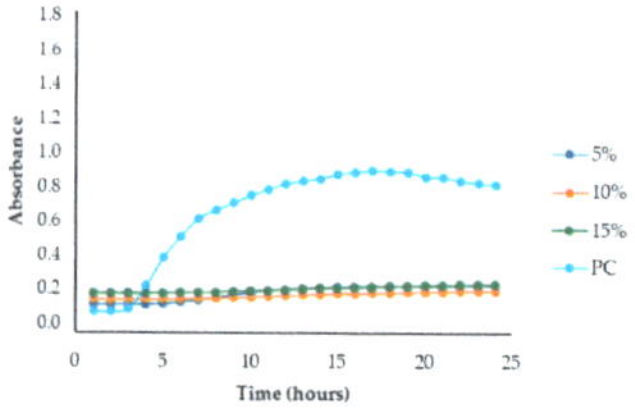	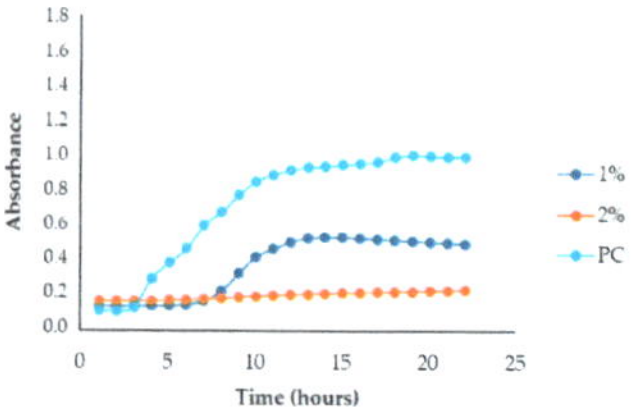

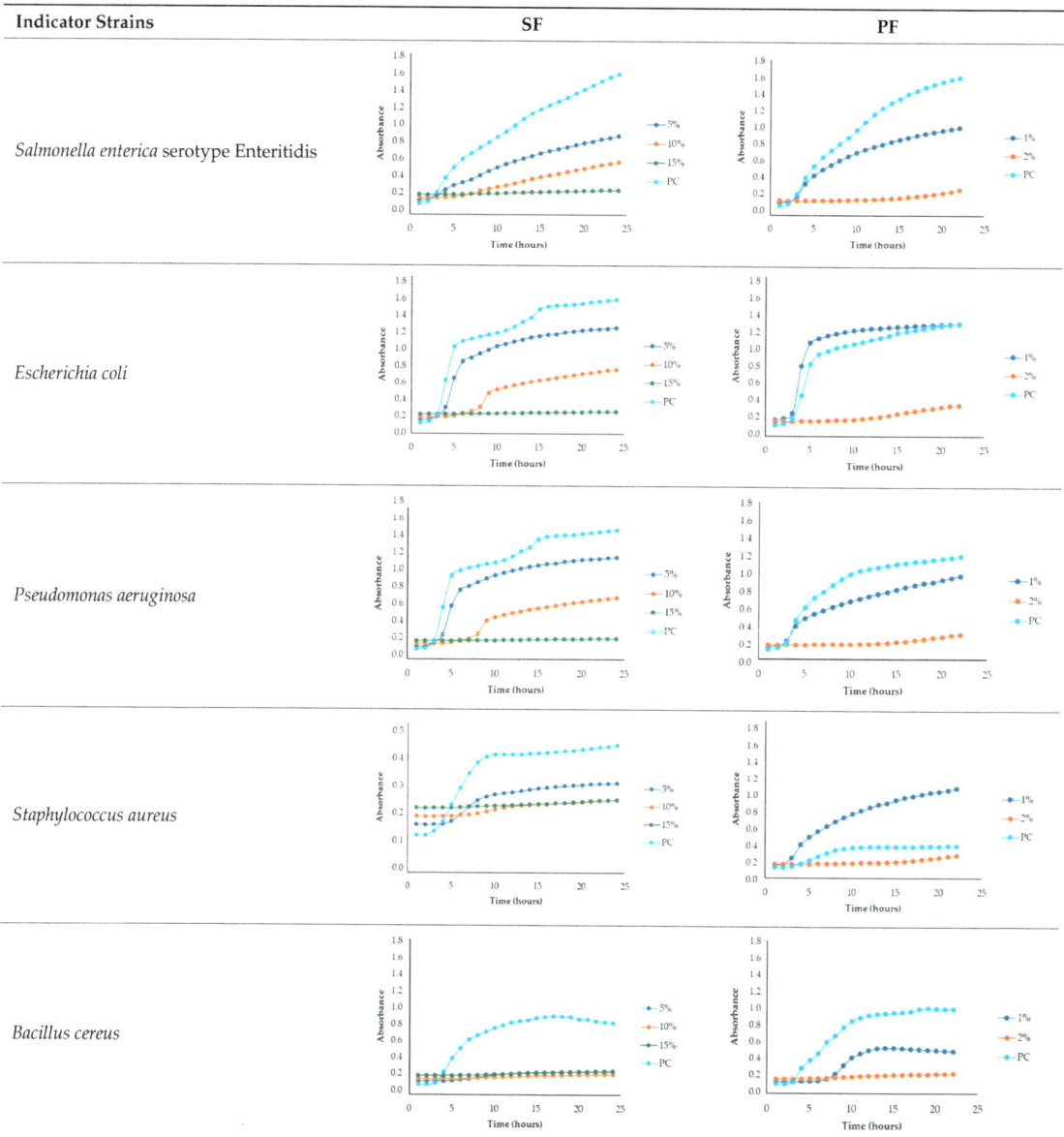

Specifically, for *Salmonella enterica* serotype Enteritidis, an increase in the inhibitory index (%) was observed, directly proportional to the concentration of fennel extracts. The inhibition ranged from 44% to 82% for the SF extract and from 37% to 81% for the PF extract (Table 4). These findings are consistent with several studies investigating the antimicrobial effects of fennel plant extracts (notably essential oils) against *Salmonella enterica* serotype Enteritidis, demonstrating their inhibitory potentials on both the growth and virulence of this pathogen at different concentrations. For example, Pluta et al., [56] observed that fennel oils showed inhibitory effects at various concentrations, with the minimum inhibitory concentration (MIC) varying based on the specific strain and testing conditions. Moreover, Di Napoli et al., [51], in a study closely aligned with ours, revealed that fennel extracts, being rich in bioactive compounds, can significantly disrupt bacterial biofilms at low concentrations, showing strong antibacterial activity against various pathogens, including *Salmonella enterica*. Additionaly, Almuzaini [57] suggests that plant-derived compounds, including those found in fennel, such as phenolic compounds, polyphenols, flavonoid, terpenoids, and essential oils, can disrupt bacterial cell membranes, inhibit biofilm formation, and lead to bacterial death, particularly in strains like *Salmonella Typhimurium* and *Salmonella Enteritidis*.

Table 4. Inhibitory potentials of the solid fennel (SF) and pellet extract (PF) on *Salmonella enterica* serotype Enteridis.

Item	SF			PF	
	50 mg DE/mL (5%)	**100 mg DE/mL (10%)**	**150 mg DE/mL (15%)**	**10 mg DE/mL (1%)**	**20 mg DE/mL (2%)**
Absorbance	0.843	0.556	0.271	1.013	0.300
Inhibition index, %	43.82	62.92	81.92	36.57	81.21

The absorbance value and the calculation of inhibition index (%) were recorded after 22 h.

Unlike the first preliminary screening, in the second one, both fennel extracts (SF and PF) showed antimicrobial effectiveness against *Escherichia coli*. An inhibitory index (%) ranging from 21% to 83% for the SF and from 0.12% to 72% for the PF was observed (Table 5). These data are in accordance with the study by Di Napoli et al. [51], where the fennel extract effectiveness was found to be dose-dependent, with higher concentrations leading to greater bacterial inhibition. According to Pluta et al., [56], this antimicrobial activity may be attributed to the bioactive components found in the extracts, which disrupt bacterial cell membranes and interfere with essential cellular processes. Notably, the mechanism involves disruption of the bacterium's outer membrane, which, although more resistant due to its lipopolysaccharide layer, can still be permeated by the extract's hydrophobic compounds under appropriate conditions [56]. Additionally, it is important to highlight that among all the Gram-negative bacteria analyzed in the present study, *Escherichia coli* was the most resistant to fennel extracts, as also observed Shabnam et al. [58]. However, unlike the study carried out by Manonmani and Khadir [59], which found that the ethanolic fennel extracts did not exhibit antibacterial effects against *Escherichia coli*, we found antimicrobial activity of the extracts even at a concentration of 20 mg DE/mL (MIC = 2%). The obtained MIC values are comparable to those reported by Gheorghita et al., (5–10%) [60].

Table 5. Inhibitory potentials of the solid fennel (SF) and pellet extract (PF) on *Escherichia coli*.

Item	SF			PF	
	50 mg DE/mL (5%)	**100 mg DE/mL (10%)**	**150 mg DE/mL (15%)**	**10 mg DE/mL (1%)**	**20 mg DE/mL (2%)**
Absorbance	1.199	0.717	0.253	1.013	0.300
Inhibition index, %	21.01	52.74	83.31	0.117	72.189

The absorbance value and the calculation of inhibition index (%) were recorded after 22 h.

Regarding the solid fraction, a similar pattern to that observed for *Escherichia coli* was noted for *Pseudomonas aeruginosa,* with an inhibitory index (%) ranging from 33% to 79% (Table 6). However, this Gram-negative bacterium proved to be more sensitive than *Escherichia coli* to the PF extract, with an inhibitory index (%) ranging from 19% to 75% (Table 5). This finding agrees with the study carried out by Daniela Gheorghita et al., [60] which reported the greater susceptibility of *Pseudomonas aeruginosa* compared to *Escherichia coli* regarding certain essential oils derived from various plants, including fennel, particularly in biofilm formation inhibition and growth reduction. In fact, the fennel's ability to prevent biofilm formation, which is a key factor in *Pseudomonas aeruginosa's* pathogenicity, disrupts its defense mechanism even at low concentrations, allowing for better control of bacterial growth [61]. Furthermore, generally, the minimum inhibitory concentrations (MIC) of essential oils for antibacterial activity may vary depending on the specific bacterial strain and the oil's composition [61]. In our study, the MIC for PF (which is the best extract) was even lower (2 mg DE/mL) than that reported by Diao et al. [35] (>10 mg DE/mL), despite their research focusing on the antimicrobial activity of essential oils from fennel seeds (*Foeniculum vulgare* Mill.) and not on fennel scrap extracts.

Table 6. Inhibitory potentials of the solid fennel (SF) and pellet extract (PF) on *Pseudomonas aeruginosa.*

Item	SF			PF	
	50 mg DE/mL (5%)	100 mg DE/mL (10%)	150 mg DE/mL (15%)	10 mg DE/mL (1%)	20 mg DE/mL (2%)
Absorbance	0.820	0.475	0.253	0.956	0.291
Inhibition index, %	33.44	61.42	79.44	18.91	75.27

The absorbance value and the calculation of inhibition index (%) were recorded after 22 h.

Unlike the previous trends, the PF at a concentration of 10 mg DE/mL was unable to inhibit or reduce the proliferation of *Staphylococcus aureus.* In fact, for this microorganism, higher growth was observed compared to the untreated positive control (PC) (Table 3). At a concentration of 20 mg DE/mL, the PF showed slight efficacy, with an inhibitory index (%) of 28% (Table 7). For the SF (Table 3), the best result was observed at a concentration of 100 mg DE/mL (10%) with an inhibition of 43%, confirming the data obtained in the first preliminary screening (Figure 1 and Table 1). These results differ from those of Rafiejan et al., [47] and Barrahi et al., [44] who found that *Staphylococcus aureus* was among the bacteria most sensitive to fennel extracts. Conversely, Manonmani and Khadir reported that fennel ethanolic extracts showed no antibacterial activity against *Staphylococcus aureus.* In our study, however, the most effective ethanolic extract (PF) reduced microbial growth by 28%. According to Shahat et al., [62], these variations are likely due to differences in the concentrations used, extraction methods, and the specific parts of the plant analyzed. In fact, fennel extracts (notably essential oils) are effective in concentrations ranging from low to moderate, depending on the strain, the conditions, and the type of extract used.

Table 7. Inhibitory potentials of the solid fennel (SF) and pellet extract (PF) on *Staphylococcus aureus.*

Item	SF			PF	
	50 mg DE/mL (5%)	100 mg DE/mL (10%)	150 mg DE/mL (15%)	10 mg DE/mL (1%)	20 mg DE/mL (2%)
Absorbance	0.298	0.240	0.242	1.027	0.266
Inhibition index, %	29.30	42.90	42.58	-	28.30

The absorbance value and the calculation of inhibition index (%) were recorded after 22 h.

For *Bacillus cereus*, the SF at a concentration of 100 mg DE/mL (10%) exhibited greater effectiveness than the SF at a concentration of 150 mg DE/mL (15%), with an inhibitory index of 72% (Table 8). For the PF, an increase in antibacterial activity directly proportional to the fennel extract concentrations was observed (Table 3), with an inhibitory index (%) ranging from 50% to 76% (Table 8). These findings are in accordance with Lemiasheuski et al., [45], who observed an inhibitory effect of approximately 50% on *Bacillus cereus* growth. The variation in antimicrobial efficacy may be attributed to differences in the type of extract used (essential oil versus ethanolic extract) and the concentration applied.

Table 8. Inhibitory potentials of the solid fennel (SF) and pellet extract (PF) on *Bacillus cereus*.

Item	SF			PF	
	50 mg DE/mL (5%)	100 mg DE/mL (10%)	150 mg DE/mL (15%)	10 mg DE/mL (1%)	20 mg DE/mL (2%)
Absorbance	0.266	0.236	0.274	0.498	0.236
Inhibition index, %	68.92	72.42	67.99	50.17	76.36

The absorbance value and the calculation of inhibition index (%) were recorded after 22 h.

Overall, the data show that the maximum antibacterial activity was detected at a concentration of 20 mg DE/mL (2%) for the PF and 150 mg DE/mL (15%) for the SF, except for the Gram-positive bacteria *Bacillus cereus* and *Staphylococcus aureus*, where the best results were obtained at a concentration of 100 mg DE/mL (10%). Additionally, it is important to note that, unlike in the first preliminary screening, both fennel extracts demonstrated antibacterial effectiveness against *Escherichia coli* in this case, successfully reducing its proliferation.

2.4. *Effects of Selected Fennel Extract on Beef Burgers*

The preliminary in vitro screenings showed that the fennel leave extract with the best antimicrobial properties was the hydroalcoholic extract obtained from the semi-solid pellet. Based on these initial results, a shelf-life test was conducted in an Italian meat cutting and processing company to confirm its antimicrobial activity on beef burgers packaged in a skin pack for 18 days.

Microbiological Analysis

The results of the microbiological analysis of treated (TRT) and untreated (CTR) beef burgers are shown in Table 9. Overall, the data showed that all the analyzed parameters of the TRT beef burgers were not only recorded below the limits (m and M) set by the EC Regulation 1441/07 [63] but also exhibited lower values compared to those of CTR group (Table 9). However, no significant differences were found between CTR and TRT beef burgers (<1.0 Log (CFU/g)), likely because these products originated from the production facility and were prepared for commercial distribution, resulting in low contamination levels.

The TAB 30 °C of the TRT group exhibited an increasing trend during the storage until reaching a concentration equal to 5.32 Log (CFU/g) at 15 and 18 d. Despite this, it always remained lower than the TAB 30 °C registered for the CTR group (5.52 Log (CFU/g) at 18 d) and lower than 5.69 Log (CFU/g), which is the m limit set by EC Regulation 1441/07.

From 0 to 8 days, yeasts and molds showed a constant trend, with a concentration always below 1.0 Log (CFU/g). An increase in microbial proliferation was observed only at the end of the storage (from 15 to 18 d), but with concentrations (1.56 Log (CFU/g)) always lower than those obtained for the CTR group, which instead exhibited an increasing trend starting from day 8 (with 1.1 Log (CFU/g) at 8 d until 1.63 Log (CFU/g) at 18 d).

Table 9. Microbiological ((Log (CFU/g)) results of treated (TRT) and untreated (CTR) beef burgers.

Item		0 d	5 d	8 d	12 d	15 d	18 d	Limits	RE
TAB 30 °C	TRT	2.36	2.80	3.26	3.71	5.32	5.32	m = 5.69	Reg. CE 2073/05
	CTR	2.43	2.84	3.34	3.81	5.50	5.52	M = 6.69	s.m.i. 1441/07
Total Coliforms	TRT	<1.0	<1.0	<1.0	<1.0	<1.0	<1.0		
	CTR	<1.0	<1.0	<1.0	<1.0	<1.0	<1.0		
β-glucuronidase-positive *Escherichia coli*	TRT	<1.0	<1.0	<1.0	<1.0	<1.0	<1.0	m = 2.69	Reg. CE 2073/05
	CTR	<1.0	<1.0	<1.0	<1.0	<1.0	<1.0	M = 3.69	s.m.i. 1441/07
Yeasts and molds	TRT	<1.0	<1.0	<1.0	<1.0	1.56	1.56		
	CTR	<1.0	<1.0	1.1	1.32	1.61	1.63		
Coagulase-positive Staphylococci	TRT	<1.0	<1.0	<1.0	<1.0	<1.0	<1.0		
	CTR	<1.0	<1.0	<1.0	<1.0	<1.0	<1.0		
Listeria monocytogenes	TRT	<1.0	<1.0	<1.0	<1.0	<1.0	<1.0	2.00	Reg. CE 2073/05
	CTR	<1.0	<1.0	<1.0	<1.0	<1.0	<1.0		s.m.i. 1441/07
pH	TRT	5.81 ± 0.02	5.70 ± 0.03	5.59 ± 0.02	5.48 ± 0.01	5.45 ± 0.01	5.40 ± 0.03		
	CTR	5.81 ± 0.02	5.73 ± 0.03	5.65 ± 0.02	5.53 ± 0.01	5.45 ± 0.02	5.37 ± 0.02		
a_w	TRT	0.979 ± 0.000	0.980 ± 0.002	0.981 ± 0.001	0.975 ± 0.006	0.976 ± 0.006	0.975 ± 0.004		
	CTR	0.979 ± 0.000	0.979 ± 0.001	0.980 ± 0.002	0.980 ± 0.000	0.975 ± 0.009	0.974 ± 0.003		

TRT, treated beef burgers during cold storage; CTR, untreated beef burgers during cold storage; TAB 30 °C, total aerobic plate count 30 °C.

The other microbial determinations (coliforms, Beta-glucuronidase-positive *Escherichia coli*, coagulase-positive staphylococci, and *Listeria monocytogenes*) had a constant trend over time in both groups (CTR and TRT), always remaining below 1.0 Log (CFU/g) (respecting the m limits set by the EC Regulation 1441/07 [63] of 2.69 Log (CFU/g) for Beta-glucuronidase-positive *Escherichia coli* and of 2.0 Log (CFU/g) for *Listeria monocytogenes*). *Salmonella* spp. was not found at any time during the storage period. Similar findings were also reported by others [51,64,65] who evaluated the antimicrobial activity of extracts obtained from different parts of the fennel plant on meat products. Minor differences compared to other studies may be attributed to our use of extracts from fennel waste and their use of extracts or essential oils obtained directly from the plant.

With regard to the chemical analyses, pH and activity water (a_w) are essential quality indicators for meat and meat products, as they impact the microbiological safety, water-holding capacity, and tenderness of the final products [26,66].

Throughout the storage period, the water activity (a_w) values for the TRT group, ranging in intervals from 0.975 to 0.981, increased and dropped (from 0.979 at day 0 to 0.981 at day 8, 0.975 at day 12, 0.976 at day 15, and 0.975 at day 18) in narrow intervals, showing a fluctuating trend (Table 9). In comparison, the water activity (a_w) values for the CTR group, ranging in intervals from 0.974 to 0.980, exhibited a gradual increase until day 12 before declining by day 18 (from 0.979 at day 0 to 0.980 at days 8 and 12, 0.975 at day 15, 0.974 at day 18).

Regarding pH values, both groups showed a decreasing trend over time. According to Huang et al., [67] this decline could be attributed to the activity of muscle and microbial enzymes, as well as the formation of organic acids. However, the TRT group showed a more pronounced pH reduction compared to the CTR group, except on the last two days (day 15 and day 18). Notably, on day 15, both groups had identical pH values (5.45), while on day 18, the CTR group had a slightly lower pH value (5.37) compared to the TRT group (5.40). This greater decrease in pH values observed in the TRT group may be due to a reduced rate of ammonia production, possibly influenced by the terpenoids present in natural extracts [67].

3. Materials and Methods

3.1. Reagents and Standards

Anhydrous sodium carbonate (Na_2CO_3) and Folin–Ciocalteu's reagent were purchase from Merck (Algés, Portugal). Standards of gallic acid, caffeic acid, chlorogenic acid, p-coumaric acid, dihydroxicaffeic acid, ferulic acid, myricetin, quercetin, resveratrol, and rutin were acquired from Sigma-Aldrich (St. Louis, MA, USA), whereas catechin, (−epicatechin, epicatechin gallate, epigallocatechin gallate, epicatechin-3-O-gallate, hydroxytyrosol, kaempferol, and luteolin-7-glycosidewere purchased from Extrasyn these (Lyon, France). Formic acid and methanol were purchased from Fischer Scientific (Oeiras, Portugal).

3.2. Preparation of Vegetable Extracts

The vegetable extracts used in this study were provided by the Benefit Company Evra S.r.l (Lauria, PZ, Italy) and obtained according to the Patent No. 102021000007460. Briefly, fresh fennel (*Foeniculum vulgare* Mill.) scraps were centrifuged at 7000 rpm for 15 min at room temperature to obtain a liquid fraction and a solid fraction. The solid fraction was dried in a vacuum oven type M40-VT (MPM Instruments s.r.l.—Bernareggio, Italy) at 40 °C for 48 h and then ground into a fine powder using a grinder model WSG30E (Waring Commercial—Torrington, CT, USA). The liquid fraction was further centrifuged to obtain a semi-solid pellet and a liquid supernatant (LF). Subsequently, the powdered solid fraction and the semi-solid pellet were extracted with ethanol (30–70%) in a raw material to solvent ratio of 1:5 (g/mL) using a magnetic stirrer for 45 min at temperature ranging from 40 to 70 °C, resulting in extracts of the fennel solid (SF) and pellet fraction (PF). The obtained extracts were filtered and dried. For the experimental tests, the dried extracts (SF and PF) were reconstituted in distilled water to obtain hydroalcoholic fennel extracts at the required concentrations.

3.3. Total Phenolic Content (TPC)

The total phenolic content of the three extracts (SF, LF, and PF) was determined using the Folin–Ciocalteu method with some modifications, as reported by Gómez-García et al., [68]. In a 96-well plate, 20 µL aliquots of each sample were mixed with 80 µL of Folin–Ciocalteu reagent previously diluted 1:10 (*v*/*v*) in water and 100 µL of 7.5% (*w*/*v*) sodium carbonate. After incubating the mixture in the dark at room temperature for 1h, the absorbance was measured at 750 nm using a microplate reader (Synergy H1, Winooski, VT, USA). Gallic acid (Sigma-Aldrich, St. Louis, MA, USA) was used as the standard, and the results were expressed as mg gallic acid equivalents (GAE) per 100 g of dry extract (DE). Gallic acid was used as a calibration curve standard (0.05–0.50 mg/mL). All measurements were performed in triplicate for each experiment.

3.4. Analysis of Phenolic Compounds Using HPLC

The polyphenolic profile of fennel extracts (SF, LF, and PF) was obtained using high-performance liquid chromatography coupled to a diode-array detector (HPLC-DAD) according to the method described by Campos et al., [69] with some modifications. The samples were injected into a Waters Series e2695 Separation Module System (Mildford, MA, USA) interfaced with a UV/Vis photodiode array detector (PDA 190–600 nm). Separation was performed using a reverse-phase column (COSMOSIL 5C1 8-AR-II Packed Column—4.6 mm I.D. × 250 mm: Dartford, UK). The chromatographic separation of phenolic compounds was carried out with mobile phase A—water/methanol/formic acid (92.5:5:2.5, *v*/*v*/*v*) and mobile phase B—methanol/water/formic acid (92.5:5:2.5, *v*/*v*/*v*) under the following conditions: 50 µL of sample was injected at a continuous flow rate of 0.5 mL/min, with gradient elution starting at 100% mobile phase A for 50 min. From 50 to 55 min, the gradient was adjusted to 45% A and 55% B, then mobile phase A was returned to 100% for 4 min (until 59 min). Data acquisition and analysis were carried out using Empower 3 software. Measurements were taken at wavelengths ranging from 200 to 600 nm. Phenolic compounds were identified and quantified using an external calibration curve of each specific phenolic compound at concentrations ranging from 0.008 to 0.125 mg/mL by comparing retention times, UV absorption spectra, and peak areas with pure standards. All measurements were performed in triplicate for each experiment, and the results were expressed as mg of phenolic compounds per 100 g of dry extract (DE).

3.5. Antioxidant Activity of Fennel Fraction Extracts

The DPPH-radical scavenging activity of the fennel extracts (LF, SF, and PF) was determined according to the method described by Gómez-García et al., [68]. Briefly, 1.75 mL of DPPH solution (60 µM) was mixed with 250 µL of each extract, and the mixture was incubated at room temperature for 30 min in the dark. Then, the absorbance was measured at 515 nm using a UV spectrophotometer (Shimadzu UV-2401PC) (Kyoto, Japan). Trolox was used as a standard for the preparation of a calibration curve (0.005–0.08 mg/mL), and the results were expressed in µM of Trolox equivalents (TE) per 100 g of dry extract (DE). All measurements were performed in triplicate for each experiment.

The $ABTS^+$ radical method was performed following the method described by Gómez-García et al., [68] mixing 10 µL of the sample extract with 1 mL of ABTS radical solution. After 6 min in the dark, the absorbance was recorded at 734 nm using a UV spectrophotometer. The stock solution was prepared by combining ABTS+ (7 mM) with potassium persulfate (2.5 mM) in ultra-pure water, and the mixture was stirred at room temperature for 16 h. The ABTS+ radical solution was subsequently diluted with water to achieve an absorbance of 0.700 ± 0.020 at 734 nm. The results were expressed in mg ascorbic acid (EAA) per 100 g of dry extract (DE) equivalents. The standard curve was made with l-ascorbic acid (0.05–0.5 mg/mL). All measurements were performed in triplicate for each experiment.

The antioxidant activity was calculated as follows:

$$\text{Antioxidant activity} = (A_{blank} - A_{sample}/A_{blank}) \times 100$$

3.6. *Antimicrobial Screening of Fennel Fraction Extracts*

The antimicrobial properties of fennel leaves extract were evaluated in vitro against pathogenic and spoilage microorganisms commonly found in the meat industry. Two preliminary screenings (agar-well diffusion assay and microplate-based assay) were carried out to test the antibacterial activity of the fennel extracts (LF, PF, and SF) against the following microorganisms: *Salmonella enterica* serotype Enteritidis ATCC 13076, *Escherichia coli* ATCC 25922, *Staphylococcus aureus* ATCC 25923, *Bacillus cereus* NCTC 2599, and *Pseudomonas aeruginosa* from the CBQF-ESB collection. Moreover, a fourth extract (LF*), obtained by concentrating the LF using a rotary evaporator under a vacuum at a temperature below 40 °C, was tested using the agar-well diffusion method.

3.6.1. Inoculum Preparation

All microbial strains used as indicators, stored in a freezer at −80 °C, were reactivated on a solid non-selective nutrient medium (Nutrient Agar, provided by Biolife, Milan, Italy) to assess their purity and vitality. According to ISO 4833-1:2013 [70], sterilized Petri dishes (9 cm in diameter) containing Plate Count Agar (PCA) were inoculated with the microbial indicator strains and incubated at 37 °C for 24 h. Bacterial cells were then picked from the colonies on the plates and suspended in phosphate buffer (pH 7.4) via agitation, aiming to achieve a final concentration of 1.5×10^8 CFU/mL (0.5 McFarland standard), measured using a densitometer (Sensititre—Nephelometer, Thermo Fisher, Roskilde, Denmark).

3.6.2. Agar-Well Diffusion Assay

A preliminary screening was performed using the agar-well diffusion method to test the antimicrobial properties of four fennel hydroalcoholic extracts (liquid fraction used directly, LF; liquid fraction concentrated, LF*; solid fraction, SF; and pellet fraction, PF) at different concentrations (38.7 mg DE/mL for LF, 160 mg DE/mL for LF*, 110 mg DE/mL for SF, and 20 mg DE/mL for PF) against the Gram-positive bacterium *Staphylococcus aureus* and Gram-negative *Escherichia coli*. The aim of this laboratory assay was to identify the most effective extracts and concentrations. Briefly, a standardized inoculum suspension of each bacterial strain, with a final concentration of 1.5×10^8 CFU/mL, was swabbed onto PCA agar plates. Subsequently, four-millimeter wells were then created and inoculated with 50 µL of each hydroalcoholic extract. The plates were incubated for 24 h at 37 °C. The effectiveness of these extracts was evaluated by measuring the diameter of the inhibition zones.

3.6.3. Microplate-Based Assay

After identifying the two most effective fennel extracts (SF and PF) and their respective concentrations (50, 100, and 150 mg/mL for the SF; 10 and 20 mg of dry extract DE/mL for PF), a second screening using a microplate-based assay was conducted to evaluate their antibacterial activity against pathogenic and spoilage microorganisms commonly found in the meat industry: *Salmonella enterica* serotype Enteritidis ATCC 13076, *Escherichia coli* ATCC 25922, *Pseudomonas aeruginosa*, *Staphylococcus aureus* ATCC 25923, and *Bacillus cereus* NCTC 2599. Briefly, for each fennel extract, various solutions at different concentrations were prepared using distilled water as follows: SF at concentrations of 50 mg DE/mL = 5%, 100 mg DE/mL = 10%, and 150 mg DE/mL = 15%, and PF at concentrations of 10 mg DE/mL = 1% and 20 mg DE/mL = 2%; the best inhibitory concentration being considered as the minimum inhibitory concentration (MIC). The MIC value corresponded to the lowest extract concentration that inhibited visible bacterial growth. The sterilized samples were added to the wells of sterile polystyrene 96-well microtiter plates (Sarstedt, Wexford, Ireland). Mueller–Hinton broth with the inoculum (without any extract solution) was used as the positive control, while Mueller–Hinton broth without the inoculum was used as the negative control (blank). The plates were incubated at 37 °C for 24 h. Absorbance was measured at 660 nm at 1 h intervals using a Thermo Scientific™ Multiskan™ GO Microplate Spectrophotometer (Roskilde, Denmark). The laboratory test was conducted in triplicate. According to Sahar Roshanak et al., [71] and María-Leonor Pla et al., [72], the increase in turbidity is a sign of microorganism growth. Thus, based on these studies and in accordance with Ju-Sung Kim et al., [66], we calculated the percent inhibition (%) of each fennel extract at the different concentrations (SF at concentrations of 50 mg DE/mL = 5%, 100 mg DE/mL = 10%, and 150 mg DE/mL = 15% and PF at concentrations of 10 mg DE/mL = 1% and 20 mg DE/mL = 2%) using the measured absorbance as follows:

$$\text{Inhibition rate (\%)} = (1 - (Abs_{sample} - Abs_{blank})/Abs_{control})) \times 100$$

where Abs_{sample} is the absorbance of the experimental sample, Abs_{blank} is the absorbance of the blank, and $Abs_{control}$ is the absorbance of the positive control.

3.7. *Evaluation of Effects of Selected Fennel Extracts on Beef Burgers*

The preliminary in vitro screenings showed that the fennel extract with the best antimicrobial properties was the hydroalcoholic extract obtained from the semi-solid pellet represented by the pellet fraction (PF). Consequently, to evaluate the effects of this extract on beef burgers, microbiological and physicochemical analyses were conducted. The results of the treated beef burger (TRT) analyses were compared to those of an untreated beef burger (control, CTR) commonly produced by the meat industry involved in this study.

3.7.1. Beef Burger Processing and Analyses

The tested beef burgers were taken from a batch produced by the meat product company of interest. The following ingredients were used for the preparation of the dough: minced beef (84%), water (8.5%), salt (1.5%), and a powdered mixture of aromas, vegetable fibers, plant extracts, and potato starch (6%). In the TRT group, 2% (20 mg DE/mL) of the best fennel extract (PF) was added to the beef burgers and thoroughly mixed. The percentage refers to final product concentration. Beef burgers with no additions were used as controls (CTR). The mixture was then manually blended to obtain a homogeneous mass. The samples were divided into treated (TRT) and untreated (control, CTR) groups. All burgers were formed using a conventional burger maker, packaged in skin packs, and stored for 18 days (shelf life determined by the company based on preliminary studies) in a relative vacuum at a refrigerator temperature of 4 ± 1 °C (in a refrigerator with digital control of air temperature and relative humidity). When opening each package, the product's acceptability was evaluated based on the color and smell to determine its suitability for analysis. Analyses were performed at 0, 5, 8, 12, 15, and 18 days from production. Sampling times were chosen based on preliminary shelf-life studies (data not published). At the laboratory, during when each package was opened from 0 to 18 days, microbiological and chemical analyses were carried out in triplicate.

3.7.2. Microbiological and Chemical Analysis

In the microbiological laboratory, the following meat hygiene indicator microorganisms were determined: total aerobic plate count at 30 °C (TAB 30 °C), total coliforms, beta-glucuronidase-positive *Escherichia coli*, coagulase-positive staphylococci, yeasts, and molds. Briefly, 10 g of each sample was added to 90 mL (1:10, *w*/*v*) of sterilized peptone water (PW, CM0009, OXOID, Basingstoke, UK) in a sterile stomacher bag and homogenized at 230 rpm for three minutes in a peristaltic homogenizer (BagMixer®400 P, Interscience, Saint Nom, France). Subsequently, ten-fold serial dilutions were prepared from each homogenate to isolate and enumerate the following: the total aerobic plate count at 30 °C on plate count agar (PCA; CM0325, Oxoid, Madrid, Spain) incubated at 30 °C for 48–72 h according to ISO 4833-1:2013 [70]; the total coliforms on violet–red bile lactose agar (VRBL,

Oxoid, Madrid, Spain) incubated at 37 °C for 48 h in agreement with ISO 4831:2006 [73]; the β-glucuronidase-positive *Escherichia coli* on tryptone bile x-glucuronide agar (TBX, CM0945, Oxoid, Basingstoke, Hampshire, UK) incubated at 44 °C for 24–48 h in agreement with ISO 16649-2:2001 [74]; the coagulase-positive staphylococci on Baird–Parker agar (Oxoid, Madrid, Spain) at 37 °C for 24–48 h in agreement with ISO 6888-1:1999 [75]; and yeasts and molds on dichloran rose–Bengal chloramphenicol agar (DRBC, Oxoid, Madrid, Spain) incubated at 25 °C for 120/168 h according to ISO 21527-2:2008 [76].

In addition, the potential presence of two pathogens most commonly associated with meat products was also evaluated: *Listeria monocytogenes*, according to ISO 11290-1:2017 [77], and *Salmonella* spp., according to ISO 6579-1:2017 [78]. For this, 25 g of each meat sample was homogenized in 225 mL (1:10 *w*/*v*) of buffer peptone water (BPW, CM0509, Oxoid, Basingstoke, UK) and incubated for 24 h at 37 °C for the detection of *Salmonella* spp. and in 225 mL of half Fraser broth (HF, CM1053, Oxoid, Basingstoke, Hampshire, UK), incubated for 24 h at 30 °C for the detection of *Listeria monocytogenes*.

During the storage period, water activity (a_w) and pH were evaluated using Aqualab 4 TE—Decagon Devices (METER Group, Inc., Pullman, WA, USA) and a pH-meter (Crison-Micro TT 2022, Crison Instruments, Barcelona, Spain) equipped with a glass insertion electrode, respectively. The pH values were measured by inserting the glass electrode 2 cm deep into the beef burgers. Each measurement was performed in triplicate (n = 3), and the mean value was recorded as the result. Moreover, the overall acceptability was assessed based on visual and odor evaluation of the samples.

3.8. Statistical Analysis

Statistical analyses were performed using SAS software version 6. 3rd ed. One-way ANOVAs were performed for the results of the fennel fraction extracts by comparing the inhibition halos of each extract against the indicator microorganisms tested. For the analysis of total phenols, individual phenolic content, and DPPH and ABTS activity, Student's t-test was applied. Physicochemical and microbiological data of beef burger analyses were statistically analyzed using generalized linear mixed models (GLMMs), including the fixed effects of treatment (CTR and TRT) and time (T0, T1, T2, T3, T4, and T5). Each measurement was performed in triplicate (n = 3), taking the mean value as the result. The means were compared using the Tukey test ($p < 0.05$; $p < 0.01$). All data were presented as the mean (M) ± standard error (SE).

4. Conclusions

Overall, as observed in other studies [64], the preliminary results from the present work suggest that fennel scraps and their extracts have significant potential as natural antioxidant and antimicrobial agents in minced meat products such as beef burgers.

The growing interest in optimizing the use of food waste and by-products from agriculture and other food sectors presents substantial opportunities for recovering high-value compounds such as bioactive compounds. The fact that fennel extracts obtained from these by-products were shown to have very high antioxidant activity and inhibitory potentials against microorganisms commonly found in meat products opens new possibilities for using vegetable waste in the meat industry. These extracts could serve as natural antimicrobial agents that are healthier and more sustainable than the chemical additives currently used in industrial markets. In fact, they have the potential to extend the shelf life of minced meat products, enhance food safety, and transform waste into valuable resources, aligning with the principles of the circular economy. This approach could reduce food waste while generating novel, safe, and natural value-added products.

However, to fully realize this potential, given the relatively limited antimicrobial effects of fennel extracts (notably the pellet fraction) observed in the beef burger experiments of this study, optimization of the extraction process and incorporation methods in minced meat is necessary to improve their antimicrobial efficacy. Further studies, such as challenge tests, are required to support these preliminary findings and will be the focus of future research.

5. Patents

Patent no. 102021000007460 was filed on 26 March 2021.

Author Contributions: Conceptualization, M.D.P. and A.S.; methodology, F.D.B. and L.C.; software, M.D.P. and M.E.; validation, F.D.B., M.P.; formal analysis, F.D.B. and R.G.-G.; investigation, F.D.B. and M.E.; resources, L.C., A.S. and M.D.P.; data curation, M.E., M.D.P. and F.D.B.; writing—original draft preparation, M.E. and L.C.; writing—review and editing, M.D.P., R.M. and F.D.B.; visualization, A.S., M.P. and R.G.-G.; supervision, R.M.; project administration, R.M.; funding acquisition, M.P. All authors have read and agreed to the published version of the manuscript.

Funding: This research was supported in part by the H2020 EU FODIAC project (reference number 778388).

Institutional Review Board Statement: Not applicable.

Informed Consent Statement: Not applicable.

Data Availability Statement: The data presented in this study are available upon request from the corresponding authors.

Acknowledgments: The authors thank the "Gruppo Loma S.r.l", Giuseppe Ruocco, Maddalena Stompanato, and Alessandro Di Luca, who provided both material and technical support for this research. Moreover, the authors acknowledge the scientific collaboration of the Escola Superior de Biotecnologia–Universidade Católica Portuguesa through CBQF and the program Marie Skłodowska-Curie grant (MSCA-RISE; FODIAC; 778388).

Conflicts of Interest: Author Filomena De Biasio was employed by the company EVRA S.r.l. Società Benefit. The remaining authors declare that the research was conducted in the absence of any commercial or financial relationships that could be construed as a potential conflict of interest.

References

1. Meinilä, J.; Virtanen, J.K. Meat and Meat Products—A Scoping Review for Nordic Nutrition Recommendations 2023. *Food Nutr. Res.* **2024**, *68*, 333–342. [CrossRef] [PubMed]
2. Morsy, M.K.; Elsabagh, R. Quality Parameters and Oxidative Stability of Functional Beef Burgers Fortified with Microencapsulated Cod Liver Oil. *LWT-Food Sci. Technol.* **2021**, *142*, 110959. [CrossRef]
3. Abril, M.; Campo, M.; Onenc, A.; Sanudo, C.; Albertì, P.; Negueruela, A. Beef Colour Evolution as a Function of Ultimate PH. *Meat Sci.* **2001**, *58*, 69–78. [CrossRef] [PubMed]
4. Bantawa, K.; Rai, K.; Subba Limbu, D.; Khanal, H. Food-Borne Bacterial Pathogens in Marketed Raw Meat of Dharan, Eastern Nepal. *BMC Res. Notes* **2018**, *11*, 618. [CrossRef]
5. Del Nobile, M.A.; Conte, A.; Cannarsi, M.; Sinigaglia, M. Strategies for Prolonging the Shelf Life of Minced Beef Patties. *J. Food Saf.* **2009**, *29*, 14–25. [CrossRef]
6. Bhandare, S.G.; Sherikar, A.T.; Paturkar, A.M.; Waskar, V.S.; Zende, R.J. A Comparison of Microbial Contamination on Sheep/Goat Carcasses in a Modern Indian Abattoir and Traditional Meat Shops. *Food Control* **2007**, *18*, 854–858. [CrossRef]
7. Addis, M. Major Causes Of Meat Spoilage and Preservation Techniques: A Review. *Food Sci. Qual. Manag.* **2015**, *41*, 106–110.
8. Shahbazi, Y.; Shavisi, N.; Mohebi, E. Effects of Ziziphora Clinopodioides Essential Oil and Nisin, Both Separately and in Combination, to Extend Shelf Life and Control Escherichia ColiO157: H7 and Staphylococcus Aureus in Raw Beef Patty during Refrigerated Storage. *J. Food Saf.* **2016**, *36*, 227–236. [CrossRef]
9. Regulation (EC). Regulation (EC) No 1333/2008 of the European Parliament and of the Council of 16 December 2008 on Food Additives. *Off. J. Eur. Union* **2008**, *L354*, 16–33.
10. Zhang, Y.; Zhang, Y.; Jia, J.; Peng, H.; Qian, Q.; Pan, Z.; Liu, D. Nitrite and Nitrate in Meat Processing: Functions and Alternatives. *Curr. Res. Food Sci.* **2023**, *6*, 100470. [CrossRef]
11. Costagliola, A.; Roperto, F.; Benedetto, D.; Anastasio, A.; Marrone, R.; Perillo, A.; Russo, V.; Papparella, S.; Paciello, O. Outbreak of Fatal Nitrate Toxicosis Associated with Consumption of Fennels (Foeniculum Vulgare) in Cattle Farmed in Campania Region (Southern Italy). *Environ Sci Pollut R.* **2014**, *21*, 6252–6257. [CrossRef] [PubMed]
12. Nair, B. Food Science and Technology Final Report on the Safety Assessment of Benzyl Alcohol, Benzoic Acid, and Sodium Benzoate. *Int. J. Toxicol.* **2001**, *20*, 23–50. [PubMed]
13. Honorato, T.C.; Batista, E.; de O. do Nascimento, K.; Pires, T. Food Additives: Applications and Toxicology. *Rev. Verde De Agroecol. E Desenvolv. Sustentável* **2013**, *8*, 1–11.
14. Silva, M.M.; Lidon, F.C. Food Preservatives - An Overview on Applications and Side Effects. *Emir. J. Food Agric.* **2016**, *28*, 366–373. [CrossRef]

15. Rangel Guimarães, R.; Cristina, M.; De Freitas, J.; Lucia, V.; Da Silva, M. Bolos Simples Elaborados Com Farinha Da Entrecasca de Melancia (Citrullus Vulgaris, Sobral): Avaliação Química, Fisica e Sensorial Simple Cakes Elaborated with Flour of Watermelon Inner Skin (Citrullus Vulgaris, Sobral): Chemical, Physical, and Sensory Evaluation. *Ciênc. Tecnol. Aliment.* **2010**, *30*, 354.
16. Ramos, E.M.; de Miranda Gomide, L.A. *Avaliação Da Qualidade de Carnes: Fundamentos e Metodologias*, 1st ed.; Editora UFV: Viçosa, Brazil, 2007.
17. Sales, P.V.G.; Sales, V.H.G.; Oliveira, E.M. Avaliação Sensorial de Duas Formulações de Hambúrguer de Peixe. *Rev. Bras. De Prod. Agroindustriais* **2015**, *17*, 17–23. [CrossRef]
18. Varela, P.; Fiszman, S.M. Exploring Consumers' Knowledge and Perceptions of Hydrocolloids Used as Food Additives and Ingredients. *Food Hydrocoll.* **2013**, *30*, 477–484. [CrossRef]
19. Ge, H.; Fu, S.; Guo, H.; Hu, M.; Xu, Z.; Zhou, X.; Chen, X.; Jiao, X. Application and Challenge of Bacteriophage in the Food Protection. *Int. J. Food Microbiol.* **2022**, *380*, 109872. [CrossRef]
20. Elois, M.A.; da Silva, R.; Pilati, G.V.T.; Rodríguez-Lázaro, D.; Fongaro, G. Bacteriophages as Biotechnological Tools. *Viruses* **2023**, *15*, 349. [CrossRef]
21. Bhattacharya, D.; Nanda, P.K.; Pateiro, M.; Lorenzo, J.M.; Dhar, P.; Das, A.K. Lactic Acid Bacteria and Bacteriocins: Novel Biotechnological Approach for Biopreservation of Meat and Meat Products. *Microorganisms* **2022**, *10*, 2058. [CrossRef]
22. Kalogianni, A.I.; Lazou, T.; Bossis, I.; Gelasakis, A.I. Natural Phenolic Compounds for the Control of Oxidation, Bacterial Spoilage, and Foodborne Pathogens in Meat. *Foods* **2020**, *9*, 794. [CrossRef] [PubMed]
23. Yu, H.H.; Chin, Y.W.; Paik, H.D. Application of Natural Preservatives for Meat and Meat Products against Food-borne Pathogens and Spoilage Bacteria: A Review. *Foods* **2021**, *10*, 2418. [CrossRef] [PubMed]
24. Danilović, B.; Đorđević, N.; Milićević, B.; Šojić, B.; Pavlić, B.; Tomović, V.; Savić, D. Application of Sage Herbal Dust Essential Oils and Supercritical Fluid Extract for the Growth Control of Escherichia Coli in Minced Pork during Storage. *LWT-Food Sci. Technol.* **2021**, *141*, 110935. [CrossRef]
25. de Alencar, M.G.; de Quadros, C.P.; Luna, A.L.L.P.; Neto, A.F.; da Costa, M.M.; Queiroz, M.A.Á.; de Carvalho, F.A.L.; da Silva Araújo, D.H.; Gois, G.C.; dos Anjos Santos, V.L.; et al. Grape Skin Flour Obtained from Wine Processing as an Antioxidant in Beef Burgers. *Meat Sci.* **2022**, *194*, 108963. [CrossRef]
26. Elhadef, K.; Smaoui, S.; Ben Hlima, H.; Ennouri, K.; Fourati, M.; Chakchouk Mtibaa, A.; Ennouri, M.; Mellouli, L. Effects of Ephedra Alata Extract on the Quality of Minced Beef Meat during Refrigerated Storage: A Chemometric Approach. *Meat Sci.* **2020**, *170*, 108246. [CrossRef]
27. Kalleli, F.; Bettaieb Rebey, I.; Wannes, W.A.; Boughalleb, F.; Hammami, M.; Saidani Tounsi, M.; M'hamdi, M. Chemical Composition and Antioxidant Potential of Essential Oil and Methanol Extract from Tunisian and French Fennel (*Foeniculum Vulgare* Mill.) Seeds. *J. Food Biochem.* **2019**, *43*, e12935. [CrossRef]
28. Noreen, S.; Rehman, H.U.; Tufail, T.; Badar Ul Ain, H.; Awuchi, C.G. Secoisolariciresinol Diglucoside and Anethole Ameliorate Lipid Abnormalities, Oxidative Injury, Hypercholesterolemia, Heart, and Liver Conditions. *Food Sci. Nutr.* **2023**, *11*, 2620–2630. [CrossRef]
29. Saddiqi H., A.; Iqbal, Z. *Chapter 55—Usage and Significance of Fennel (Foeniculum Vulgare Mill.) Seeds in Eastern Medicine*; Elsevier: Amsterdam, The Netherlands,, 2011; pp. 461–467.
30. Barros, L.; Carvalho, A.M.; Ferreira, I.C.F.R. The Nutritional Composition of Fennel (Foeniculum Vulgare): Shoots, Leaves, Stems and Inflorescences. *LWT-Food Sci. Technol.* **2010**, *43*, 814–818. [CrossRef]
31. Badgujar, S.B.; Patel, V.V.; Bandivdekar, A.H. *Foeniculum vulgare* Mill: A Review of Its Botany, Phytochemistry, Pharmacology, Contemporary Application, and Toxicology. *Biomed. Res. Int.* **2014**, *2014*, 842674. [CrossRef]
32. Choi, E.M.; Hwang, J.K. Antiinflammatory, Analgesic and Antioxidant Activities of the Fruit of Foeniculum Vulgare. *Fitoterapia* **2004**, *75*, 557–565. [CrossRef]
33. Agarwal Scholar, D.; Agarwal, D.; Sharma, L.; Saxena, S. Anti-Microbial Properties of Fennel (*Foeniculum vulgare* Mill.) Seed Extract. *J. Pharmacogn. Phytochem.* **2017**, *6*, 479–482.
34. Abdellaoui, M.; Derouich, M.; El-Rhaffari, L. Essential Oil and Chemical Composition of Wild and Cultivated Fennel (*Foeniculum Vulgare* Mill.): A Comparative Study. *S. Afr. J. Bot.* **2020**, *135*, 93–100. [CrossRef]
35. Diao, W.R.; Hu, Q.P.; Zhang, H.; Xu, J.G. Chemical Composition, Antibacterial Activity and Mechanism of Action of Essential Oil from Seeds of Fennel *(Foeniculum vulgare* Mill.). *Food Control* **2014**, *35*, 109–116. [CrossRef]
36. Sun, Y.; Zhang, M.; Bhandari, B.; Bai, B. Fennel Essential Oil Loaded Porous Starch-Based Microencapsulation as an Efficient Delivery System for the Quality Improvement of Ground Pork. *Int. J. Biol. Macromol.* **2021**, *172*, 464–474. [CrossRef] [PubMed]
37. Gómez-García, R.; Campos, D.A.; Aguilar, C.N.; Madureira, A.R.; Pintado, M. Valorisation of Food Agro-Industrial by-Products: From the Past to the Present and Perspectives. *J. Environ. Manag.* **2021**, *299*, 113571. [CrossRef] [PubMed]
38. Roby, M.H.H.; Sarhan, M.A.; Selim, K.A.H.; Khalel, K.I. Antioxidant and Antimicrobial Activities of Essential Oil and Extracts of Fennel (*Foeniculum vulgare* L.) and Chamomile (*Matricaria chamomilla* L.). *Ind. Crops Prod.* **2013**, *44*, 437–445. [CrossRef]
39. Salami, M.; Rahimmalek, M.; Ehtemam, M.H. Inhibitory Effect of Different Fennel (Foeniculum Vulgare) Samples and Their Phenolic Compounds on Formation of Advanced Glycation Products and Comparison of Antimicrobial and Antioxidant Activities. *Food Chem.* **2016**, *213*, 196–205. [CrossRef]
40. Ahmed, A.F.; Shi, M.; Liu, C.; Kang, W. Comparative Analysis of Antioxidant Activities of Essential Oils and Extracts of Fennel (Foeniculum Vulgare Mill.)Seeds from Egypt and China. *Food Science and Human Wellness* **2019**, *8*, 67–72. [CrossRef]

41. Tlili, N.; Elfalleh, W.; Hannachi, H.; Yahia, Y.; Khaldi, A.; Ferchichi, A.; Nasri, N. Screening of Natural Antioxidants from Selected Medicinal Plants. *Int J Food Prop* **2013**, *16*, 1117–1126. [CrossRef]
42. Oktay, M.; Gülçin, I.; Küfrevioğlu, Ö.I. Determination of in Vitro Antioxidant Activity of Fennel (Foeniculum Vulgare) Seed Extracts. *LWT* **2003**, *36*, 263–271. [CrossRef]
43. Ghasemian, A.; Al-Marzoqi, A.H.; Mostafavi, S.K.S.; Alghanimi, Y.K.; Teimouri, M. Chemical Composition and Antimicrobial and Cytotoxic Activities of Foeniculum Vulgare Mill Essential Oils. *J. Gastrointest. Cancer* **2020**, *51*, 260–266. [CrossRef] [PubMed]
44. Barrahi, M.; Esmail, A.; Elhartiti, H.; Chahboun, N.; Benali, A.; Amiyare, R.; Lakhrissi, B.; Rhaiem, N.; Zarrouk, A.; Ouhssine, M. Chemical Composition and Evaluation of Antibacterial Activity of Fennel (Foeniculum Vulgare Mill) Seed Essential Oil against Some Pathogenic Bacterial Strains. *Casp. J. Environ. Sci.* **2020**, *18*, 295–307. [CrossRef]
45. Lemiasheuski, V.; Ji, Y.; Buchenkov, I.; Gritskevitch, E.; Sysa, A. Evaluation of the Antibacterial Effect of Foeniculum Vulgare Mill. Essential Oil on Opportunistic Microflora: Growth and Enzymatic Activity Indicators. *Asian J. Res. Biochem.* **2024**, *14*, 138–148. [CrossRef]
46. Ghafarizadeh, A.; Seyyednejad, S.M.; Motamedi, H.; Shahbazi, F. Evaluation of the antibacterial activity of different parts of Foeniculum vulgare (fennel) extracts. *J. Med. Herbs* **2018**, *9*, 33–38.
47. Rafieian, F.; Amani, R.; Rezaei, A.; Karaça, A.C.; Jafari, S.M. Exploring Fennel (*Foeniculum vulgare*): Composition, Functional Properties, Potential Health Benefits, and Safety. *Crit. Rev. Food Sci. Nutr.* **2024**, *64*, 6924–6941. [CrossRef]
48. Elbaz, M.; Abdesslem, S.B.; St-Gelais, A.; Boulares, M.; Moussa, O.B.; Timoumi, M.; Hassouna, M.; Aider, M. Essential Oils Profile, Antioxidant and Antibacterial Potency of Tunisian Fennel (*Foeniculum vulgare* Mill.) Leaves Grown under Conventional and Organic Conditions. *Food Chem. Adv.* **2024**, *4*, 100734. [CrossRef]
49. Joshi, R.K. Chemical Constituents and Antibacterial Property of the Essential Oil of the Roots of *Cyathocline Purpurea*. *J. Ethnopharmacol.* **2013**, *145*, 621–625. [CrossRef]
50. Shakeri, A.; Khakdan, F.; Soheili, V.; Sahebkar, A.; Rassam, G.; Asili, J. Chemical Composition, Antibacterial Activity, and Cytotoxicity of *Essential oil* from *Nepeta ucrainica* L. Spp. Kopetdaghensis. *Ind. Crops Prod.* **2014**, *58*, 315–321. [CrossRef]
51. Di Napoli, M.; Castagliuolo, G.; Badalamenti, N.; Maresca, V.; Basile, A.; Bruno, M.; Varcamonti, M.; Zanfardino, A. Antimicrobial, Antibiofilm, and Antioxidant Properties of Essential Oil of Foeniculum Vulgare Mill. Leaves. *Plants* **2022**, *11*, 3573. [CrossRef]
52. Lo Cantore, P.; Iacobellis, N.S.; De Marco, A.; Capasso, F.; Senatore, F. Antibacterial Activity of *Coriandrum sativum* L. and *Foeniculum vulgare* Miller Var. vulgare (Miller) Essential Oils. *J. Agric. Food Chem.* **2004**, *52*, 7862–7866. [CrossRef]
53. Chiboub, W.; Sassi, A.B.; Amina, C.M.H.; Souilem, F.; El Ayeb, A.; Djlassi, B.; Ascrizzi, R.; Flamini, G.; Harzallah-Skhiri, F. Valorization of the Green Waste from Two Varieties of Fennel and Carrot Cultivated in Tunisia by Identification of the Phytochemical Profile and Evaluation of the Antimicrobial Activities of Their Essentials Oils. *Chem. Biodivers.* **2019**, *16*, e1800546. [CrossRef] [PubMed]
54. Özcan, M.M.; Sağdıç, O.; Özkan, G. Short Communication Inhibitory Effects of Spice Essential Oils on the Growth of *Bacillus* Species. *J. Med. Food* **2006**, *9*, 418–421. [CrossRef] [PubMed]
55. Górniak, I.; Bartoszewski, R.; Króliczewski, J. Comprehensive Review of Antimicrobial Activities of Plant Flavonoids. *Phytochem. Rev.* **2019**, *18*, 241–272. [CrossRef]
56. Berthold-Pluta, A.; Stasiak-Różańska, L.; Pluta, A.; Garbowska, M. Antibacterial Activities of Plant-Derived Compounds and Essential Oils against Cronobacter Strains. *Eur. Food Res. Technol.* **2019**, *245*, 1137–1147. [CrossRef]
57. Almuzaini, A.M. Phytochemicals: Potential Alternative Strategy to Fight Salmonella Enterica Serovar Typhimurium. *Front. Vet. Sci.* **2023**, *10*, 1188752. [CrossRef]
58. Shabnam, J.; Sobia, M.; Ibatsam, K.; Rauf, A.; M, S.H. Comparative Antimicrobial Activity of Clove and Fennel Essential Oils against Food Borne Pathogenic Fungi and Food Spoilage Bacteria. *Afr. J. Biotechnol.* **2012**, *11*, 16065–16070. [CrossRef]
59. Manonmani, R.; Khadir, V.A. Khadir Antibacterial Screening on Foeniculum Vulgare Mill. *Int. J. Pharma Bio Sci.* **2011**, *2*, 390–394.
60. Gheorghita, D.; Robu, A.; Antoniac, A.; Antoniac, I.; Ditu, L.M.; Raiciu, A.D.; Tomescu, J.; Grosu, E.; Saceleanu, A. In Vitro Antibacterial Activity of Some Plant Essential Oils against Four Different Microbial Strains. *Appl. Sci.* **2022**, *12*, 9482. [CrossRef]
61. Yi, Q.; Zhang, J.; Ai, D. Antimicrobial Effects of Essential Oils on Pseudomonas Aeruginosa. *J. Pathol. Res. Rev. Rep.* **2022**, *152*, 2–6. [CrossRef]
62. Shahat, A.A.; Ibrahim, A.Y.; Hendawy, S.F.; Omer, E.A.; Hammouda, F.M.; Abdel-Rahman, F.H.; Saleh, M.A. Chemical Composition, Antimicrobial and Antioxidant Activities of Essential Oils from Organically Cultivated Fennel Cultivars. *Molecules* **2011**, *16*, 1366–1377. [CrossRef]
63. Commission Regulation (EC) Commission Regulation (EC) No 1441/2007 of 5 December 2007 Amending Regulation (EC) No 2073/2005 on Microbiological Criteria for Foodstuffs. *Off. J. Eur. Communities* **2007**, *L322*.
64. Mujović, M.; Šojić, B.; Danilović, B.; Kocić-Tanackov, S.; Ikonić, P.; Đurović, S.; Milošević, S.; Bulut, S.; Đorđević, N.; Savanović, J.; et al. Fennel (Foeniculum Vulgare) Essential Oil and Supercritical Fluid Extracts as Novel Antioxidants and Antimicrobial Agents in Beef Burger Processing. *Food Biosci.* **2023**, *56*, 103283. [CrossRef]
65. Saleh, S.M.; Kassab H., A.; Mariam, A.R.; El-Bana M., A. Fennel and Turmeric Powders' Effectiveness as Natural in Beef Burgers. *Rom. Biotechnol. Lett.* **2022**, *27*, 3186–3192. [CrossRef]
66. Kim, J.S.; Kwon, Y.S.; Chun, W.J.; Kim, T.Y.; Sun, J.; Yu, C.Y.; Kim, M.J. Rhus Verniciflua Stokes Flavonoid Extracts Have Anti-Oxidant, Anti-Microbial and α-Glucosidase Inhibitory Effect. *Food Chem.* **2010**, *120*, 539–543. [CrossRef]

67. Huang, L.; Wang, Y.; Li, R.; Wang, Q.; Dong, J.; Wang, J.; Lu, S. Thyme Essential Oil and Sausage Diameter Effects on Biogenic Amine Formation and Microbiological Load in Smoked Horse Meat Sausage. *Food Biosci.* **2021**, *40*, 100885. [CrossRef]
68. Gómez-García, R.; Campos, D.A.; Oliveira, A.; Aguilar, C.N.; Madureira, A.R.; Pintado, M. A Chemical Valorisation of Melon Peels towards Functional Food Ingredients: Bioactives Profile and Antioxidant Properties. *Food Chem.* **2021**, *335*, 127579. [CrossRef]
69. Campos, D.A.; Ribeiro, T.B.; Teixeira, J.A.; Pastrana, L.; Pintado, M.M. Integral Valorization of Pineapple (*Ananas Comosus* L.) By-Products through a Green Chemistry Approach towards Added Value Ingredients. *Foods* **2020**, *9*, 60. [CrossRef]
70. *ISO 4833-1:2013*; Microbiology of the Food Chain: Horizontal Method for the Enumeration of Microorganisms—Part 1: Colony Count at 30 °C by the Pour Plate Technique. International Organization for Standardization: Geneva, Switzerland, 2013.
71. Roshanak, S.; Shahidi, F.; Yazdi, F.T.; Javadmanesh, A.; Movaffagh, J. Evaluation of Antimicrobial Activity of Buforin I and Nisin and the Synergistic Effect of Their Combination as a Novel Antimicrobial Preservative. *J. Food Prot.* **2020**, *83*, 2018–2025. [CrossRef]
72. Pla, M.L.; Oltra, S.; Esteban, M.D.; Andreu, S.; Palop, A. Comparison of Primary Models to Predict Microbial Growth by the Plate Count and Absorbance Methods. *Biomed. Res. Int.* **2015**, *2015*, 365025. [CrossRef]
73. *ISO 4831:2006*; Microbiology of Food and Animal Feeding STUFFS—Horizontal Method for the Detection and Enumeration of Coliforms—Most Probable Number Technique. International Organization for Standardization: Geneva, Switzerland, 2006.
74. *ISO 16649-2:2001*; Microbiology of Food and Animal Feeding Stuffs Horizontal Method for the Enumeration of Beta-Glucuronidase-Positive Escherichia Coli Part 2: Colony-Count Technique at 44 Degrees C Using 5-Bromo-4-Chloro-3-Indolyl Beta-D-Glucuronide. International Organization for Standardization: Geneva, Switzerland, 2001.
75. *ISO 6888-1999*; Microbiology of Food and Animal Feeding Stuffs—Horizontal Method for Enumeration of Coagulase-Positive Staphylococci (Staphylococcus aureus and Other Species)—Part 1: Technique Using Baird-Parker Agar. ISO: Geneva, Switzerland, 1999; pp. 1–11.
76. *ISO 21527-2:2008*; Microbiology of Food and Animal Feeding Stuffs-Horizontal Method for the Enumeration of Yeast and Moulds-Part 2: Colony Count Technique in Products with Water Activity Less than or Equal to 0.95. International Organization for Standardization: Geneva, Switzerland, 2008.
77. *ISO 11290-1:2017*; Microbiology of the Food Chain Horizontal Method for the Detection and Enumeration of Listeria Monocytogenes and of *Listeria* spp. Part 1: Detection Method. International Organization for Standardization: Geneva, Switzerland, 2017.
78. *ISO 6579-1:2017*; Microbiology of the Food Chain–Horizontal Method for the Detection, Enumeration and Serotyping of Salmonella-Part 1: Detection of Salmonella spp. International Organization for Standardization: Geneva, Switzerland, 2017.

MDPI AG
Grosspeteranlage 5
4052 Basel
Switzerland
Tel.: +41 61 683 77 34

Antibiotics Editorial Office
E-mail: antibiotics@mdpi.com
www.mdpi.com/journal/antibiotics

www.ingramcontent.com/pod-product-compliance
Lightning Source LLC
LaVergne TN
LVHW072354090526
838202LV00019B/2544

* 9 7 8 3 7 2 5 8 3 6 1 6 1 *